THE GEOGRAPHY OF
THE WORLD ECONOMY

AN INTRODUCTION TO ECONOMIC GEOGRAPHY

Third edition

PAUL KNOX AND JOHN AGNEW

A member of the Hodder Headline Group
LONDON
Co-published in the United States of America by
Oxford University Press Inc., New York

First published in Great Britain in 1989
Second Edition 1994
Third Edition 1998 by
Arnold, a member of the Hodder Headline Group
338 Euston Road, London NW1 3BH
http://www.arnoldpublishers.com

Copublished in the United States of America by
Oxford University Press, Inc.,
198 Madison Avenue, New York, NY 10016

Whilst the advice and information in this book are believed to be true and accurate at the date of going to press, neither the authors nor the publishers can accept any legal responsibility or liability for any errors or omissions.

British Library Cataloguing in Publication Data
A catalogue entry for this book is available from the British Library

Library of Congress Cataloging-in-Publication Data
A catalog entry for this book is available from the Library of Congress

ISBN 0 340 70612 0

3 4 5 6 7 8 9 10

Production Editor: Liz Gooster
Production Controller: Priya Gohil
Cover design: Terry Griffiths

Composition by Photoprint, Torquay, Devon
Printed and bound in Great Britain by The Bath Press, Bath

CONTENTS

ACKNOWLEDGEMENTS

The American Geographical Association for figures 5.5 and 5.6 from 'The urban and industrial transformation of Japan' by Harris published in *Geographical Review* 72 (1982); The American Sociological Association for table 3.4 from 'Incorporation in the world-system: toward a critique' by Hall published in *American Sociological Review* 15 (1986); Edward Arnold for figure 4.1 from *An Introduction to Urban Historical Geography*; The Association of American Geographers for figure 2.3 from 'The disparity between the level of economic development and human welfare' by Holloway and Pandit published in *Professional Geographer* 44, figure 7.1 from 'Japan's direct manufacturing investment in the US' by Chang published in *Professional Geographer* 41, figure 7.2 from 'US employment in foreign-owned high-technology firms' by Warf published in *Professional Geographer* 42, and figure 7.4 from 'The location and growth of business and professional services in American metropolitan areas, 1976–1986' by Ó hUalacháin and Reid published in *Annals, Association of American Geographers*, 81; Cambridge University Press for figure 3.1 from *The United States in the World-Economy: A Regional Geography* by Agnew, figure 9.1 from *Farm Labour* by Swindell, and figure 10.5 from 'The semi-conductor in South-East Asia: organization, location, and the international division of labour' by Scott published in *Regional Studies* 21(1987); Basil Blackwell for figure 8.11 from O'Loughlin in Johnston and Taylor (eds) *A World in Crisis?*; The Economist for figure 3.1 from *The Economist* 24 September 1983 and figure 12.2 from *The Economist* 5 December 1992; Harper & Row and Paul Chapman Publishing Ltd for figure 7.13 from *Global Shift* by Peter Dicken; Jessica Kingsley Publishers for tables 6.2, 7.4 and 7.1 from Albrechts and Swyngedouw in Albrechts *et al.* (eds) *Regional Policy at the Crossroads*; The Johns Hopkins Press for figure 1.4 from *Long Wave Rhythms in Economic Development and Political Change* by Berry; Longman Group UK for figure 5.3 from *The US: A Contemporary Human Geography* by Knox *et al.*, and figures 8.1 and 8.2 from *Political Geography: World-Economy, Nation-State and Locality* by Taylor; Macmillan Publishers Ltd for figure 12.1 from Martin and Rowthorn (eds) *The Geography of De-Industrialization*; Methuen & Co for figure 8.8 from 'The River Plate Countries' by Crossley published in Blakemoor and Smith (eds) *Latin America: Geographical Perspectives* 2e, and figures 3.5, 8.9 and 8.10 from *The Fragmented World: Competing Perspectives on Trade, Money and Crisis* by Edwards; MIT Press for figure 3.3 from *The Age of Diminished Expectations* by Krugman; Open University Press for figure 10.4 from *The Third World Atlas*; Oxford University Press for figure 5.1 from *Peaceful Conquest: The Industrialization of Europe, 1760–1970* by Pollard; Penguin Books Ltd for figure 8.4 from *Industry and Empire* by E J Hobsbawn (Penguin Books, 1969) © E J Hobsbawn, 1968, 1969; Pergamon Press for figure 6.1 from 'Industrial restructuring: an international problem' by Hamilton published in *Geoforum* 15 (1984); Pion Ltd for figures 7.9 and 7.12 from 'Local economic performance in Britain during the 1980s: the results of the Booming Towns Study' by Champion and Green published in *Environment and Planning A* 24 (1992); Stanford University Press for table 3.3 from *The Evolution of Human Societies* by Johnson and Earle; Unwin Hyman for figure 1.3 from *The World Economy in Transition* by Beenstock; Westview Press for figure 10.6 from *Urbanization and Urban Politics in Pacific Asia* by Fuchs and Pernia.

Every effort has been made to trace copyright holders of material reproduced in this book. Any rights not acknowledged here will be acknowledged in subsequent printings if notice is given to the publisher.

We are indebted to many of our colleagues for their advice at various stages in the conception and preparation of this book, and in its current revision. We would particularly like to recognize Stuart Corbridge, Raymundo Cota, Bob Dyck, Larry Grossman, Richard Grant, Naeem Inayatullah, D. Michael Kirchoff, Chase Langford, Soo-Seong Lee, Andrew Leyshon, Ragnhild Lund, Sallie Marston, Ezzeddine Moudoud, Pritti Ramamurthy, Bon Richardson, Susan Roberts, Freddy Robles, David J. Robinson, Mark Rupert, David Short and Colin Warren for their contributions.

PART 1

ECONOMIC PATTERNS AND THE SEARCH FOR EXPLANATION

In this first part of the book, we introduce the scope and complexity of our subject matter, establish the salient patterns in the world's economic landscapes, and review alternative theoretical approaches to understanding the development of these patterns. Chapter 1 provides an orientation for the whole book by outlining the relationships between economic organization and spatial change. In Chapter 2 the major dimensions of the world's contemporary landscapes are described. The objective here is to identify dominant and recurring patterns and to note the major exceptions to these patterns. Both the patterns and the exceptions raise a number of critical questions about process and theory in economic geography. For example: how should the development process be conceptualized? What are the processes that initiate and sustain spatial inequalities? Such questions are pursued in Chapter 3, where we outline a broad theoretical framework, that enables us to understand the interdependence of the entire world economy and its spatial components.

Picture credit: Paul Knox

CHAPTER 1

THE CHANGING WORLD ECONOMY

Picture credit: Städelsches Kunstinstitut, Frankfurt

As its title suggests, the perspective of this book is global. There is a very compelling reason for this. The rapidly increasing interdependence of the world economy means that the economic and social well-being of nations, regions and cities everywhere depends increasingly on complex interactions that are framed at the global scale. Although local, regional and national circumstances remain very important, what happens in any given country or locality is broadly determined by its role in systems of production, trade and consumption which have become global in scope. Most of the world's population now lives in countries that are either integrated into global markets for goods and finance, or rapidly becoming so. As recently as the late 1970s, only a few less-developed countries (LDCs) had opened their borders to flows of trade and investment capital. About a third of the world's labour force lived in countries like the Soviet Union and China with centrally planned economies, and at least another third lived in countries insulated from international markets by prohibitive trade barriers and currency controls. Today, three giant population blocs – China, the republics of the former Soviet Union, and India – with nearly half the world's labour force among them, are being drawn into the global market. Many other countries, from Brazil to Taiwan, have already become involved in deep linkages. According to World Bank estimates, fewer than 10 per cent of the world's labour force will remain isolated from the global economy by the year 2000. (The World Bank, properly called the International Bank for Reconstruction and Development, is a United Nations affiliate established in 1948 to finance productive projects that further the economic development of member nations.) Robert Reich, former US Secretary of Labor, was unequivocal in his estimation of the significance of economic globalization:

> We are living through a transformation that will rearrange the politics and economics of the coming century. There will be no national products or technologies, no national corporations, no national industries. There will no longer be national economies, at least as we have come to understand that concept. . . . As almost every factor of production – money, technology, factories, and equipment – moves effortlessly across borders, the very idea of an American economy is becoming meaningless, as are the

notions of an American corporation, American capital, American products, and American technology. A similar transformation is affecting every other nation, some faster and more profoundly than others; witness Europe, hurtling toward economic union.

(Reich, 1991, pp. 3, 8)

The World Bank, on page 1 of the 1995 edition of its *World Development Report*, noted that 'These are revolutionary times in the global economy.' The report shows how globalization now affects the lives of three very different people in very different places:

Joe lives in a small town in southern Texas. His old job as an accounts clerk in a textile firm, where he had worked for many years, was not very secure. He earned $50 a day, but promises of promotion never came through, and the firm eventually went out of business as cheap imports from Mexico forced textile prices down. Joe went back to college to study business administration and was recently hired by one of the new banks in the area. He enjoys a comfortable living even after making the monthly payments on his government-subsidized student loan.

Maria recently moved from her central Mexican village and now works in a US-owned firm in Mexico's maquiladora sector. Her husband, Juan, runs a small car upholstery business and sometimes crosses the border during the harvest season to work illegally on farms in California. Maria, Juan, and their son have improved their standard of living since moving out of subsistence agriculture, but Maria's wage has not increased in years: she still earns about $10 a day.

Xiao Zhi is an industrial worker in Shenzhen, a Special Economic Zone in China. After three difficult years on the road as part of China's floating population, fleeing the poverty of nearby Sichuan province, he has finally settled with a new firm from Hong Kong that produces garments for the U.S. market. He can now afford more than a bowl of rice for his daily meal. He makes $2 a day and is hopeful for the future.

These examples begin to reveal a complex and volatile interdependence that would have been unthinkable just 15 or 20 years ago. Joe lost his job because of competition from poor Mexicans like Maria, and now her wage is held down by cheaper exports from China. But Joe now has a better job, and the American economy has gained from expanding exports to Mexico. Maria's standard of living has improved and her son can hope for a better future. Joe's pension fund is earning higher returns through investments in growing enterprises around the world; and Xiao Zhi is looking forward to higher wages and the chance to buy consumer goods. But not everyone has benefited, and economic globalization has come under attack by some in industrial countries where rising unemployment and wage inequality are making people feel less secure about the future. In particular, production-line workers in affluent countries are fearful of losing their jobs because of cheap exports from lower-cost producers. Others worry about their employers relocating abroad in search of low wages and lax labour laws.

STUDYING ECONOMIC GEOGRAPHY

The task of the student of economic geography is to make sense of the world – the real world – and the ways in which its economic landscapes are changing. *But how* can we cope, intellectually, with what is happening to the likes of Joe, Maria, and Xiao Zhi? How can we cope with the local, regional and national implications of a succession of what are literally headline-making events? Acute unemployment in Detroit, Liverpool and Bochum; a chronic budget deficit in the United States; bitter trade disputes between developed and underdeveloped countries, between the United States and Europe; and between Japan and nearly every other trading nation; three-digit inflation in parts of Africa and Latin America; violent labour disputes in South Korea and Taiwan; famine in Somalia; and so on.

Furthermore, how should we approach the local, regional and national implications of less newsworthy but equally profound changes in the world economy, such as the remarkable developments that have taken place in international finance and banking? Most of all, how should we interpret the significance of specific changes that have been occurring in the world's economic landscapes: the **deindustrialization** of traditional manufacturing regions (e.g. northern England, the Ruhr), the economic revival of formerly 'lagging' regions (e.g. New England, Bavaria), the spread of branch factories in the countryside of industrialized nations (e.g. East Anglia, Jutland) and in the towns and cities of some newly industrializing nations (e.g. Taipei, Seoul), the emergence of high technology complexes (e.g. Silicon Valley in California, Research Triangle in North Carolina), and the consolidation of global financial and corporate control functions in a few cities (London, New York, Tokyo)?

In attempting to answer such questions, many of the methods, models and theories of 'traditional' economic geography seem to fall short. The traditional approach to economic regions has been undermined by events: the very constitution of 'urban' and 'regional' scales has been radically redefined through advances in telecommunications and the speed and volatility of economic restructuring. On the other hand, traditional approaches to systematic economic geography (locational analysis, etc.), have become less convincing. This is because they rest so heavily on the assumptions of neo-classical economics (and in particular the assumption of rational behaviour on the part of firms and individuals). In reality, the influence of monopolistic and oligopolistic elements and the powers of national and regional governments have become significant factors in shaping economic landscapes. Furthermore, the normative approaches of traditional neo-classical economic models tend to make for an unfortunate bias towards the general and away from the variability that characterizes the 'real world'.

The task, as Johnston (1984) pointed out, is to develop an understanding both of the general economic forces and socio-economic relationships within the world economy *and* of the unique features that represent local and historical variability. Following Johnston, we should clarify here the use of 'general' and 'unique'. By 'general' we mean something that is universally applicable within the domain to which it refers. By 'unique' we mean something which is distinctive, because there is no other instance of it, but whose distinctiveness *can be accounted for* by a particular combination of general processes and individual responses. For those

5

deindustrialization
This term is used rather loosely by many authors. At the heart of the concept is a *relative* decline in industrial *employment* in a nation or region where industry has traditionally been a significant component of the economy. It may be the result of climacteric changes or of secular shifts in an economy that are related to technological change and/or the globalization of the economy. In some instances, such trends may involve not just a relative decline but an *absolute* one; and may involve declining industrial *output* as well as employment.

phenomena that are distinctive but entirely remarkable because no general state-ments can be made in reference to them, we can use the term 'singular'. With this perspective, as Johnston observes, 'the world is our oyster':

> There is but one world-economy, to which all places are linked, to a greater or lesser degree. That economy and the ways in which it operates – almost independently now of the human societies that created it – provide the framework within which regional differences have evolved and are evolving. What is done, where and how, reflect human interpretations of how land should be used. These interpretations are shaped through cultural lenses (which may be locally created, or may be imported); they reflect reactions to both the local physical environment and the international economic situation; they are mediated by local institutional structures; they are influenced by the historical context; and they change that context, hence the environment for future interpretations. Unique they certainly are. Singular they are not.

(Johnston, 1984, p. 446)

From such a perspective we can begin to establish some of the central interrelation-ships surrounding economic organization and spatial change. Fig. 1.1 shows that economic organization, while critical to spatial change, is itself implicated with demographic, political, cultural, social and technological change, as both cause and effect. The diagram also shows that there are many interactions between, for example, political change and cultural change, and between locally contingent factors and spatial change. The point to emphasize at the moment is that all these direct, indirect and interaction effects are important to an understanding of spatial change. *They are all implicated, in other words, in accounting for both the general and the unique.* The task of the economic geographer is to unravel these relation-ships within a coherent and comprehensive framework. In order to do this, we must have a clear perspective on the central relationship between economic organization and spatial change. In the next section, we outline the most important aspects of this relationship, introducing several important concepts that we shall refer to throughout the book.

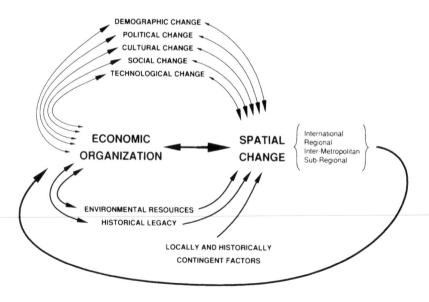

Figure 1.1
The inter-relationships surrounding economic organization and spatial change

ECONOMIC ORGANIZATION AND SPATIAL CHANGE

At the most fundamental level of abstraction, the idea of 'economic organization' approximates to the concept of **mode of production**: the way in which human societies organize their productive activities and thereby reproduce their socio-economic life. The theoretical and historical identification of different modes of production is a difficult and controversial matter, but there are five major modes of production, or forms of economic organization, that are commonly recognized:

- subsistence
- slavery
- feudalism (or rank redistribution)
- capitalism
- socialism.

These are all broad categories, however, and each can be broken down into more specific forms of economic organization. It is often useful, for example, to differentiate between merchant capitalism (or mercantilism), industrial capitalism (or competitive capitalism), and advanced capitalism; and it is sometimes useful, as we shall see in subsequent chapters, to subdivide each of these still further.

What most distinguishes one mode of production from another are differences in the relations between the **factors of production** (land, resources, labour, capital, enterprise). With the slave mode of production, for example, the labourer is bought and sold, along with other instruments of production, by the slave owner. Under feudalism, the peasant labourer may own some of the instruments of production, but the land and a certain amount of the product are the property of the feudal lord and the peasant is legally tied to a specific tract of land. Under capitalism, the labourer owns no instruments of production but is free to sell his or her labour power. It should be noted, however, that different modes of production are also characterized by different *forces of production* (technology, machinery, means of transportation) and by different *social formations* (made up of specific proportions of different social classes).

The significance of all this to spatial change is that the economic 'logic' of different modes of production brings about substantially different forms of spatial organization. Thus, for example, whereas feudalism brings a patchwork of self-sufficient domains with little trade and, therefore, few market centres, merchant capitalism requires a highly developed system of market towns and brings a built-in tendency for the colonization of new territories (in order to furnish new resources and bigger markets). Industrial capitalism requires spatial restructuring in order to exploit new energy sources, new production techniques and new forms of corporate organization. New mining and manufacturing towns appear, and whole regions become specialized in production. It is, of course, much more complicated than this, as we show in subsequent chapters. In addition, there are no definitive forms of spatial organization that can be associated with particular modes of production. Just as each mode of production is a dynamic, evolving set of relationships, so economic landscapes are restructured and reorganized to reflect, articulate and, sometimes, constrain the evolution of the economic dynamo.

mode of production
A fundamental form of economic organization with distinctive relationships between the main factors of production (land, labour and capital). The concept is derived from Marxian economics but now has much wider use.

7

factors of production
The fundamental components of any economic system: land, labour and capital. Land includes not only space or territory but also the associated soils and natural resources. Labour includes not only the size of the available work-force but also its skills, experience, and discipline. Capital includes not only money capital but also everything deliberately created for the purpose of production, such as factories and machinery (i.e. 'fixed' capital). A fourth factor, enterprise, is often recognized, though it may be missing from some forms of economic organization (e.g. subsistence); while it may legitimately be regarded as one aspect of labour.

8

It should also be stressed that there is no set sequence of transformation from one mode of production to another, although the 'classic' sequence, as experienced in much of western Europe, runs from subsistence economies through slavery, feudalism, merchant capitalism and industrial capitalism to advanced capitalism. Largely because capitalism first developed in Europe and because the 'logic' of capitalism requires ever-expanding markets, the sequence elsewhere has been different. In North America, capitalism was imposed directly on subsistence (i.e. native American) economies. In Japan, feudalism was displaced very suddenly by state-sponsored industrial capitalism. In Russia, an embryonic industrial capitalism was displaced by a socialism which soon gave way to state capitalism; and so on. As a result of these variations, important regional differences have come about within the world economy.

More spatial change, and further regional differentiation, occur with the evolution of a particular mode of production. At this more detailed level of resolution, we can see that regional differences develop as different social, cultural and political contexts affect the ways in which people react to the economic imperatives of particular modes of production. Thus, for example, different socio-cultural contexts make for different interpretations of environmental possibilities, desirable products and marketable opportunities. Thus a particular regional agricultural landscape must be seen as *just one of a number of possible realizations*, not as a straightforward reflection of a particular mode of production. Each economic landscape should be interpreted, therefore, as the product of the combination of broad economic forces interacting with local social, cultural, political and environmental factors: a product of both the 'general' and the 'unique'.

The evolution of capitalism

competitive capitalism
The early phase of industrial capitalism, lasting until *c.* 1890, characterized by comparatively high levels of free-market competition, with many small-scale producers, consumers and workers who acted almost completely independently, and relatively little government intervention.

imperialism
The extension of the power of a nation through direct or indirect control of the economic and political life of other territories.

organized capitalism
The phase of industrial capitalism that was characterized by comparatively highly structured relationships between labour, government and corporate enterprise. These relationships were mediated through legal and legislative instruments, formal agreements, and public institutions.

The evolution of the capitalist mode of development within the world's developed economies has been a particularly important influence on the development of the world's economic landscapes. There have already been two broad phases in the nature of capitalism, and we are now entering a third phase. The earliest phase, which took place in the United Kingdom and spread to much of the rest of northwestern Europe and North America, lasted from the late 1700s until the end of the nineteenth century. It was a phase of **competitive capitalism**: the heyday of free enterprise, *laissez-faire* economic development, with markets characterized by competition between small family businesses and with few constraints or controls imposed by governments or public authorities (Fig. 1.2). In the earlier years of this phase, the dynamism of the whole system rested on the profitability of agriculture and, increasingly, manufacture and 'machinofacture' (industrial production that was based less on handicraft and direct labour power than on mechanization, automation, and intensively used skilled labour). It was toward the end of this phase that the first climacteric (see p. 11) took place, the result of which was that the United States took over from the United Kingdom as the leading industrial nation.

The collective prosperity of the industrial nations was meanwhile consolidated by their **imperialism** which ensured both supplies of raw materials and markets for

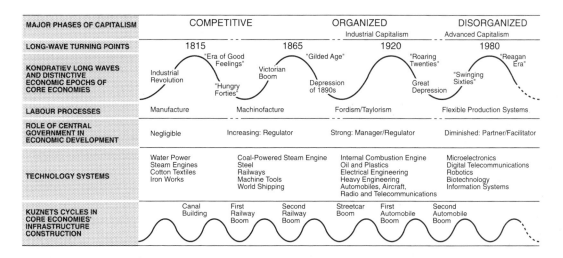

MAJOR PHASES OF CAPITALISM	COMPETITIVE	ORGANIZED	DISORGANIZED
		Industrial Capitalism	Advanced Capitalism
LONG-WAVE TURNING POINTS	1815 1865	1920	1980
KONDRATIEV LONG WAVES AND DISTINCTIVE ECONOMIC EPOCHS OF CORE ECONOMIES	"Era of Good Feelings" "Gilded Age" Industrial Victorian Revolution Boom "Hungry Depression Forties" of 1890s	"Roaring Twenties" Great Depression	"Reagan Era" "Swinging Sixties"
LABOUR PROCESSES	Manufacture Machinofacture	Fordism/Taylorism	Flexible Production Systems
ROLE OF CENTRAL GOVERNMENT IN ECONOMIC DEVELOPMENT	Negligible Increasing: Regulator	Strong: Manager/Regulator	Diminished: Partner/Facilitator
TECHNOLOGY SYSTEMS	Water Power Coal-Powered Steam Engine Steam Engines Steel Cotton Textiles Railways Iron Works Machine Tools World Shipping	Internal Combustion Engine Oil and Plastics Electrical Engineering Heavy Engineering Automobiles, Aircraft, Radio and Telecommunications	Microelectronics Digital Telecommunications Robotics Biotechnology Information Systems
KUZNETS CYCLES IN CORE ECONOMIES' INFRASTRUCTURE CONSTRUCTION	Canal First Second Building Railway Railway Boom Boom	Streetcar First Boom Automobile Boom	Second Automobile Boom

Figure 1.2
Major features of economic change in the world's core economies, 1790–1990

9

their manufactures. Gradually, the most successful family businesses within the industrial economies grew bigger and began to take over their competitors. Business became more organized as companies set out to serve regional or national consumer markets rather than local ones. Labour markets became more organized as wage norms spread, and governments began to be more organized as the need for *regulation* in public affairs became increasingly apparent.

By the turn of the century, these trends had reached the point where the nature of capitalist enterprise had changed significantly. It could now be characterized as **organized capitalism** – a label that came to be increasingly appropriate with the evolution of the economy over the next 75 years or so. In the early decades of the twentieth century, the dynamism of the system (that is, the basis of profitability) shifted away from industrial manufacture and machinofacture as a new labour process took hold. This process was **Fordism**, named after Henry Ford, the automobile manufacturer who was a pioneer of the principles involved: mass production, based on assembly-line techniques and 'scientific' management (known as **Taylorism**) together with mass consumption, based on higher wages and sophisticated advertising techniques.

The success of Fordism was associated with the development of a workable relationship between business interests and the labour unions, whose new strength was in itself another important element of 'organization'. The role of government, meanwhile, also expanded – partly to regulate the unwanted side effects of free-enterprise capitalism, and partly to mediate the relationship between organized business and organized labour. After the Great Depression of 1929–34, the role of government expanded dramatically to include responsibility for full employment, the management of the national economy, and the organization of various dimensions of social well-being.

Fordism
A regime of accumulation that centres on the mutual reinforcement of mass production and mass consumption. Named after Henry Ford because of his innovations and philosophy concerning automobile manufacture, it features a highly specialized and differentiated division of labour with assembly-line production geared to the availability of standardized, affordable goods for mass markets.

Taylorism
The name given (after analyst F. W. Taylor) to forms of organization in manufacturing industries wherein the planning and control of work are given over entirely to management, leaving production workers to be allocated specialized tasks that are subject to careful analysis – 'scientific management' using techniques such as time-and-motion studies.

10

flexible production systems
Various practices whereby manufacturing operations achieve flexibility in what they produce, when they produce it, how they produce it, and where they produce it. These practices include the exploitation of various kinds of enabling technologies, greater use of sub-contracting, the exploitation of different labour markets, the exploitation of different market niches for products, and the development of new labour processes using flexible working hours, part-time workers, etc.

disorganized capitalism
The label given to the most recent phase of the political economy of capitalism. Under disorganized capitalism, the relationships between capital, labour and government are more flexible, largely because a great deal of corporate activity has escaped the framework of nation–states and their institutions that still constrain organized labour and most governmental functions.

informational economy
A new mode of economic production and management in which productivity and competitiveness rely heavily on the generation of new knowledge and on the access to, and processing of, appropriate information.

After the Second World War, another important transformation in the nature of capitalism became evident. There was a shift away from industrial production and toward services, particularly sophisticated business and financial services, as the basis for profitability within the more developed economies. It is denoted on Fig. 1.2 as an evolution from 'industrial capitalism' to 'advanced capitalism'. With this shift, the decline in manufacturing jobs (but *not* in manufacturing production) combined with a second climacteric shift and an increasing globalization of the economy (in which huge *trans*national corporations were able to outmanoeuvre the national scope of both governments and labour unions), contributed to a destabilization of the relationship between business, labour, and government.

Meanwhile, Fordism had begun to be a victim of its own success, with mass markets for many products becoming saturated. As it became increasingly difficult to extract profits from mass production and mass consumption, many enterprises sought profitability through serving specialized market niches. Such specialization required variability and, above all, **flexible production systems**. The overall result has been labelled **disorganized capitalism** not so much because of the lack of organization or purpose in business, government, or labour but because of the contrast with the orderly interdependence of all three during the previous phase of organized capitalism. The driving force behind economic growth in this new phase of capitalism is a global **informational economy**, a new mode of economic production and management in which productivity and competitiveness rely heavily on the generation of new knowledge and on the access to, and processing of, appropriate information. The most important economic sectors in this informational economy are:

- high-technology manufacturing
- design-intensive consumer goods, ranging from high-fashion footwear to entertainment products, selling in market niches around the world
- business and financial services (Scott, 1996).

Long-wave economic fluctuations and spatial change

In general terms, this book is devoted to exploring in detail the **geographical path dependence** of economic activities: how over time the patterns of economic activities shaped by the historical contingencies of one period meet with changed organizational and technological conditions to produce new patterns of economic activities. Changes in technology are crucial in this perspective. As old technologies are eclipsed by new ones, 'old' industries – and sometimes entire old industrial regions – have to be 'dismantled' (or, at least, neglected) in order to provide the capital to fund the creation of new centres of profitability and employment. This process is often referred to as **creative destruction**, something that is inherent to the dynamics of capitalism. Creative destruction provides us with a powerful image to understand entrepreneurs' need, from time to time, to withdraw investments from activities (and regions) that yield low rates of profit, in order to reinvest in new activities (and, often, in new places). One of the most fascinating aspects of this, for geographers and economists, is that the see-saw of investment and disinvestment seems to follow distinctive cyclical fluctuations.

Within the dynamics of the capitalist mode of production there have been several kinds of cyclical fluctuation, each with different amplitudes, that have had profound effects on economic geography. Four of these are particularly important:

- logistic cycles
- climacteric cycles
- Kondratiev cycles
- Kuznets cycles.

The longest of these are the **logistic cycles** identified by Cameron (1973), with an amplitude of 150–300 years. They are called logistics because, rather than conforming to wavelike curves of expansion and recession, they conform to the shape of a statistical logistic curve, with an expansion phase followed by a phase of stagnation. Two such cycles have been completed: 1100–1450 and 1450–1750. The industrial era, representing the third, is still in progress. In the first cycle, the expansion phase (between 1100 and 1300) was the period of expanding population and territorial colonization of the late Middle Ages; the second phase, between 1300 and 1450, saw the crisis of feudalism and the stagnation, throughout Europe, of economic development.

The expansion phase of the second cycle, from 1450 to 1600, saw a further increase in population and economic growth with the transition to merchant capitalism. The second phase, as Wallerstein (1979) points out, was a period of relative stagnation only in an *overall* sense. It was 'a vector of several curves: some zones expanding, others staying level and still other declining' (p. 75). The reason for this was that the global economic landscape had become sharply differentiated, with different regions of the world experiencing asymmetrical trajectories as a reflection of the overall consolidation of the European world economy. The mechanisms of change associated with these two cyclical periods will be outlined in Chapter 3. The asymmetries underlying the expansion phase of the third cycle are the focus of Chapter 4; and the emergent asymmetries of the second phase of the third cycle – which seems to have begun between 1960 and 1970 – are the subject of Chapters 5 and 6.

International interdependence within the third logistic has been associated with a second kind of long-term cycle: the **climacteric** eclipse of industrial economies by emergent rivals from other world regions. The first large-scale climacteric involved the relative decline of Britain and the ascent of the United States, Germany, France and Russia between 1870 and 1900. Beenstock (1983, pp. 162–3) summarizes the British climacteric as follows (see also Fig. 1.3):

1. Industrial growth in less-developed competitors adversely affects the relative price of manufactures and/or threatens British trading power.
2. The new set of relative prices and/or the trade threat to Britain on the part of competitors causes deindustrialization in Britain as the share of industrial production in GDP declines from what it otherwise would have been.
3. Meanwhile, reduced profits from manufacturing adversely affect the return on capital in Britain.
4. The lower returns to capital cause a reduction of investment in Britain and increased foreign investment by British investors in the competing countries.

geographical path dependence
The historical relationship between the present economic activities associated with a place and its past experience.

11

creative destruction
The withdrawal of investments from activities (and regions) that yield low rates of profit, in order to reinvest in new activities (and new places).

logistic cycles
Long-term (150–300 years) cycles of growth and stagnation in overall levels of economic development.

climacteric cycles
Long-term changes in relative levels of national economic development based on the exploitation of cheaper factor costs (particularly land and labour) in peripheral or semi-peripheral countries and the concurrent failure of formerly dominant core economies to successfully adjust to changing market, technologies, and international conditions.

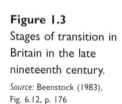

Figure 1.3

Stages of transition in Britain in the late nineteenth century.

Source: Beenstock (1983), Fig. 6.12, p. 176

newly industrializing countries (NICs)
Countries, formerly peripheral within the world-system, that have acquired a significant industrial sector, usually through direct foreign investment.

Kondratiev cycles
Long waves of 50–55 years in duration that have characterized the rate of change in price inflation within the capitalist world economy for the past 250 years. Their origins and significance remain controversial, but in recent years they have been widely recognized to be closely tied in to distinctive phases of political-economic development.

5. While deindustrialization takes place there is a temporary if protracted period of 'mismatch' unemployment and slower economic growth in Britain.

It has been suggested that while the climacteric was triggered by foreign competition, this alone was not a sufficient condition for decline, since Britain could have moved up-market by taking advantage of the new science-based industries that emerged during the latter part of the nineteenth century. According to Lewis (1978), Britain's failure to do so reflected (1) rigidities imposed on the labour market by growing trade unionization and (2) snobbery on the part of the bourgeoisie who did not understand the need for technological and vocational education upon which the new industries would be based. Beenstock (1983) suggests that a second major climacteric cycle began around 1970, with an expanded core of advanced industrial economies (represented by the OECD countries) experiencing deindustrialization, while parts of what were formerly peripheral regions – South Korea, Taiwan, Mexico, Brazil, Hong Kong and Singapore – begin to ascend rapidly within the world economy, creating a new category of **newly industrializing countries** (NICs). These events are explored in more detail in Chapters 6 and 7.

Interwoven with the long-wave fluctuations described above has been a series of cyclical movements in the overall rate of change of prices in the economy. These fluctuations are known as **Kondratiev cycles** after the Russian economist who first identified them. There have been four complete Kondratiev cycles and we are now in the fifth. Each one has been marked by a progressive acceleration in the rate of price increases for about twenty years, followed by a rapid inflationary spiral. After the peak, prices collapse, eventually reaching a trough some 50–55 years after the start of the cycle. Although the mechanics of these cycles remain in dispute (see Chapter 3), it is clear that each of the cycles has been associated with the development and exploitation of a distinctive **technology system,** a cluster of energy sources, transportation technologies and key industries (Fig. 1.2). The inflationary phase of Kondratiev cycles is tied into the initial exploitation of these technology systems, when the new technologies are relatively expensive; the deflationary phase is associated with the maturation of the technology systems, when the technologies have become routine and widely available. The five technology systems can be summarized as follows:

- First was the early mechanization based on water power and steam engines, the development of cotton textiles, pottery and iron working, and the development of river systems, canals, and turnpike roads for the assembly of raw materials and the distribution of finished products (the 'water' style of technology system).
- Second was the development of coal-powered steam engines, steel products, railroads, world shipping, and machine tools (the 'steam transport' style of technology system).
- Third was the development of the internal combustion engine, oil and plastics, electrical and heavy engineering, automobiles, aircraft, radio, and telecommunications (the 'steel and electricity' style of technology system).
- Fourth was the exploitation of nuclear power, the development of limited-access highways, durable goods consumer industries, aerospace industries, electronics, and petrochemicals (the 'Fordist' style of technology system).
- The most recent (and still incomplete) technology system is based on microelectronics, digital telecommunications, biotechnology, robotics, fine chemicals, and information systems (the 'microelectronics and biotechnology' style of technology system).

Superimposed on the Kondratiev fluctuations have been a variety of *business cycles* of different amplitudes. These can be regarded as the result of the fundamental dynamics of capitalist markets, in which expansion (production and supply increasing to meet increased consumer demand) is followed inevitably by overshoot (when overly optimistic producers and suppliers overestimate the rate of increase in demand) and then collapse (as overcapacity and excess inventory lead to falling profits, reduced asset values, pessimism, and reduced levels of investment). The most significant of these cycles is the **Kuznets cycle.** This is a cycle of regular changes in the rate of economic growth, as measured by indicators such as per capita Gross National Product (GNP). As indicated on Fig. 1.2, these are characterized by an expansion phase of 11–15 years, followed by a collapse phase of similar duration. They have affected many aspects of economic development, including the rhythm of investment in transport infrastructure, in city building, and in migration.

technology systems
Distinctive 'packages' of technologies, energy sources, and political-economic structures that represent the most efficient means for the organization of production at any given phase of economic development. Based on key sets of interdependent technologies, they represent the underpinnings of successive regimes of accumulation and modes of regulation.

13

Kuznets cycles
Long waves of approximately 25 years in duration that have characterized the pattern of acceleration and deceleration in economic growth. Named after Ukrainian-born economist Simon Kuznets, who established their existence in the 1920s, they are cycles of activity in investment and building.

Overaccumulation

stagflation
Episodes of economic recession accompanied by comparatively high rates of price inflation.

14

overaccumulation
A distinctive phase in the long-term dynamics of capitalist economies, characterized by unused or underutilized capital and labour. It is an inevitable outcome of the difficulty of matching supply to demand under changing conditions, and it represents a critical moment for the political economy of capitalism. It can be recognized by the appearance of idle productive capacity, excess inventories, gluts of commodities, surplus money capital, and high levels of unemployment.

The relationship between the rhythms of Kondratiev cycles and Kuznets cycles has been highlighted by B. J. L. Berry (1991) as being of particular significance for patterns of economic development. Fig. 1.4 summarizes this relationship, showing how two Kuznets cycles are embedded within each Kondratiev cycle. As the economy moves out of recession and price collapse, there is a 'post depression' Kuznets cycle that begins with what Berry calls Type A growth – growth that coincides with a price recovery. Prices continue to accelerate, however, even after the overshoot and collapse of Type A economic growth. This produces a combination of economic slowdown or stagnation combined with price inflation: **stagflation**. Stagflation represents a crisis point in economic development that is usually marked by banking crises. There has always followed a second, Type B, Kuznets cycle, during which a cycle of economic expansion, overshoot, and collapse takes place under conditions of price deflation.

As we look back at economic history, Berry points out, we can identify the rhythm of these synchronized long waves in distinctive epochs of economic, political, social, and urban development. The 'Roaring Twenties' (1920s), 'Swinging Sixties' (1960s), and the 'Reagan Era' (1980s), for example, were all episodes of Type B economic growth during price deflation, marked by significant bursts of growth and development. Similarly, the 'Hungry Forties' (1840s), the 'Great Depression' (1929–34) and the recession of the early

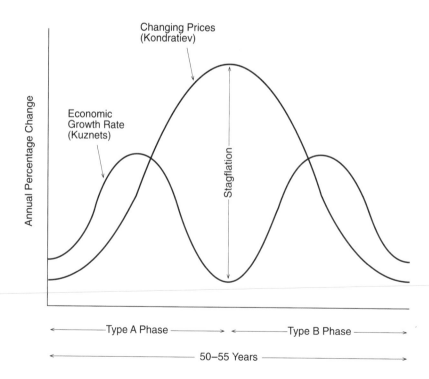

Figure 1.4
The relationship between Kuznets cycles and Kondratiev cycles.

Source: Berry (1991), Fig. 5.1, p. 100

1990s can all be seen as episodes of Type B economic decline during price deflation, marked by economic stagnation, acute social distress, and pressure for economic and social reform.

Such episodes are known by some theorists as **overaccumulation** crises, because three surpluses exist simultaneously: surplus labour (the unemployed and underemployed), surplus productive capacity (idle factories and machinery), and surplus capital (because owners of profits, interest and dividends cannot find enough reasonably safe, profitable investment opportunities in productive enterprises). These epochs make convenient markers that help in understanding the relationship between economic development and spatial change, the point being that *the ups and downs of each cycle have been imprinted differentially on the landscapes of different regions*. We must bear in mind, however, that economic history cannot always be framed into neat periodizations – there are many leading or lagging aspects of economic change that are not captured by fixed chronologies.

SPATIAL DIVISIONS OF LABOUR

The unfolding evolution of capitalism and its accompanying long waves have brought a changing **spatial division of labour**. This is the idea that the division of labour within and between firms and over space is not fixed but responds to changes in the historical–structural context in which firms must operate. In the Fordist period in national economies such as Britain and the United States, the basic division of labour was organized within the national economy or, even more typically, within regional parts of the national economy. In terms of production, plant, firm and industry were national phenomena, organized around national markets and national industries and creating national social (class) divisions. Although capital, labour and technology were often imported and exported, they were subject to intensive regulation by national governments.

The internal geography of a national economy such as Britain's reflected its position in the international division of labour. For example, in the 1930s a small group of industries – coal-mining and exporting, iron and steel manufacture, shipbuilding – owed their significance to the role of Britain in an international division of labour in which Britain specialized in certain key manufacturing industries. British trading patterns were shaped by the economic implications of previous investment and the increasing returns to scale and external economies that this produced. Elsewhere, different industries, often newer, mass production ones based on larger firms, took root. Trade was a result of cumulative competitive advantage in sectors where each had come to have a 'headstart' (e.g. Elbaum, 1990).

The locational consequences in the British case are laid out by Massey (1984, pp.128–9) as follows:

> It was the United Kingdom's position as an imperial power, its early lead in the growth of modern industry, and its consequent commitment to free trade and its own specialization in manufacturing *within* this international division of labour, which

spatial division of labour
Regional economic specialization, based on the distribution of resources and markets and on the exploitation of economies of scale, agglomeration economies, and localization economies.

15

enabled the rapid growth, up to the First World War, of these major exporting industries. The spatial structures that were established by those industries were those where all the stages of production of the commodity are concentrated within single geographical areas. The comparatively low level of separation of functions within the process of production, and the relatively small variation in locational requirements between such potentially separable functions, were not sufficient to make geographical differentiation a major attraction.

In other words, the spatial division of labour of key industries within 'national economies' was based on different regional industrial specializations. National economies were as a consequence regionally differentiated. Physical accessibility of functions to one another (directional, component production, assembly) and **agglomeration** were major features of economic organization across a large number of manufacturing industries. For example, in the United States in this period the northeast contained a vast array of specialized manufacturing clusters – steel in Pittsburgh and vicinity, automobiles in Detroit, chemicals in Wilmington, photographic equipment in Rochester, NY – and regions of agricultural and raw materials specialization elsewhere. Places and regions could be readily associated with specific products.

agglomeration
The clustering together of functionally related activities.

Globalization and changing spatial divisions of labour

Under the new conditions of post-Fordist, disorganized capitalism, however, such regional specialization has been challenged and, to a considerable degree, undermined. Spatial divisions of labour are now structured in a variety of ways depending on the needs and characteristics of particular industries. In addition to (1) *regional specialization* and (2) *regional dispersal* that has long characterized **consumer services** (stores, restaurants, hospitals, etc.) and some manufacturing industries (e.g. shoe production, food processing), four other spatial divisions of labour can be identified. These are:

consumer services
Personal services, including retailing, medical care, personal grooming, leisure and recreation, culture and entertainment.

3. *functional separation* between management/research activities in major metropolitan regions, skilled labour used in 'old' manufacturing areas, and unskilled labour used in a regional 'peripheries' to take advantage of low wages/disorganized labour force;
4. *functional separation* between management/research in major metropolitan regions and semi-skilled and unskilled labour used in a regional peripheries (the pattern in the British electrical engineering industry);
5. *functional separation* between management/research and skilled labour in more advanced industrial regions and unskilled labour in the 'global periphery'; and
6 *division* between areas with investment, technical change and job expansion, and other areas with stagnant and progressively less competitive production and job loss (Urry, 1985).

These new spatial divisions of labour have been possible because of a set of transportation and communications technologies that have provided a 'permissive' environment in which firms could decentralize manufacturing and primary production activities yet maintain central control, e.g. computers, telecommunications, air travel (see p. 242). There is now the possibility of intensive interaction and diffusion without geographical proximity. Firms can remain headquartered in, or relocate to, New York, Zürich, or Hamburg, but locate manufacturing facilities in places with isolated and disorganized labour forces, with particular combinations of labour force skills, costs, militance, and captivity, or close to highly concentrated regional markets. The main push for *restructuring* corporate operations has come from the increasingly competitive environment faced by large firms resulting from a less regulated and more internationalized market-place.

Under this **New International Division of Labour** (NIDL), investment, and production are no longer organized primarily around national economies. The actual process of production, most obviously in the automobile and electronics industries, is now global. Components are 'sourced' or obtained from multiple suppliers in different countries and assembled in several. Increasing numbers of products have no obvious nationality; it is difficult to distinguish some 'American' from some 'Japanese' cars, for example, now that American car companies import vehicles under 'their' names from Japan and Japanese companies now manufacture cars in the United States (e.g. Honda in Marysville, Ohio). For many multinational firms, national markets for capital, labour and plant and office location exist only as parts of global ones. Recent German and American evidence suggests that even small firms have now acquired the propensity and capability to operate globally. So the 'new' conditions cannot be solely identified with giant multinational or global corporations (Fröbel *et al.*, 1977).

The contemporary world economy is constituted through the myriad **commodity chains** that criss-cross global space as a result of this global reach of transnational corporations. Commodity chains are networks of labour and production processes whose origin is in the extraction or production of raw materials and whose end result is the delivery and consumption of a finished commodity. These networks often span countries and continents, linking the production and supply of raw materials, the processing of raw materials, the production of components, the assembly of finished products, and the distribution of finished products into vast global assembly lines. As we shall see in subsequent chapters, these global assembly lines are increasingly important in shaping economic landscapes everywhere.

The advantages to manufacturers of a global assembly line are several. First, a standardized global product for a global market allows them to maximize economies of scale. Second, a global assembly line allows production and assembly to take greater advantage of the full range of geographical variations in costs. Basic wages in manufacturing industries, for example, are between 25 and 75 times higher in advanced industrial countries than in some LDCs. With a global assembly line, labour-intensive work can be done where labour is cheap, raw materials can be processed near their source of supply, final assembly can be done close to major markets, and so on. Third, a global assembly line means that a company is no longer dependent on a single source of supply for a specific component, thus reducing its vulnerability to industrial troubles and other disturbances.

The pace of this economic globalization has accelerated since the late 1960s. Between 1961 and 1976, for example, the number of employees of German firms outside of Germany increased tenfold. The number of firms with foreign operations doubled during the same period of time. German firms have generally been less

New International Division of Labour
The idea of the reorganization of spatial divisions of labour, formerly organized principally the national scale, to a global scale based on international production and marketing systems.

commodity chain
A network of labour and production processes beginning with the extraction or production of raw materials and ending with the delivery of a finished commodity.

willing to expand foreign operations compared to American and British firms, so these figures indicate something of a lower bound among countries with long histories of industrialization.

Paralleling and stimulating this trend has been the emergence of international devices for coordinating and steering capital beyond national control (e.g. the Eurodollar market of dollars in circulation outside the USA; see p. 50) and the 'offshore' banking and financial centres which, rather like some old city states did in their day, now service the new international division of labour (e.g., Hong Kong, Singapore). The circulation of capital through international circuits and in the form of investment in commercial real estate in the major financial centres (especially Tokyo, London and New York) became important features of the new world economy in their own right in the late 1970s and 1980s. Indeed, some small states such as Switzerland have successfully cashed in on the new world economy to the extent that they now have median income levels higher than those of the 'old' national manufacturing economies such as Britain and the United States.

The national economy, therefore, is no longer the sole building block of the world economy. For an increasing proportion of agricultural and manufactured commodities and for some services, production and markets have become world-wide. This shift has had important consequences for the spatial distribution of economic activities both globally and within countries. Globally it has given rise to the growth of the Newly Industrializing Countries (NICs) such as South Korea, Taiwan, Hong Kong, Singapore, Brazil and Mexico. It has also contributed to a significant polarization of income and wealth. According to the United Nations Development Program, the differential between the wealthiest 20 per cent of the world's countries and the poorest 20 per cent increased from a factor of 30 in 1960 to a factor of 60 in 1990 (UNDP, 1993). Within the national economies of the 'core', the NIDL has led to both a reorientation in employment away from manufacturing to services and massive restructuring of regional economies. In Britain, for example, three sorts of local area have fallen victim to the loss of 'traditional' manufacturing industries and the failure of new ones to replace them.

1. The centres of nineteenth-century industrialization in the north of England, South Wales and Central Scotland.
2. The 'inner cities' of London and other large metropolitan areas with concentrations of poor people and few of the unskilled jobs that they used to fill.
3. The centres of the growth industries of the 1950s and 1960s (vehicles and engineering) in the West Midlands and Northwest of England.

We will draw on this broad framework throughout the remainder of this book as we analyse and describe the geography of the world economy. We begin, in Chapter 2, by establishing the major dimensions of the contemporary economic landscapes within the world economy. In Chapter 3 we establish a comprehensive global historical framework which serves as the context for the rest of the book. In Part 2, we trace the emergence of the world's core economies – Europe, North America and Japan – and follow their different paths towards increasing scale and complexity. Part 3 deals with the world outside the core of the world-economy, paying special attention to the spatial transformations that have occurred as a consequence of the colonialism and global capitalism emanating from the core economies, and to the role of agriculture and manufacturing industry in economic development and spatial change. Finally, in Part 4, we examine some of the reactions to the emergence of ever-larger and more powerful economic forces that have come to characterize the world economy, describing the spatial consequences of transnational political and economic integration and of decentralist reactions: nationalism, regionalism and grassroots movements towards economic democracy.

Key Sources and Suggested Reading

Beenstock, M. 1983. *The World Economy in Transition*. London: Allen & Unwin.

Berry, B. J. L. 1991. *Long-wave Rhythms in Economic Development and Political Change*. Baltimore: Johns Hopkins University Press.

Cameron, R. 1973. The logistics of European economic growth: A note on historical periodization. *Journal of European Economic History*, **2**, 145–58.

Elbaum, B. 1990. Cumulative or comparative advantage? British competitiveness in the early 20th century. *World Development*, **18**, 1255–72.

Fröbel, F. *et al.* 1977. *Die Neue Internationale Arbeitsteilung*. Reinbeck: Rowhohlt.

Johnston, R. J. 1984. The world is our oyster. *Transactions, Institute of British Geographers*, **9**, 443–59.

Johnston, R. J., Taylor, P. J. and Watts, M. J. (eds) 1995. *Geographies of Global Change*. Oxford: Blackwell.

Lewis, W. A. 1978. *Growth and Fluctuations 1870–1913*. London: Allen & Unwin.

Massey, D. 1984. *Spatial Divisions of Labour*. London: Methuen.

Reich, R. 1991. *The Work of Nations. Preparing Ourselves for 21st Century Capitalism*. New York: Vintage Books.

Scott, A. J. 1996. Regional motors of the global economy, *Futures*, **28**, 391–411.

Storper, M. and Scott, A. J. (eds) 1992. *Pathways to Industrialization and Regional Development*. London: Routledge.

Tylecote, A. 1992. *The Long Wave in the World Economy*. London: Routledge.

United Nations Development Programme 1993. *Human Development Report 1993*. Oxford: Oxford University Press.

Urry, J. 1985. Social relations, space and time. In D. Gregory and J. Urry (eds), *Social Relations and Spatial Structures*. London: Macmillan.

Wallerstein, I. 1979. Underdevelopment and Phase-B. In W. Goldfrank (ed.), *The World-System of Capitalism: Past and Present*. Beverly Hills: Sage, 73–84.

Wallerstein, I. 1991. *Geopolitics and Geoculture: Essays on the Changing World-System*. Cambridge: Cambridge University Press.

World Bank 1995. *Workers in an Integrating World. World Development Report 1995*. Oxford: Oxford University Press.

CHAPTER 2

Picture credit: World Bank

GLOBAL PATTERNS AND TRENDS

Geography is about local variability within a general context.

R. J. Johnston (1984, p. 444)

In this chapter, we describe the major dimensions of the contemporary economic landscape. Space does not permit anything like a full coverage of patterns of economic activity or of the quilt of economic development, let alone a systematic review, resource by resource, industry by industry, flow by flow of commodities, services and capital. Such a catalogue, in any case, is not our purpose. Rather, our objective is to identify dominant and recurring patterns and to note the major exceptions to these patterns. We are, in other words, concerned primarily with characterizing the *general context* referred to by Johnston in the quote above. To the extent that we identify exceptions and contradictions, we are also concerned to some degree with *local variability*. In subsequent chapters our objective will be to uncover the processes that have contributed to these patterns – both the general and the locally distinctive or unique. As we shall see, *it is the interaction of the unique with the general that produces distinctive economic regions*.

It has been widely recognized for several decades that the dominant components of economic geography at the global scale are cast in terms of core–periphery differences. Meier and Baldwin (1957) were perhaps the earliest writers to attempt a conceptual description of this core–periphery structure on a global scale. According to them, a country is at the centre of the world economy

> if it plays a dominant, active role in world trade. Usually such a country is a rich, market-type economy of the primarily industrial or agricultural–industrial variety. Foreign trade revolves around it: it is a large exporter and importer, and the international movement of capital normally occurs from it to other countries.

In contrast, they argued, a country could be considered peripheral

> if it plays a secondary or passive role in world trade. In terms of their domestic characteristics, peripheral countries may be market-type economies or subsistence-type economies. The common feature of a peripheral economy is its external dependence on

the centre as the source of a large proportion of imports, as the destination for a large proportion of exports, and as a lender of capital.

(Meier and Baldwin, 1957, p. 147)

This cleavage of the economic world into two interdependent but highly unequal camps was echoed by the Brandt Report (Independent Commission on International Development Issues, 1980, 1983), where core–periphery contrasts were cast in terms of 'North' and 'South'. The North includes the industrialized countries of North America, Europe (East and West), the former Soviet Union, Japan, Australia and New Zealand; the South includes China and all the countries of Latin America, Africa, the Middle East, South Asia and Southeast Asia. By this definition the South has over 4.35 billion people – three-quarters of the world's population – living on one-fifth of the world's income. But, as the report noted,

> It is not just that the North is so much richer than the South. Over 90 per cent of the world's manufacturing industry is in the North. Most patents and new technology are the property of multinational corporations of the North, which conduct a large share of world investment and world trade in raw materials and manufactures. Because of this economic power Northern countries dominate the international economic system – its rules and regulations, and its international institutions of trade, money and finance (p. 32).

Figure 2.1 shows the North–South divide of the 'Brandt line' superimposed on a map of Gross Domestic Product (GDP) per capita: one of the most widely used international economic indicators. Gross Domestic Product is an estimate of the total value of all materials, foodstuffs, goods and services that are produced by a country in a particular year. To standardize for countries' varying sizes, the statistic is normally divided by total population, which gives an indicator, per capita GDP, that provides a good yardstick of relative levels of economic development. (**Gross National Product**, a similar measure, includes the value of income from abroad – flows of profits or losses from overseas investments, for example.) In making international comparisons, GDP and GNP can be problematic, because they are based on each nation's currency. Recently, it has become possible to compare national currencies based on purchasing power parity (PPP). In effect, PPP measures how much of a common 'market basket' of goods and services each currency can purchase locally, including goods and services that are not traded internationally. Using PPP-based currency values to compare levels of economic prosperity usually produces lower GDP figures in wealthy countries and higher GDP figures in poorer nations, compared with market-based exchange rates. Nevertheless, even with this compression between rich and poor, economic prosperity is very unevenly distributed across nations.

As Fig. 2.1 shows, the burden of poverty lies heavily across the peripheral regions of Asia, Africa and Latin America. The extreme cases include countries such as Chad, Ethiopia, Guinea, Malawi, Sudan, Nepal and Vietnam, where in 1996 per capita GDP (measured in the 'international dollars' of PPP) was less than US$1000. In all, over 1.3 billion people have to subsist on less than $US1 a day. Yet not all of the South exhibits low levels of per capita income. Oil-rich countries like Saudi Arabia, Libya and Venezuela, together with one or two small export-processing

Gross National Product
A measure of the market value of the production of a given economy in a given period (usually a year). It is based on the market price of finished products and includes the value of subsidies; it does not take into account the costs of replacing fixed capital. When GNP is adjusted to remove the value of profits from overseas investments and the 'leakage' of profits accruing to foreign investors, the result is a measure of Gross Domestic Product (GDP).

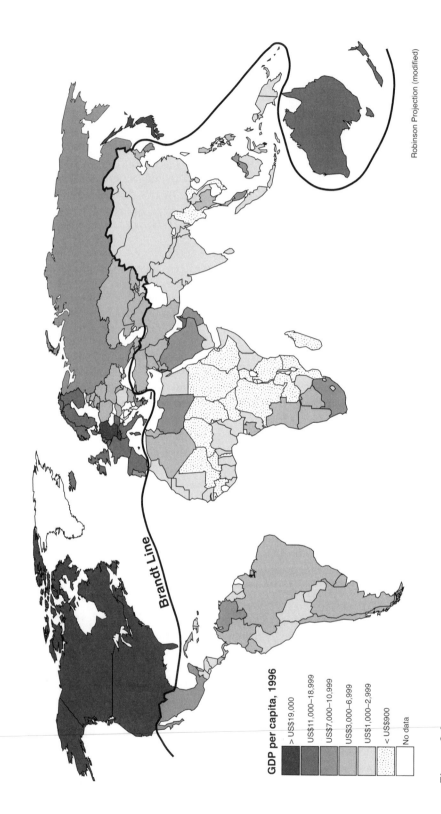

Robinson Projection (modified)

GDP per capita, 1996

> US$19,000

US$11,000–18,999

US$7,000–10,999

US$3,000–6,999

US$1,000–2,999

< US$900

No data

Brandt Line

Figure 2.1

International variations in Gross Domestic Product (GDP) per capita, 1996

countries like Hong Kong and Singapore, are comparable, in terms of GDP per capita, with some of the less affluent nations of the North. And countries like Chile, Brazil, Malaysia, Uruguay and South Africa, with favourable resource endowments, vigorous export industries and a degree of economic diversification, generate a GDP per capita that is significantly higher than that of their regional neighbours. Meanwhile, it should be acknowledged that there is also considerable variability within the North: GDP per capita is less than US$6000, for example, in Poland and Hungary; around US$18 500 in France and Australia; and over US$23 000 in the United States.

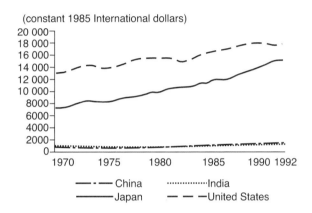

Figure 2.2

Per capita Gross Domestic Product in constant 1985 international dollars.

Source: World Resources Institute (1996), Fig 7.2, p. 164.

We should also note here that the gap between the world's rich and poor is also getting wider rather than narrower. In 1970, the average GDP per capita of the 10 poorest countries in the world was just one-fiftieth of the average GDP per capita of the 10 most prosperous countries. By 1990, the relative gap had doubled: the average of the top 10 was 100 times greater than the average of the bottom 10. Consider the differences between China and India, the largest LDCs, and Japan and the United States, the largest developed economies (Fig. 2.2). Between 1970 and 1992, the absolute difference in per capita GDP between China and Japan (measured in constant 1985 'international dollars') more than doubled, while the gap between China and the United States increased by more than 30 per cent. India, meanwhile, fell behind Japan and the United States in almost exactly the same way.

WHAT 'ECONOMIC DEVELOPMENT' MEANS

Major international economic cleavages not only reflect differences in prosperity but also reflect different forms of economic organization, different kinds of resource bases, different demographic characteristics, different political systems, and different roles in the system of international specialization and trade. Defining and measuring 'economic development' are therefore problematic. As we shall see, there are strong grounds for thinking in terms of *under*-development rather than development as far as the LDCs are concerned, since the term 'development' implies a trajectory of improvement, in both relative and absolute terms.

In addition, it is now widely accepted that 'development' must be conceived in broad terms of social well-being. Narrowly economic definitions, while admirably precise, provide only part of the picture. They encompass changes in the amount, composition, rate of growth, distribution and consumption of resources but they do not extend to the effects these changes have on people's lives. Fig. 2.3 shows a scatterplot of per capita GNP against a simple measure of human welfare ('PQLI' – the physical quality of life index) that is based on national rates of infant mortality, literacy, and life expectancy at age one (Holloway and Pandit, 1992). As we might expect, the overall relationship between GNP and welfare is not linear. Once per capita GNP reaches about US$4000, increments to basic human welfare tend to be marginal. But there are many exceptions to this overall relationship. Positive deviations, reflecting nations whose levels of human welfare are *higher* than might be expected from their level of economic development, are recorded by the long-affluent nations of North America, Europe and Australasia, though the largest positive deviations are recorded by socialist countries: China, Cuba, and Vietnam. The largest negative deviations, on the other hand, are recorded by two very different groups of countries: oil-rich Middle Eastern countries (including Iraq, Iran, and Saudi Arabia) and many African countries (including Algeria, Angola, Libya, Namibia, and Niger).

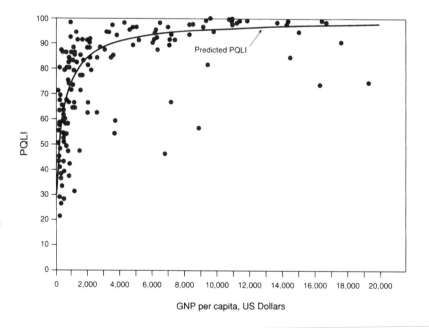

Figure 2.3

The relationship between economic development (measured here by GNP per capita) and the quality of life (measured here by an index, PQLI, based on infant mortality rates, life expectancy, and literacy).

Source: Holloway and Pandit (1992), Fig. I, p. 60

Development, then, should really be thought of not only in terms of income and consumption but also in terms of people's health, education, housing conditions, security, civil rights, and so on. Seen in this light, development is clearly a 'normative' concept, i.e. it involves values, goals and standards that make it possible to compare a particular situation against a preferred one. Development can properly be evaluated only in the context of the human needs and values as perceived by the very societies undergoing change. It also follows that although 'development' implies economic, social, political and cultural transformations, these should be seen not as ends in themselves but as means for enhancing social well-being and the quality of human life. In this book we shall be concerned with both the means and the ends, addressing our subject matter from a broad perspective that sees economic geography as the dynamic core of human geography. In the present context, this brings us first to an examination of some of the international patterns that reflect the 'means' of transformation: global patterns of resources, population, manufacturing, trade, investment, aid and debt. We shall then summarize the 'ends', or net outcomes, in terms of an overall typology of socio-economic development.

An index of international development

Recognizing the limitations of measures of national income (i.e. GDP and GNP) the United Nations Development Programme (UNDP) has established a Human Development Index (UNDP, 1994). This index combines countries' scores on three basic components of development:

- physical well-being, as measured by life expectancy;
- education, as measured by a combination of adult literacy rates (two-thirds weight) and mean years of schooling (one-third weight); and
- standard of living, as measured by GDP per capita, adjusted to PPP.

These components are rated against normative minimum and maximum values, and scaled into an index with a theoretical range from zero to 1.0. A country unable to meet the minimum values on all three components would score zero on the index; a country that met the maximum achievable values on all three would score 1.0. In 1994 the top score was 0.932, achieved by Canada; the bottom score, 0.191, was recorded for Guinea.

Figure 2.4 shows the global pattern of development according to this index. Many countries in Latin America and East Asia have clearly moved beyond the basic threshold of development, with index scores that are comparable to those of the advanced industrial economies of northwest Europe and North America. In contrast, most countries in South Asia and Sub-Saharan Africa still have very low levels of development.

The UNDP has also established a gender-sensitive development index, based on employment levels, wage rates, adult literacy, years of schooling, and life expectancy. In no country are women better off than men, according to this gender-sensitive index. In many of the advanced industrial countries, the score for women is between 85 and 95 per cent of the score for men, with the affluent and liberal countries of Scandinavia and Australasia coming closest to gender equality. Canada, ranked first on the overall index of development, slips to eighth place in a gender-sensitive ranking because Canadian women

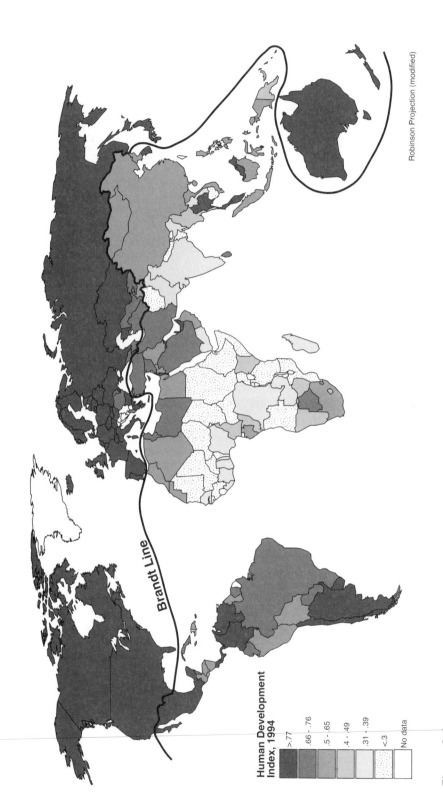

Robinson Projection (modified)

Human Development
Index, 1994

>.77
.66 - .76
.5 - .65
.4 - .49
.31 - .39
<.3
No data

Brandt Line

Figure 2.4
UNDP Human Development Index, 1994

have lower employment and wage rates than men. In affluent but less progressive core countries (such as Italy, Japan, and Switzerland), the index score for women is barely 80 per cent of the score for men. As with so many patterns in economic geography, however, the most striking difference is between developed countries and LDCs. In much of Africa and Latin America, the scores for women are only between 55 and 65 per cent of those for men.

These patterns reflect deep-seated cultural positions rather than women's actual contributions to economic development. In many peripheral countries, women constitute the majority of workers in the new manufacturing sector created by the new international division of labour (Christopherson, 1995). On average, women earn 30 to 40 per cent less than men for the same work. They also tend to work longer hours than men: 12 to 13 hours a week more (counting both paid and unpaid work) in Africa and Asia.

INTERNATIONAL PATTERNS OF RESOURCES AND POPULATION

The distribution of natural resources has a very important influence on patterns of international economic activity and development. Not only are key resources – energy, minerals, cultivable land – unevenly distributed, but the *combination* of particular resources in particular nations and regions makes for a complex mosaic of opportunities and constraints. A lack of resources can, of course, be remedied through international trade (Japan is the prime example here: see p. 171); but for most countries the resource base is an important determinant of development.

In overall terms, a very high proportion of the world's key non-renewable natural resources are concentrated in Russia, the USA, Canada, South Africa and Australia. The United States, for example, has 44 per cent of the world's known resources of hydrocarbons (oil, natural gas, oil shales, etc.), 38 per cent of the lignite, 38 per cent of the molybdenum, 21 per cent of the lead, 19 per cent of the copper, 18 per cent of the bituminous coal, and 15 per cent of the zinc. Russia has 75 per cent of the vanadium, 50 per cent of the lignite, 40 per cent of the bituminous coal, 38 per cent of the manganese, 30 per cent of the iron, 20 per cent of the cultivable land, and 18 per cent of the hydrocarbons. This concentration is largely a function of geology and physical geography. But it is also partly a function of political instability in much of ex-colonial Africa, Asia and Latin America: instability that has been a serious hindrance to resource exploration and exploitation. Current estimates of the distribution of resources will inevitably understate, therefore, the relative size of the resource base of many LDCs.

It should also be noted that the significance of particular resources is sometimes very much a function of prevailing technologies. *As technologies change, so resource requirements change*: the switch from coal to oil, gas and electricity early in the twentieth century, for example; and the switch from natural to synthetic fibres for mass-produced textiles. This also means that regions and countries that are heavily dependent on one particular resource are very open to the consequences of technological change. This is particularly important for countries such as Bolivia, Chile, Guinea, Guyana, Liberia, Mauritania, Sierra Leone, Surinam and Zambia, whose economies are heavily dependent on non-fuel minerals.

Meanwhile, the rate of exploitation of some of the world's natural resources can also be a cause for concern. Large-scale commercial agriculture has contributed to the 'desertification' of marginal environments such as the Sahel, for example. But the most dramatic consequences are to be found in relation to commercial forestry. The indiscriminate logging of sub-tropical forests means that global deforestation is now of the order of between 10 and 40 hectares per *minute* (Blaikie, 1989) – about half the size of Finland every year. Apart from the overall loss of timber resources, this has also resulted in other economic and ecological problems: the loss of livelihood of local inhabitants, the silting of reservoirs, damage to hydroelectric plants, and an increase in flash floods with consequent damage to property, crops and livestock. In addition, deforestation results in a serious loss of genetic diversity. Tropical forests support millions of plant, insect and animal species, some of which may prove invaluable to human welfare. The Madagascan Rosy Periwinkle, for example, has been discovered to be the source of two powerful drugs that can be successfully used to treat leukaemia and Hodgkin's Disease.

The idea of sustainable development

In 1972, the publication of a report on the limits to growth (Meadows *et al.*, 1972), sponsored by the Club of Rome, began an international debate on resources and development. The following year, a major energy crisis (caused by a fourfold increase in the price charged by the Organization of Petroleum Exporting Countries (OPEC) for crude oil) emphasized the resource issue in very dramatic terms. Within a few years, however, public concern had fallen away, leaving only a residual worry over specific 'green' issues (saving whales and pandas, doing without nuclear energy, and conserving 'wilderness' areas, for example).

Meanwhile, the 'limits to growth' thesis – that population growth and resource consumption will inevitably lead to a global economic/ecological/ demographic crisis – lost a great deal of its force in academic circles because the computer simulations on which the argument was based were quickly shown to be unsophisticated. They failed, for example, to take account of the probability of major technological advances, and of the multiple relations and feedback loops between resources, populations, production and pollution. Unfortunately, the technical critique of the argument was strongly linked to an ideological perspective that sought to defend a free-enterprise system with no constraints on the exploitation of scarce and non-renewable resources.

More recently, the World Commission on Environment and Development, chaired by former Norwegian Prime Minister Gro Harlem Brundtland, has revisited the issues within the context of a globalizing world economy. The Commission's report, *Our Common Future* (the 'Brundtland Report,' 1987), stressed the intensification and interdependence of ecological and economic crises and made a strong plea for the principle of **sustainable development** – economic development that seeks to meet the needs and aspirations of the present without compromising the ability to meet those of the future. Sustainable development means using renewable natural resources in a manner that does not eliminate or degrade them – by making greater use, for example, of solar and geothermal energy, and by greater use of recycled materials. It means managing economic systems so that all resources – physical and human – are used optimally. It means regulating economic systems so that the benefits of development are distributed more equitably (if only to prevent poverty from causing environmental degradation). It also means organizing societies so that improved education, health care, and social welfare can contribute to environmental awareness and sensitivity and an improved quality of life. A final and more radical aspect of sustainable development is to move away from wholesale globalization toward increased 'localization': a desire to return to a more locally based economy where production, consumption, and decision-making can be oriented to local needs and conditions.

Put this way, sustainable development sounds eminently sensible yet impossibly Utopian. A widespread discussion of sustainability took place in the early 1990s, and focused on the 'Earth Summit' (the United Nations Conference on Environment and Development) meeting in Rio de Janeiro in 1992. Attended by 128 heads of state, it attracted intense media attention. At the conference, many examples were described of successful sustainable development programs at the local level. Most of these centred on sustainable agricultural practices for LDCs, including the use of intensive agricultural features such as raised fields and terraces in Peru's Titicaca Basin: techniques that had been successfully used in this difficult agricultural environment for centuries, before European colonization. After the United Nations conference, however, many observers commented bitterly on the deep conflict of interest between core countries and peripheral countries that was exposed by the summit.

The Brundtland Report showed that one of the most serious obstacles to prospects for sustainable development is continued heavy reliance on fossil fuels as the fundamental source of energy for economic development. This not only perpetuates international inequalities but also leads to transnational problems such as acid rain, global warming, climatic changes, deforestation, and health hazards. The sustainable alternative – renewable energy resources such as solar energy, tides, waves, winds, geothermal and hydroelectric energy – has been pursued half-heartedly, the report argues, because of the vested interests of the powerful corporations and governments that control fossil fuel resources.

Demographic growth in LDCs was recognized by the Brundtland Report as a second important challenge to the possibility of sustainable development. As the report pointed out, sustainable development is feasible only if population size and growth are in harmony with the changing productive capacity of the ecosystem. About 80 per cent of all deforestation, for example, can be attributed to population pressure. Even more startling is the contrast in trends

sustainable development
A pattern of resource use and economic development that does not jeopardize non-renewable resources, damage existing ecosystems, or harm individual species.

in population and food production. Between 1985 and 1988, for example, world population increased by 5 per cent while world food output declined by 5 per cent. The obstacle here is not the inertia of powerful governments and corporations but the complexity of interactions between demographic change and economic development, education, culture, and resources.

But the greatest obstacle to sustainable development, according to the Brundtland Report, is the inadequacy of the institutional framework. Sustainable development requires economic, financial and fiscal decisions to be fully integrated with environmental and ecological decisions. In practice, national and local governments everywhere have evolved institutional structures that make for the separation of decision-making about what is economically rational and environmentally desirable. Supranational organizations, while better placed to integrate policy across these sectors and better able to address economic and environmental 'spillovers' from one nation to another, have (with the notable exception of the European Union) never acquired sufficient power to promote integrated, harmonized policies. Without radical and widespread changes in value systems and unprecedented changes in political will, 'sustainable development' will remain an embarrassing contradiction in terms.

Two key resources: energy and cultivable land

Two particularly important resources in terms of the world's economic geography are energy and cultivable land. The major sources of commercial energy are oil, natural gas and coal, all of which are very unevenly distributed across the globe. Most of the world's developed economies are reasonably well off in terms of energy *production*, the major exceptions being Japan and parts of Europe. Most LDCs, on the other hand, are energy-poor. The major exceptions are Algeria, Ecuador, Gabon, Indonesia, Libya, Nigeria, Venezuela and the Persian Gulf states – all major oil producers. Because of this unevenness, energy has come to be an important component of world trade. Oil is in fact the most important single commodity in world trade, making up around 13 per cent of the total by value.

For many LDCs, the costs of energy imports represent a huge burden. Consider, for example, the predicament of countries like India, Ethiopia, Paraguay, Jordan and Turkey, where in the mid-1990s the cost of energy imports amounted to more than one-quarter of the total value of exported merchandise. Nevertheless, few LDCs can afford to consume energy on the scale of the developed economies, so that patterns of commercial energy *consumption* tend to mirror the fundamental core–periphery cleavage of the world economy (Table 2.1). In 1993, energy consumption in North America was 30 times higher than in India and 60 times higher than in sub-Saharan Africa.

It should be noted that these figures do not reflect the use of firewood and other traditional fuels for cooking, lighting, heating, and, sometimes, industrial needs. In

Table 2.1: Energy consumption per capita (kilograms of oil equivalent, 1994)

Low-income economies	339
Lower-middle income economies	1531
Upper-middle income economies	1632
High-income economies	5245

Source: World Bank (1995, Table 5).

total, such forms of energy probably account for around 20 per cent of total world energy consumption. In parts of Africa and Asia they account for up to 90 per cent of energy consumption. This points us to yet another North–South contrast. Whereas massive investments in exploration and exploitation are enabling more of the developed, energy-consuming countries to become self-sufficient through various combinations of coal, oil, natural gas, hydroelectric power and nuclear power, *1.5 billion of the people who depend on fuelwood as their principal source of energy are cutting wood faster than they can grow it* (World Resources Institute, 1996). The problem is most serious in arid and semiarid areas and in cooler mountainous areas, where the regeneration of shrubs, woodlands and forests is particularly slow. Nearly 100 million people in 22 countries (16 of them in Africa) cannot meet their minimum needs even by overcutting remaining forests. Even if consumption rates can be pegged to the level of the early 1980s, the fuelwood deficit will have doubled by the year 2000 (Table 2.2).

The distribution of cultivable land represents another important environmental influence on international economic differentiation. Much more than half of the earth's land surface is unsuitable for any productive form of agriculture, as suggested by Figure 2.5. This map gives an approximation of the world's cultivable land by excluding regions that have too short a growing season (less than 90 days), are too dry (less than 25 cm annual rainfall), or too mountainous (elevations over 500 metres). This does not mean that agriculture is absent from the unshaded areas of the map – rather, that agriculture in these regions is likely to be marginal. By this measure, we should note, the distribution of the world's cultivable land is highly uneven, being concentrated in Europe, west-central Russia, eastern North America, the Australian littoral, Latin America, West Africa, India, and eastern China. In detail, of course, some of these regions may be marginal for agriculture because of marshy soils or other adverse conditions; while irrigation, for example, sometimes extends the local frontier of productive agriculture.

We also have to bear in mind that not all cultivable land is of the same quality. This leads us to the concept of the **carrying capacity** of agricultural land: the maximum population that could be fed a minimum daily diet, given the particular soils and climate. The UN Food and Agriculture Organization (FAO) has attempted to measure the carrying capacity of developing nations, drawing on UNESCO (UN Educational, Scientific and Cultural Organization) soil maps and climate data from their own Agroecological Zones project, and using a computerized routine to determine the particular food crops that would provide the greatest amounts of calories for each land/climate unit. The theoretical carrying capacity of each region

carrying capacity
Used in the context of food and agriculture, this term refers to the maximum population that can be supported within a given territory on a minimum daily diet, given the quality of local soils and local climatic conditions, and assuming the availability of appropriate forms of mechanization.

Table 2.2: Fuelwood deficits by region, 1980 and 2000

Fuelwood situation	Region	Populations involved and fuelwood deficit in 1980	Countries mainly concerned[d]
Acute scarcity[a]	Africa	13 million people 6 million m^3	Burkina Faso, Cape Verde, Chad, Djibouti, Mali, Mauritania, Niger, Sudan, Kenya, Ethiopia, Somalia, Botswana, Namibia
Arid and semi-arid areas	Asia	9.5 million people 3.6 million m^3	Afghanistan, Pakistan
	Latin America	6.8 million people 3.5 million m^3	Chile, Peru
Mountainous areas	Africa	36 million people 40 million m^3	Burundi, Rwanda, Lesotho, Swaziland
	Asia	29 million people 34 million m^3	Nepal
	Latin America	2 million people 2 million m^3	Bolivia, Peru
	Total	96.3 million people 89.1 million m^3	23 countries
Deficit[b]	Africa	131 million people 66 million m^3	Cameroon, Congo, Zaire, Malawi, Kenya, Madagascar, Uganda, Tanzania, Gambia, Guinea, Benir, Togo, Senegal, Sierra Leone, Nigeria, Mozambique
Areas with rapidly increasing population and agriculture	Asia	288 million people 75 million m^3	India, Nepal, Pakistan
	Latin America	143 million people 36 million m^3	Brazil, Colombia, Peru, Cuba, Domincan Republic, Guatemala, Mexico, Trinidad and Tobago
Densely populated lowlands	Asia	412 million people 120 million m^3	Bangladesh, India, Sri Lanka, Thailand, Indonesia (Java), Philippines, Vietnam
	Latin America	9 million people 6 million m^3	El Salvador, Haiti, Jamaica
	Total	983 million people 303 million m^3	37 countries
Prospective deficit[c]	Africa	(in year 2000: 175 million people facing a 40 million m^3 deficit)	Ghana, Ivory Coast, Central African Republic, Angola, Zimbabwe, Guinea-Bissau
	Asia	(in year 2000: 239 million people facing a 50 million m^3 deficit)	Burma, India, Indonesia, Philippines, Vietnam
	Latin America	(in year 2000: 50 million people facing substantial degradation of fuelwood supplies)	Ecuador, Paraguay, Uruguay, Venezuela
	Total	464 million people	15 countries
Surplus potential for wood-based energy	Africa	Surplus potential 50 million m^3	Cameroon, Congo, Equatorial Guinea, Angola, Zaire, Central African Republic
Low population tropical forest areas	Asia	Surplus potential 200 million m^3	Bhutan, Laos, Democratic Kampuchea, Indonesia (except Java)
	Latin America	Surplus potential 200 million m^3	Amazon Basin

Notes:
[a] Acute scarcity: available supplies of fuelwood are insufficient to meet minimum requirements, even with overcutting.
[b] Deficit: fuelwood supplies are being consumed faster than they are replenished by natural regeneration and forest growth.
[c] Prospective deficit: fuelwood supplies will be in a deficit situation by the year 2000, if present trends continue.
[d] Data not available for China.

Source: World Resources Institute (1996), Table 5.8, p. 70.

Figure 2.5
The world's cultivable land. *Source:* After Berry, Conkling and Ray (1976), Fig. 2.2, p. 16

Table 2.3: Potential population-supporting capacities, 2000

Input level	Africa	Southwest Asia	South America	Central America	Southeast Asia	Average
Low	1.6	0.7	3.5	1.4	1.3	1.6
Intermediate	5.8	0.9	13.3	2.6	2.3	4.2
High	16.5	1.2	31.5	6.0	3.3	9.3

was then calculated by dividing the total potential calorie production of each by the FAO's minimum daily requirement (United Nations, 1984). The results were expressed as ratios by dividing this figure by the projected population for the year 2000. Because agricultural output depends so heavily on technological inputs, the exercise was run for low, medium and high input levels, involving different assumptions about the use of fertilizers, biocides, improved crop varieties, and soil conservation.

The results, summarized in Table 2.3, suggest that the developing world as a whole is in reasonable shape, potentially able to support 1.6 times its population even at low input levels. On a country-by-country basis, however, the results are less comforting. By 2000, 64 countries are projected to be in critical condition, including the entire region of Southwest Asia. In fact, the FAO model is overly optimistic, since it assumes that all potentially cultivable land is used to grow nothing but staple food crops or to provide pasture for livestock. In reality, land has to be set aside for lumber and fuelwood production and for growing fibres and other non-food crops; and cultivable land is continually lost to deforestation and overgrazing – a total of about 1.2 billion hectares (almost 11 per cent of the earth's vegetated surface) over the past 45 years.

Agricultural patterns and the food question

land tenure
A system of land use rights and transfer mechanisms. The major types of land tenure include owner-occupation, cash tenancy, share-cropping (a form of tenancy in which rent is paid in kind), use rights (where there is no codified legal owner and a person or group establishes a right to land by using it) and collectivism.

These issues shift our attention from the abstract to reality, and to a consideration of the world agricultural map and the world food situation. Both of these are rather different in configuration from the patterns of cultivable land and carrying capacity described above. *The actual pattern of world agriculture is just one of a vast number of possible realizations of the world's agricultural resources.* It is the product of a variety of interpretations, at different times, of environmental possibilities, desirable products, and marketable opportunities – all influenced, in turn, by prevailing **land-tenure** systems, levels of technology, and global power politics. It is, in short, a legacy of the world's economic history. As a result, the agriculture of the developed nations has come to be dominated by corporate 'agribusiness' and strongly conditioned by government policy. The agriculture of LDCs, meanwhile, has come to be a mixture of commercial non-food crop production and peasant-based food production.

In detail, the mosaic of world agricultural regions shows a very high degree of specialization. At the same time, the broader international division of labour means that some countries depend much more on agriculture for employment and income than others. In low-income developing countries, agriculture employs roughly 70 to 80 per cent of the labour force and accounts for 35 to 45 per cent of GDP; in developed economies, agriculture employs around 7 per cent of the labour force and accounts for about 3 per cent of GDP.

One consequence of this specialization is of course a large volume of *trade* in agricultural produce. The biggest exporters of food, however, are not developing countries but a few of the more developed countries – Argentina, Australia, Canada, France and the USA – with highly productive agricultural sectors specializing in cereal production. In recent years, world trade in food has grown rapidly and there have been some significant changes in the pattern of trade. Until the 1950s the dominant flows were those of food grains into western Europe. These have now declined significantly, with the major flows currently originating in Canada and the United States with destinations in Russia, Japan and, increasingly, middle-income developing countries.

Gross inequalities in the consumption of food, one of the most basic of all human needs, are an important corollary of these patterns and flows. At least 900 million people around the world are under-nourished, 500 million of them chronically so. Directly or indirectly, malnutrition causes the death of more than 13 million under-fives each year. Measured in terms of calorie consumption per capita as a percentage of the minimum daily requirement (Fig. 2.6), there is steady gradient between low-income countries such as Ethiopia, Somalia, and Mozambique, with levels of only 75 per cent, to the advanced industrial economies, almost all of which had levels, in the early 1990s, of more than 140 per cent. Estimates of the incidence of chronic malnourishment range from 340 to 730 million people, with most of them concentrated in South and East Asia (Grigg, 1985; World Bank, 1996). The fact that almost half of them are located in countries that are net *exporters* (by value) of foods is a telling indictment of the world economic system.

Because the food question is most dramatically highlighted by localized famines, the most commonly cited causes of the food problem are climatic instability, inefficiencies in the allocation and transportation of food, and overpopulation. In fact, however, food production in developing countries has been growing faster than population. The chief source of instability in food production is not climatic or the product of unfavourable population/resource ratios but the result of profit-motivated decisions and domestic and foreign policy considerations within the major grain-exporting countries:

Thus in 1972, the crisis year of the Sahelian drought in which starvation was widespread, the US government paid farmers $3 billion to take 50 million hectares out of production . . . The intent was quite clear. In order to remedy the 'glut' of previous years, which saw the grain reserve rise to 49 million tons with a commensurate fall in the world price, the US government determined to reduce production and create an effective storage and thus raise again the world price. . . .The deliberately induced shortage was exacerbated by sales to the USSR, not, it seems, through deliberate

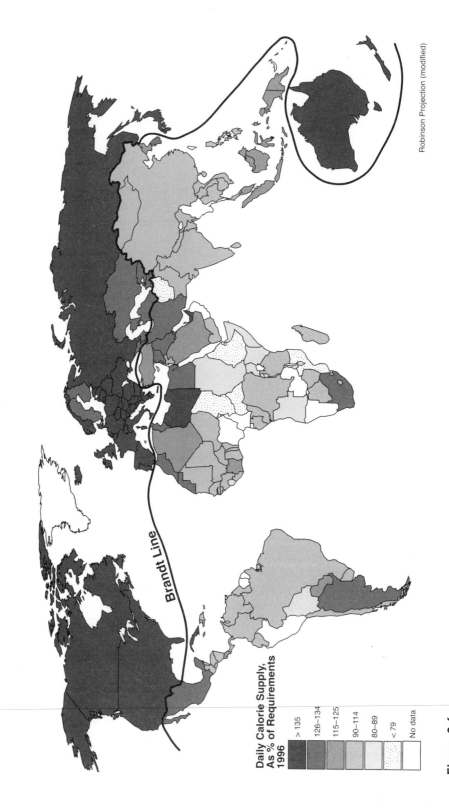

Robinson Projection (modified)

**Daily Calorie Supply,
As % of Requirements
1996**

> 135
126–134
115–125
90–114
80–89
< 79
No data

Brandt Line

Figure 2.6

The food problem: international variations in daily calorie supply per capita, 1996

government action, but via the combined efforts of a number of individual Russian buyers and the sales pitch of the controlling corporations in the USA.

(Bradley and Carter, 1989, p. 112–13)

In this context, it is not surprising that surplus food has become an important component of international aid and, therefore, an important instrument in global power politics.

International demographic patterns

The geography of population and the dynamics of population change are closely interrelated with patterns of economic development. Population density, fertility, mortality and migration are often a direct reflection of economic, social and political conditions. At the same time, they can be important determinants of economic change and social well-being. Human resources are vital to economic development in terms of both production and consumption; but at the wrong time and in the wrong place they can be more of a liability than an asset. Although it is not always easy to unravel cause and effect, it is important to understand the broad context.

In global terms, this broad context is currently dominated by the sheer growth of population. Over 1 million people are added to the population of the world every four days. The current population of nearly 5.8 billion is likely to grow to over 10 billion by the year 2050, with nine-tenths of the increase taking place in LDCs, whose populations have been growing by between 2 and 3 per cent each year. In contrast, the population of the advanced industrial countries has been growing at an annual rate of about 0.5 per cent, with some West European countries having virtually stagnant populations.

These core–periphery contrasts are the product of differences in fertility and mortality rates that are, in turn, related to differentials in the **demographic transition** that is associated with the broad sweep of economic development and social change. This transition is conventionally portrayed as involving three stages (Fig. 2.7). In the first, populations exhibit high birth rates and high, fluctuating death rates, with net growth rates of around 1 per cent. In the second stage, death rates fall sharply (largely because of improved diets, improved public health, and the availability of scientific medicine). Birth rates also fall, but the decrease in fertility is lagged (largely because it takes time for social and cultural practices concerning family size to respond to changing circumstances). The result is an explosive increase in population. This stage was experienced by most western industrial countries during the nineteenth century. In the third stage, death rates even off at a low level; while birth rates are low but fluctuating, with net growth rates once again around 1 per cent.

It is clearly important to know whether a country is just entering the critical second stage of rapid population expansion and thus has the major part of its population growth ahead of it, whether it is in the middle of the population 'explosion', or whether it is on the verge of completing the growth stage. Accordingly, the UN has suggested a threefold division of the second stage, giving five categories of population growth types (Fig. 2.7). Fig. 2.8 shows how the countries

demographic transition
The evolution of vital rates – birth rates and death rates – over time, from high to low levels. The demographic transition model posits improved diets, public health and scientific medicine as causing a steady decline in death rates with increasing levels of economic development over time. Birth rates decline later, and more slowly, as socio-cultural practices take time to adjust to these new circumstances. The result is a sharp increase in population growth, until birth rates fall to relatively low levels.

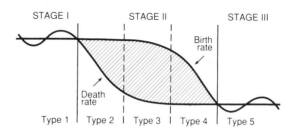

Figure 2.7
The demographic transition

of the world fit into this classification system. The economies of the North are all in the final, slow-growth stage, while most of Latin America seems to be in the final phase of the growth stage. Most of Asia, the Middle East, and much of Africa are experiencing the most explosive phase of the growth stage, while the central and western regions of Africa seem poised to enter this explosive phase. A few African countries, meanwhile – including Burkina Faso, Chad, Ethiopia, Niger and Somalia – are at the very beginning of the demographic transition, with relatively high death rates that are suppressing the rate of natural increase.

Another important aspect of population change is *migration*. International labour migration has been an important part of the world economic system ever since the industrial revolution of the nineteenth century. Current estimates of the total number of international migrant workers stand at around 25 million, with a comparable number of dependents accompanying them. About 10 million of these, including 4 or 5 million illegal immigrants, are working in the United States, which now draws most of its immigrants from Mexico. Northwestern Europe has about 5 million migrant workers, most of them from nearby countries such as Spain, Turkey and Portugal or from ex-colonial countries such as Algeria and Jamaica. Since the early 1970s, large numbers of workers have also been attracted to the oil-rich countries of the Middle East: about 4 million at present, two-thirds of them from the region itself and the rest from South and Southeast Asia. South Africa draws about 500 000 migrant workers from neighbouring countries; and there are also important flows of migrant labour between developing countries in parts of Latin America and in West Africa.

Mention should also be made of the distinctive streams of highly skilled labour – physicians, engineers, scientists, etc. – the so-called 'brain drain'. The principal recipients of these streams have been the United States, Canada, Britain and Australia. The principal countries of origin have been India, Pakistan, the Philippines, Sri Lanka and, more recently, the former Soviet Union and its satellites. Typically, the brain drain is a result of students and professionals choosing not to return home after the completion of educational courses or training programmes in developed countries. These streams are significant not because of the absolute numbers of people involved but because of the economic implications of the relative gains and losses of highly skilled personnel. The brain drain from the former Soviet Union has had added significance because of the sudden availability of scientists with key skills relevant to military technology and nuclear and biological weapons. This has added a geopolitical dimension to the issue, raising fears of a brain drain of ex-Soviet experts to LDCs. In response, the United States, Japan, and the European

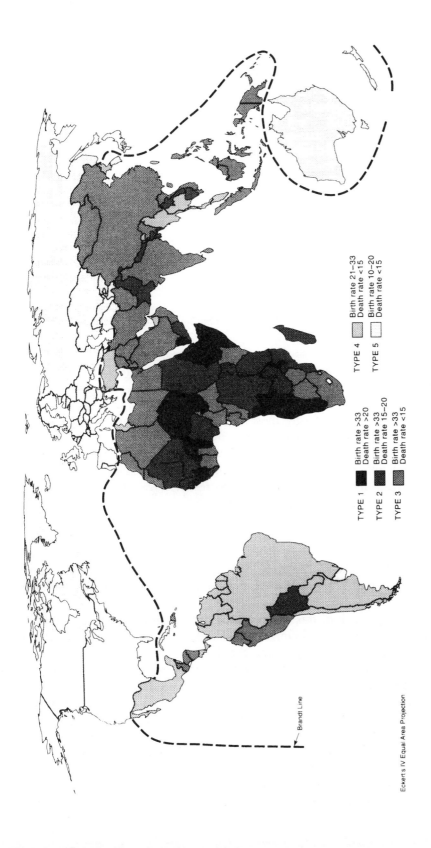

Ecken's IV Equal Area Projection

TYPE 1 ■ Birth rate >33
Death rate >20

TYPE 2 ■ Birth rate >33
Death rate 15–20

TYPE 3 ▓ Birth rate >33
Death rate <15

TYPE 4 ░ Birth rate 21–33
Death rate <15

TYPE 5 □ Birth rate 10–20
Death rate <15

Brandt Line

Figure 2.8
Demographic transition map of the world

Community established a scientific centre in Russia in 1992 with the express purpose of stemming the nuclear brain drain.

In terms of unskilled and semi-skilled workers, demand in the developed economies has been decreasing ever since the mid-1970s. Meanwhile, the demographic transition continues to flood the labour markets of most LDCs. Without the option of migration or emigration to unsettled territories – an option that represents a crucial difference in the current experience of LDCs compared to the historical experience of developed countries – the increases in population resulting from the demographic transition in LDCs have been channelled into internal, rural-to-urban migration streams. Young, reproductively active rural populations are being pushed out by shortages of land and pulled to larger cities by a combination of real and perceived advantages: employment opportunities, wages and modern amenities. As a result, the rate and scale of urbanization in developing countries represent yet another dimension of core–periphery contrasts.

INTERNATIONAL PATTERNS OF INDUSTRY AND FINANCE

As with the agricultural map of the world, the international mosaic of industrial production and employment is highly complex, with a great deal of specialization in particular activities. Once again, what we want to stress here is the overall framework. Table 2.4 provides the necessary base information. In terms of production, the United States is by far the most important source of manufactured goods, accounting for 27.7 per cent of global output in 1993. Just five countries – the United States, Japan, Germany, France and Italy – produced over two-thirds of the world total. Of the 20 biggest manufacturing countries, a number were LDCs: Argentina, Brazil, China, Mexico, and South Korea. Nevertheless, the vast majority of LDCs have a very small manufacturing output.

Another important contrast implicit in Table 2.4 concerns *productivity*. In brief, the highly capitalized manufacturing industries of the developed countries have been able to maintain high levels of productivity, with the result that (with the notable exception of the United Kingdom) their output has continued to grow appreciably even as the size of their manufacturing labour forces has tended to shrink.

Within the framework of this continuing dominance of the advanced industrial nations there are several important trends (Dicken, 1992). Although the United States has retained its leadership as the world's major producer of manufactured goods over the postwar period, its dominance has been significantly reduced. In 1963, for example, its share of world manufacturing production had been 40 per cent, compared to less than 25 per cent in 1995. The United Kingdom, meanwhile, has lost ground in both relative and absolute terms. In contrast, Japan has moved from fifth place with a share of 5.5 per cent in 1963 to second place and a share of 21 per cent in 1995. We shall examine the reasons for these shifts in Chapters 5 to 7, where we discuss in detail the evolution and transition of the world's developed economies. As we shall see, this has involved an *increasing degree of international interdependence* throughout the world economy.

Table 2.4: World manufacturing data, 1993

Rank	Country	Total Manufacturing Value Added (US$millions)	Percent of World Total	Manufacturing as share of GDP 1965	Manufacturing as share of GDP 1993	Manufacturing as share of labour force 1965	Manufacturing as share of labour force 1993
1	United States	1,481,700	27.7	29	18	36	18
2	Japan	1,023,048	19.1	32	30	30	23
3	Germany	565,603	10.5	–	23	48	26
4	France	271,133	5.1	–	20	39	17
5	Italy	250,345	4.7	–	20	40	21
6	United Kingdom	201,859	3.8	30	21	48	20
7	Russia	200,237	3.7	–	49	–	27
8	China	147,302	2.7	–	45	–	13
9	Canada	103,690	1.9	23	18	34	17
10	Spain	100,672	1.8	25	23	31	20
11	Brazil	90,062	1.7	26	25	15	16
12	South Korea	85,454	1.6	18	29	9	13
13	Mexico	67,157	1.3	21	22	20	16
14	Netherlands	58,476	1.1	–	13	42	16
15	Argentina	50,009	0.9	33	22	36	21
16	Ukraine	48,872	0.9	–	43	–	28
17	Austria	46,739	0.8	33	24	46	27
18	Belgium	45,230	0.8	30	23	48	18
19	Australia	43,679	0.8	28	15	37	16
20	Sweden	43,605	0.8	28	42	45	18

This interdependence is reflected by some important trends in manufacturing output. Of particular importance is the emergence of 10 or so LDCs as major growth points for manufacturing (Table 2.5). The most important of these 'newly industrializing countries' (NICs) in absolute terms are Spain and Brazil, followed by Mexico, but in terms of the rate of manufacturing growth it has been the four Asian NICs – South Korea, Taiwan, Hong Kong and Singapore – that have experienced the most spectacular growth. As Dicken noted:

> By far the most dramatic increase in manufacturing production was experienced by South Korea which attained *annual average* growth rates of almost 18 per cent in the 1960s and 16 per cent in the 1970s. As a result, South Korea surged up the league table of developing market economies ... The percentage of South Korea's labour force employed in manufacturing increased from 9 per cent to 29 per cent between 1960 and 1980.

(1986, p. 30)

All four Pacific Rim NICs have made spectacular progress up the world league table of exporters. In 1993 Hong Kong ranked 9th in the World (up from 26th in 1979),

Table 2.5: The growth of manufacturing production in the leading NICs, 1960–93

	Share of world manufacturing output		Average annual growth in manufacturing		Share of total labour force in manufacturing	
	1963	1993	1960–70	1980–93	1960	1993
Hong Kong	0.08	0.22	–	4.8	52	19.8
Singapore	0.05	0.25	13.0	5.1	23	26.7
South Korea	0.11	1.16	17.6	8.4	9	24.1
Taiwan	0.11	1.30	15.5	8.9	16	27.1
Brazil	1.57	1.68	–	–2.4	15	15.9
Mexico	1.04	1.25	9.4	6.7	20	15.9
Portugal	0.23	0.26	8.9	8.9	29	24.9
Greece	0.19	0.23	10.2	0.8	20	20.2

Taiwan ranked 12th (up from 22nd), South Korea ranked 13th (up from 27th), and Singapore ranked 14th (up from 30th).

These shifts are part of a globalization of economic activity that has emerged as the overarching component of the world's economic geography. As we shall see in Chapters 6 and 10, it has been corporate strategy, particularly the strategies of large transnational enterprises, that has created this globalization of economic activity.

For the moment, however, it will be sufficient to take note of the magnitude of the phenomenon. One striking measure of the importance of transnational corporations in the world economy is given by the size of their annual turnover in comparison with the GNP of entire nation states. By this yardstick, all of the top 50 transnational corporations – including the likes of Exxon, General Motors, Ford Motor Company, Matsushita Electronics, IBM, Unilever, Philips, ICI, Union Carbide, ITT, Siemens and Hitachi – carry more economic clout than many of the world's smaller LDCs; while the very biggest transnationals are comparable in size with the national economies of countries like Greece, Ireland, Portugal, and New Zealand (Fig. 2.9). Collectively, the 500 largest US corporations now employ an overseas labour force as big as their domestic labour force. Similar statistics apply to the largest Japanese corporations.

These overseas labour forces are spread between parts of both the North and the South, but it is in LDCs and, in particular, the NICs where the most rapid growth has been taking place. Thus two-thirds of the radios made by Japanese manufacturing companies are produced abroad, together with half the stereos and colour television sets – mostly in South Korea and other nearby East Asian locations. This, clearly, has had much to do with the rise of the NICs. Over 90 per cent of South Korean exports of electronic equipment, for example, are produced by affiliates of Japanese companies.

These examples also serve to emphasize the point that the most important industries involved in the globalization of economic activity are electronics, together with textiles and clothing: industries where profits are difficult to maintain

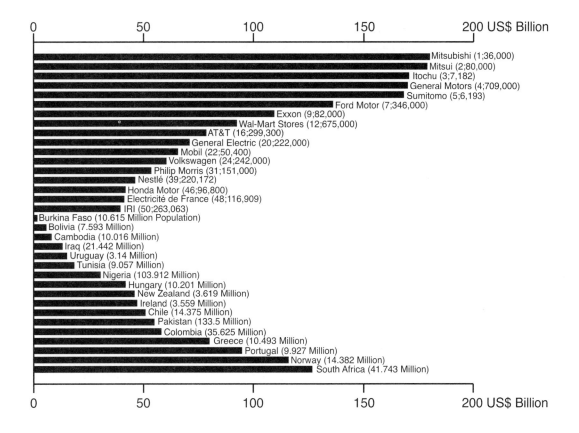

43

Figure 2.9
A comparison of nation states (GNP) and transnational corporations (sales). *Note:* Figures in parentheses indicate corporate rank-order (by sales) and number of employees; and, for nations, total population.

through increases in technological inputs but relatively easy to increase by substituting low-wage for high-wage labour. This means that the 'manufacturing' undertaken in LDCs by transnational corporations is more often than not merely *assembly*. The manufacture of components that require high levels of skill and/or technology is more commonly undertaken in developed nations – albeit in settings that are often outside traditional manufacturing regions.

We should also note here that the globalization of economic activity has, in turn, had an important impact on another dimension of economic geography: government policy. The governments of LDCs have sought to take advantage of transnational companies need for cheap labour by setting up **export-processing zones** (EPZs) (see p. 339), adaptations of free trade zones in which favourable investment and trade conditions are created by waiving excise duties on components, providing factory space and warehousing at subsidized rates, allowing tax 'holidays' of up to 5 years, and suspending foreign exchange controls. In the Philippines, for example, some 199 companies were operating within EPZs in 1996, employing about 50 000 workers. Most of these firms are foreign-owned, or owned jointly with foreigners,

export-processing zones
Small, closely definable areas within which especially favourable investment and trading conditions are created by governments in order to attract export-oriented industries, usually foreign-owned. These conditions include the absence of foreign exchange controls, the availability of factory space and warehousing at subsidized rents, low tax rates, and exemption from tariffs and export duties.

and most are engaged in the manufacture of electronics, microchips, semiconductors, toys, and garments. In 1996, the aggregate exports from these firms amounted to nearly US$5.5 billion, representing over 30 per cent of the country's exports by value. At a more general level, governments everywhere have responded to the 'global reach' of transnational corporations by intensifying their involvement with supranational economic and political organizations such as the European Union (EU), the Association of South East Asian Nations (ASEAN), and the North American Free Trade Association (NAFTA). We shall be reviewing the changing role of the state in the context of the internationalization of the world economy in Parts 2 and 3 of this book. In Part 4, we shall explore supra-national reactions to the internationalization of the world economy.

Patterns of international trade

The high degree of international specialization in agricultural and industrial production inevitably means that the geography of international trade is very complex. It is clear, however, that trade in general is of increasing importance as a component of the world economy. One significant reflection of the increased economic integration of the world-system is that global trade has grown much more rapidly over the past 25 years than global production. Between 1985 and 1995 the average annual growth rate of the value of world exports was twice that of the growth of world production and several times greater than that of world population growth. It is also clear that the fundamental structure of international trade has been based on a few **trading blocs** with most of the world's trade taking place *within* these blocs. Membership of these trading blocs is principally the result of the effects of (1) distance, (2) the legacy of colonial relationships, and (3) geopolitical alliances. For most of the period 1950–90, international trade was dominated by four trading blocs:

1. Western Europe, together with some former European colonies in Africa, South Asia, the Caribbean, and Australasia;
2. North America, together with some Latin American states;
3. the countries of the former Soviet world-empire; and
4. Japan, together with other East Asian states and the oil-exporting states of Saudi Arabia and Bahrain.

trading blocs
Groups of countries with formalized systems of trading agreements.

Meanwhile, we should note that a significant number of countries exhibit a high degree of **autarky** from the world economy. That is, they do not contribute significantly to the flows of imports and exports that constitute the geography of trade. Typically, these are smaller LDCs: examples include Bolivia, Burkina Faso, Ghana, Malawi, Samoa and Tanzania.

autarky
National economic self-sufficiency and independence.

The geography of trade has been changing rapidly, however, in response to:

1. Innovations in transport, communications and manufacturing technology. These innovations have diminished the importance of the 'classical' distance-based factors that have underpinned traditional trading blocs.
2. Shifts in global politics. The most important shift has of course been the break-up of the former Soviet world-empire. Other important changes include the trend towards political as well as economic integration in Europe, the

increasing participation of China in the world economy, and the continuing trend away from isolationism on the part of the United States.

3. The increasing internationalization and flexibility of production processes. As we have seen, the globalization of economic activity has created new flows of materials, components, information and finished products. Meanwhile, as much as 35 per cent of world trade now takes place within the 'internal markets' of transnational corporations.

These changes have already altered the fundamental pattern of international trade in that the trading bloc of the former Soviet world-empire has been dissolved, leaving a tri-polar framework for international trade: North America, the European Community, and Japan and the East Asian NICs. It is no coincidence, of course, that this tri-polar framework is focused on the countries that constitute the core regions of the world economy.

Advanced industrial countries have long dominated international trade (Fig. 2.10), while the share of world trade accounted for by LDCs has decreased (a trend that is particularly pronounced if oil-exporting countries are excluded from calculations). In general terms, then, *there has been an intensification of the long-standing domination of trade within and between developed countries at the expense of trade between developed countries and LDCs* – with the major exception of trade in oil.

In detail, however, there have been some very important shifts. Between 1950 and 1995, the United Kingdom's share of world exports shrank drastically, from 11.0 per cent to 4.8 per cent, with its share of world imports falling from 12.3 per cent to 5.4 per cent. The US share of exports fell from 17.8 per cent to 12.5 per cent, with imports falling almost imperceptibly: from 16.0 to 15.9 per cent. As the United States was acquiring this trade deficit, Japan was moving in the opposite direction and assuming a much more important role in world trade. In 1950, Japanese exports accounted for only 1.4 per cent of the world total; in 1995, they

Figure 2.10
Global trends in trade, 1974–1990

Source: International Monetary Fund (1991), p. 22

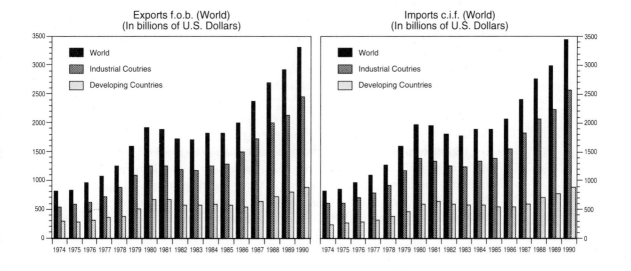

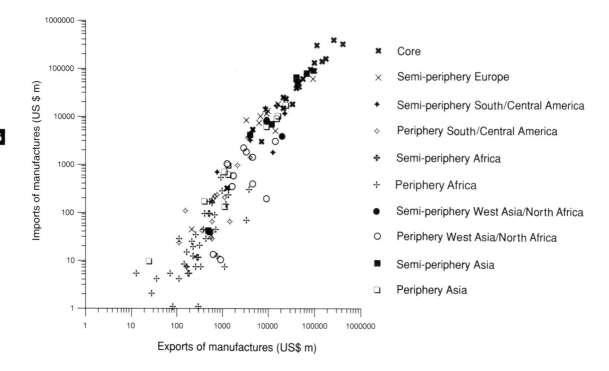

Figure 2.11
World trade in
manufactured goods,
1991

had risen to 9.8 per cent. Meanwhile, Japanese imports increased from 1.6 to 6.4 per cent of the world total. Germany has also achieved a significant improvement in its trading position; while France and Italy have increased their share of world trade but at the same time have moved into a net trade deficit. Most of the other developed countries have experienced a decrease in their share of world trade, as have the LDCs as a bloc.

The graph of world trade in manufactured products (Fig. 2.11) reflects the importance of the developed economies, but at the same time illustrates the extent to which a global system of manufacturing has emerged. Significant quantities of manufactured goods are now imported *and* exported across much of the world; no longer do developed economies export manufactures and LDCs import them. African countries are an important exception, many of them barely participating in world trade in manufactures.

Space does not permit a detailed examination of the dominant patterns of commodity flows or of the regionalization of international trade. Our purpose here is to illustrate the overall framework and note the major trends that provide the template and the context for the evolution of the world's economic landscapes. Suffice it to note, therefore, that the most striking aspect of commodity flows and the regionalization of trade is the persistence of the *dependence of LDCs on trade with developed countries that are geographically or geopolitically close.* Thus, for example, the United States is the central focus for the exports and the origin of the bulk of the imports of most Central American nations, while France is the focus for commodity flows to and from French ex-colonies such as Algeria, Cambodia,

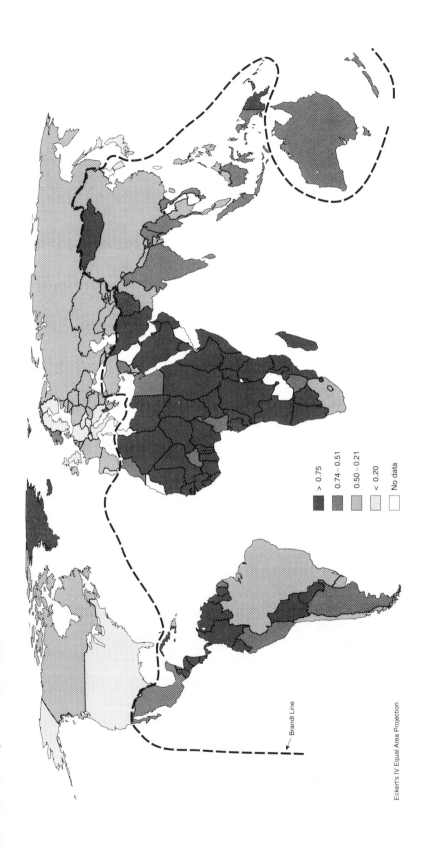

Eckert's IV Equal Area Projection

> 0.75

0.74 – 0.51

0.50 – 0.21

< 0.20

No data

Brandt Line

Figure 2.12

Index of commodity concentration of exports, 1996

Dahomey and the Ivory Coast. These flows, however, represent only part of the action for the developed economies, whose trading patterns are dominated by flows to and from other developed countries.

One of the implications of this situation is that the smaller, peripheral partners in these trading relationships are highly dependent on levels of demand and the overall economic climate in developed economies. Another aspect of dependency, in this context, is the degree to which a country's export base is diversified. Figure 2.12 shows one measure of this: the index of commodity concentration of exports. Countries with low values on this index have diversified export bases. They include Argentina, Brazil, China, India and North and South Korea as well as most of the developed countries. At the other extreme are LDCs where the manufacturing sector is poorly developed and the balancing of national accounts and the genera-tion of foreign exchange are dependent on the export of one or two agricultural or mineral resources: Bolivia, Chile, Iran, Iraq, Libya, Nigeria and Venezuela, for example.

Patterns of international finance and business services

The spatial organization of world production and trade is closely mirrored by patterns of international finance and business services. Once again, therefore, we find a tri-polar framework (Fig. 2.13). The long-standing dominance of flows of direct investment between the developed nations of what was once called the 'White North' (Buchanan, 1972) has been heavily modified by the globalization of economic activity.

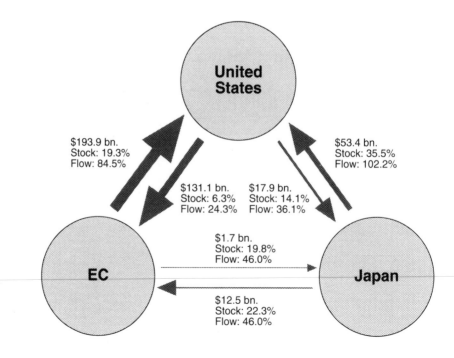

Figure 2.13

Intra-core foreign direct investment, 1988.

Source: UN Commission on Transnational Corporations (1991), *World Investment Report*, Fig. 11, p. 40

Until the early 1970s, US-based transnational corporations accounted for about two-thirds of the total outflows of foreign direct investment, and about four-fifths of this was directed towards Canada and the more advanced industrial nations of Western Europe. By the mid-1990s, US transnational corporations' share of the total had dropped to less than half, while direct foreign investment by Japanese, Canadian and West German corporations had increased significantly. Meanwhile, another source of foreign direct investment had begun to show up: transnational corporations based in the NICs. Nearly 10 per cent of the top 500 non-US-based corporations in the world are now based in NICs. Hyundai, the South Korean shipbuilding firm, is now bigger than Michelin or Rio Tinto Zinc; and Taiwan's Walsin Likwa (an electronics group) is bigger than Distillers or De Boers.

Along with these changes in the *sources* of investment have been changes in *destination*. The advanced industrial nations still absorb most of the inflows but, as we have seen, the globalization of economic activity has brought significant flows of capital into the NICs. Almost three-quarters of the total inflows to all LDCs are accounted for by just seven NICs: Argentina, Brazil, Hong Kong, Malaysia, Mexico, Singapore and South Korea.

Within the core of the world economy, meanwhile, other significant shifts have occurred. Australia, Canada, West Germany, Italy and the Netherlands have all lost ground in terms of their share of the inflow of foreign capital, while France, Spain, the United Kingdom and the United States have gained. The United States now *receives* almost one-third of the total inflow of foreign direct investment, compared to less than 10 per cent at the beginning of the 1970s. The overall result has been described as follows:

> In the 1970s the destination of foreign direct investment was neatly described as a double capital movement, towards the United States and the developing countries. . . . But in the 1980s it might more accurately be described *as a circular capital movement between the United States, Western Europe and Asia*. The developing countries have found the door to prosperity slammed shut.
>
> (Thrift, 1989, p. 31, emphasis added)

These shifts in international investment – and in particular the opening up of the United States to investment from Canadian, European and Japanese corporations – have been contingent upon other changes in the pattern of international finance. Throughout the first part of the postwar period, the pattern of international finance was set by the Bretton Woods agreement of 1944. This created what was virtually a US-run system, with fixed exchange rates and the US dollar serving as the convertible medium of currency with a fixed relationship to the price of gold. But, as the position of the United States deteriorated in terms of world manufacturing and trade, the system came under pressure.

> The result was that the Bretton Woods system crumbled. In particular, by the late 1960s fixed exchange rates effectively disappeared and every domestic currency became convertible into every other. Exchange rates 'floated' and, as a result, all domestic currencies became a medium that could be bought and sold and out of which a profit could be made.
>
> (Thrift, 1989, p. 34)

Eurodollars
United States currency that is held in banks located outside the United States, mostly in Europe. It is a pool of currency for which there is a distinctive and independent market, for it represents a fairly stable, hard currency that has been established beyond the control of the US government and its financial institutions.

Meanwhile, there had developed a pool of **Eurodollars** – US dollars held in banks located outside the United States – that was boosted after 1971 as the US government began to finance its budget deficit by paying in its own currency, flooding the world with dollars and fuelling world-wide inflation. Two years later, in 1973, the Eurodollar market was swollen still further as oil-producing countries rapidly acquired huge reserves of US dollars as a result of the quadrupling of petroleum prices in the wake of the OPEC embargo. The combined result of the floating of currencies and the creation of a large market in Eurodollars was that a new, more sophisticated system of international finance emerged. Consequently, new patterns of investment were accompanied by an expansion and international-ization of key business services such as stock exchanges, futures markets, banks, advertising agencies and business hotels. Indeed, as developed economies have lost competitiveness in industrial production, they have come to rely increasingly on these services to earn foreign currency and to balance national accounts.

The globalization of financial services has occurred in tandem with the global-ization of trade, industry, and capital itself. In large part, of course, the inter-nationalization of financial services and related business services is the result of the tremendous increase in trade, the internationalization of manufacturing, and the emergence of transnational corporate empires. And, just as technological changes have facilitated world trade, the internationalization of manufacturing and the emergence of transnational corporations, so advances in telecommunications and data processing have fostered the internationalization of financial and business services. Satellite communications systems have made it possible for firms to operate key financial and business services 24 hours a day, around the globe (though national regulatory mechanisms still impede the full exploitation of available technologies).

In addition, several factors have reinforced the globalization of financial and business services. One is the institutionalization of savings in developed countries (through pension funds and the like), which has established a large pool of capital managed by professional investors with few geographical allegiances or ties. Another is the trend towards 'disintermediation' – whereby borrowers (especially large corporations) raise capital and make investments without going through the traditional, intermediary channels of financial institutions. A third, and probably more important factor is the deregulation of financial markets that occurred in many developed countries in the 1980s. Thus:

> A series of changes in the United States since the 1970s has both eased the entry of foreign banks into the domestic market and facilitated the expansion of US banks overseas. . . . In the United Kingdom the so-called 'Big Bang' of October 1986 removed the barriers which previously existed between banks and securities houses and allowed the entry of foreign firms into the Stock Exchange. In Japan the restrictions on the entry of foreign securities houses have been relaxed (though not removed) and Japanese banks are now allowed to open international banking facilities. In France the 'Little Bang' of 1987 is gradually opening up the French Stock Exchange to outsiders and to foreign and domestic banks. In Germany foreign-owned banks are now allowed to lead-manage foreign DM issues, subject to reciprocity agreements.

(Dicken, 1992, p. 366)

Banking services provide a good example of the globalization of financial and business services in general. The expansion and internationalization of the world's banking system have been explosive. In the race to become global, US banks were the most successful:

> The number of foreign branches of US banks increased from 124 to 723 between 1960 and 1976. ... But the United States banks were soon followed by the British and French banks (with their experience of dealing with former colonial countries) and later by German, Italian, Arab and Japanese banks. Like Japanese transnational corporations, the Japanese-based banks have become a major force in global banking, backed by the enormous savings of the Japanese people. The largest Japanese banks have now overtaken many of the largest US and Western European banks.

(Thrift, 1989, p. 38)

It has been suggested that the internationalization of banking has contributed to an 'electronic colonialism' whereby large European, American and Japanese banks now exercise a great deal of control over the world economy. The increasing integration of the world economy can thus be seen as both cause and effect of the internationalization of banking. Yet, as with other dimensions of the world economy, the process is neither simple nor straightforward. For example, just as NICs have developed their own transnational corporations to compete with those of developed countries, so they have developed large and aggressive banks with an increasing interest in international opportunities. By the end of 1990, Korean banks had established 58 overseas branches, 33 overseas subsidiaries and affiliates, and 52 representative offices in overseas locations.

The growth of the global financial network has created a significant new dimension to the world economy. As Barnet and Cavanaugh (1994, p. 17) put it,

> Twenty-four hours a day, trillions of dollars flow through the world's foreign-exchange markets as bits of data traveling at split-second speed. No more than 10 per cent of this staggering sum has anything to do with trade in goods and services. International traffic in money has become an end in itself, a highly profitable game.

Not all of the needs of this 'casino economy' can be met by conventional financial and business services. The need for secrecy and the desire for shelter from taxation and regulation have resulted in the emergence of a series of **offshore financial centres**: islands and micro-States such as the Bahamas, Bahrain, the Cayman Islands, the Cook Islands, Luxembourg, Lichtenstein, and Vanuatu that have become specialized nodes in the geography of world-wide financial flows (Roberts, 1994; see Fig. 2.14).

The chief attraction of these offshore financial centres is simply that they are less regulated than financial centres elsewhere. They provide low-tax or no-tax settings for savings, havens for undeclared income and for hot money. They also provide discreet markets in which to transact currencies, bonds, loans, and other financial instruments without coming to the attention of regulating authorities or competitors. The US Internal Revenue Service estimates that about US$300 billion ends up in offshore financial centers each year as a result of tax evasion schemes. Overall, about 60 per cent of all the world's money now resides offshore.

offshore financial centre
Island or micro-state that has become a specialized node in the geography of world-wide financial flows.

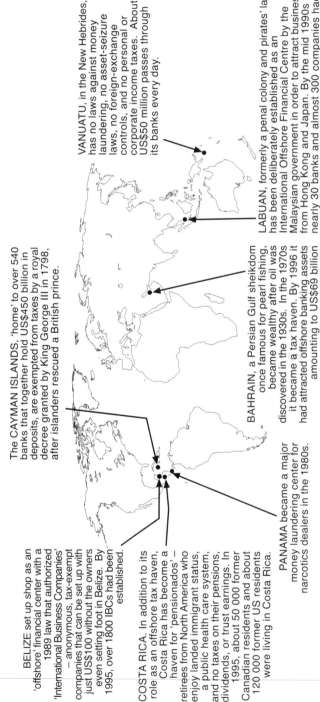

VANUATU, in the New Hebrides, has no laws against money laundering, no asset-seizure laws, no foreign-exchange controls, and no personal or corporate income taxes. About US$50 million passes through its banks every day.

The CAYMAN ISLANDS, 'home' to over 540 banks that together hold US$450 billion in deposits, are exempted from taxes by a royal decree granted by King George III in 1798, after islanders rescued a British prince.

LABUAN, formerly a penal colony and pirates' lair, has been deliberately established as an International Offshore Financial Centre by the Malaysian government in order to attract business from Hong Kong and Japan. By the mid 1990s nearly 30 banks and almost 300 companies had decided to take advantage of Labuan's zero tax on dividends, interest, and royalties and its 3 per cent tax on net profits.

BAHRAIN, a Persian Gulf sheikdom once famous for pearl fishing, became wealthy after oil was discovered in the 1930s. In the 1970s it became a tax haven. By 1996 it had attracted offshore banking assets amounting to US$69 billion

BELIZE set up shop as an 'offshore' financial center with a 1989 law that authorized 'International Business Companies' – anonymous, tax-exempt companies that can be set up with just US$100 without the owners even setting foot in Belize. By 1995, over 1800 IBCs had been established.

COSTA RICA. In addition to its role as an offshore tax haven, Costa Rica has become a haven for 'pensionados' – retirees from North America who enjoy landed immigrant status, a public health care system, and no taxes on their pensions, dividends, or trust earnings. In 1995, about 50 000 former Canadian residents and about 120 000 former US residents were living in Costa Rica.

PANAMA became a major money laundering center for narcotics dealers in the 1980s.

Figure 2.14
Offshore financial centres

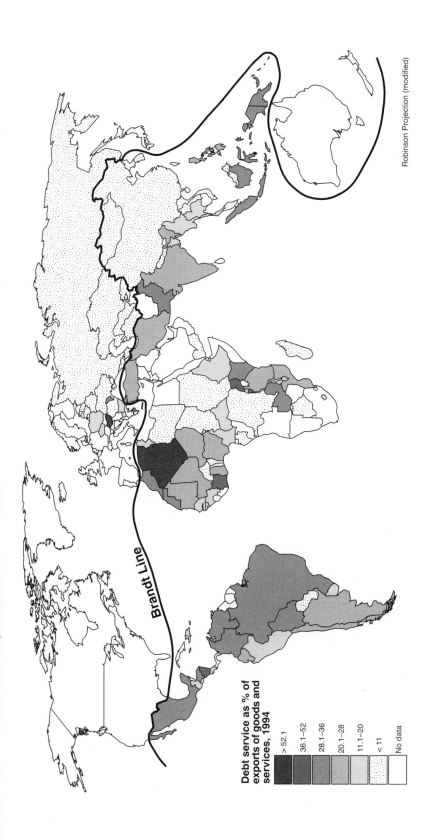

Robinson Projection (modified)

Debt service as % of exports of goods and services, 1994

> 52.1
36.1–52
28.1–36
20.1–28
11.1–20
< 11
No data

Brandt Line

Figure 2.15

The debt problem: total long-term debt service as a percentage of the value of exports of goods and services, 1994

Patterns of international debt

elasticity of demand
The degree to which levels of demand for a product or service change in response to changes in price. Where a relatively small change in price induces a significant change in demand, elasticity is high; where levels of demand remain fairly stable in spite of price changes, demand is said to be inelastic.

terms of trade
The ratio of the prices at which exports and imports are exchanged. When the price of exports rises relative to the price of imports, the terms of trade reflect an improvement for the exporting country.

import substitution
The development of domestic firms capable of producing goods or services formerly provided by foreign firms.

debt trap
The cycle of borrowing that results when the productivity gains from investments undertaken with borrowed capital are insufficient to meet interest repayments. Further loans and debt rescheduling provide temporary relief, but in the long term make it even more difficult to achieve increases in productivity sufficient for self-sustaining growth.

capital flight
The withdrawal of liquid assets from a national economy by domestic and overseas investors or transnational corporations.

Servicing long-term debts (i.e. meeting costs of both interest charges and repayments) has reached crisis proportions for some countries. The intensity of these crises is highlighted by Fig. 2.15, which shows countries' long-term debt service as a proportion of their capacity to earn foreign exchange, as reflected by the value of exports of goods and services. Some countries are clearly in serious trouble (including Hungary, and the NICs of Mexico and Brazil); but it is among the LDCs that the greatest debts and the heaviest burdens, in relation to earning capacities, have accrued: in the likes of Algeria, Nicaragua, and, where for every $100 earned through exports, more than $50 is owed in long-term debt servicing.

At the root of the international debt problem is the structured inequality of the world economy. The role inherited by most LDCs within the international division of labour has been one of producing primary goods and commodities for which both the **elasticity of demand** and price elasticity are low. That is, demand for their products in their principal markets (the more developed countries) tends to increase by relatively small amounts in response to significant increases in incomes within their principal markets. Similarly, significant reductions in the cost of their products tends to result in only a relatively small increase in demand. In contrast, the elasticity of demand and price elasticity of manufactured goods and high-order services (the specialties of developed economies) are both high. As a result, the **terms of trade** are stacked against the producers of primary goods. No matter how efficient primary producers may become, or how affluent their customers, the balance of trade will be tilted against them. Quite simply, they must run in order to stand still.

An obvious counter-strategy is to attempt to establish a new role in the international division of labour, moving away from a specialization in primary commodities towards a diversified manufacturing base. This strategy is known as **import substitution**. It is a difficult strategy to pursue, however, because establishing a diversified manufacturing base requires vast amounts of start-up capital. With the terms of trade running against them, it is extremely difficult for them to accumulate this capital. The only alternative, short of opting for self-sufficiency (e.g. Tanzania, Burma) or opting out of the capitalist world economy altogether (e.g. Cuba) is to raise the capital as loans. If the capital is invested in economic development projects that do not yield sufficient returns, further loans have to be undertaken in order to service the original debt and/or in order to finance new development projects. Hence the evolution of international debt. The syndrome of having constantly to borrow in order to fund 'development' has come to be known as the **debt trap**.

The debt problem has been seriously aggravated, however, by patterns of international capital flows that have been stimulated by changing money market conditions rather than by the geography of trade. In order to avoid the punitive 'taxation' of profits and savings by the high inflation rates caused by the combination of trade imbalances and debt, the domestic creditors of LDC governments (i.e. anyone holding its currency or its interest-bearing debt) tend to move their money abroad. The result is **capital flight**. According to estimates by the International

Monetary Fund, capital flight from the most heavily indebted nations in 1995 amounted to roughly half of the outstanding foreign debt of the countries concerned.

There is another problem. In the 1970s, world monetary reserves increased twelvefold with the availability of OPEC 'petrodollars' and as a by-product of the inflation that accompanied the break-up of the Bretton Woods system. The international banking system, awash with funds, began to 'recycle' the surpluses that oil companies had deposited with them. Bankers from developed nations suddenly found a willingness to lend to LDC governments. They found eager borrowers in LDCs desperate for capital. The result, however, was that many LDC governments committed themselves to capital projects, wage bills and debt repayments that were more than they could finance through taxes.

By 1981, LDC debt amounted to US$739 billion. When **'Reaganomics'** and **monetarism** drove interest rates up in the early 1980s, the debt burden became a debt crisis. In 1982, Mexico threatened to default on its debts, causing a widespread concern that such action might snowball to the point where international financial stability might be threatened. Recognizing this possibility (and realizing that their long-term interests depend not only on international financial stability but on reasonably healthy markets within LDCs), creditors in the developed countries allowed Mexico to reschedule its debts. This pattern of events has since been repeated several times. Brazil, for example, borrowed so much money in the 1970s and 1980s that it could no longer meet the interest payments. Between 1983 and 1989 the International Monetary Fund (IMF) bailed the country out, but on condition of austerity measures that were designed to curb imports. These included a 60 per cent increase in petroleum prices, and a reduction of the minimum wage to US$50 a month, which gave workers half the purchasing power they had in 1940. Nevertheless, by 1994, the country's debt had reached nearly US$130 billion; annual inflation rates in the 1990s have ranged between 500 per cent and 2000 per cent. The Philippine government, meanwhile, has had to reschedule its debts four times with the 'Paris Club', a group of 14 creditor nations. By 1995 the Philippines had reduced its foreign debt to $34 billion, and its debt-service ratio to less than 20 per cent of its GNP. By 1995, the accumulated debts of LDC countries had risen to US$2 trillion.

Patterns of international aid

The debt issue leads us logically to the question of aid. As with most of the other issues raised in this chapter, we shall reserve detailed discussion for subsequent chapters. For the moment, therefore, we shall merely outline the dominant patterns and the most striking trends.

Large-scale movements of aid began shortly after the Second World War with the Marshall Plan, financed by the United States to bolster war-torn European allies whose economic weakness, it was believed, made them susceptible to communism. During the 1950s and 1960s, as more LDCs gained independence, aid became a useful weapon in Western and Sino-Soviet Cold War offensives to establish and

'Reaganomics'
The application of supply-side economics to the management of the US economy in the 1980s. Supply-side economics is similar to monetarism in its disavowal of demand management; rather, the key to economic stability and well-being is seen to be the enhancement of aggregate supply. Reaganomics consisted of a set of objectives that included tax reduction, deregulation of business, increased government spending on defence, and decreased government spending on social welfare.

monetarism
A doctrine of macroeconomic management that disavows demand management and regards the money supply as the most important determinant of economic stability. Important in the USA and UK in the late 1970s and early 1980s, it reasserted the relevance of price theory and the importance of free markets.

preserve political influence throughout the world. By the late 1960s, the list of donor countries had expanded beyond the super-powers to include smaller countries such as Austria, Denmark and Sweden, whose motivation in aid-giving must be seen as more philanthropic or conscience-salving than political. In addition, there was a greater geographic dispersal of aid, thanks largely to the activities of multilateral financial agencies such as the International Monetary Fund (IMF) and the World Bank.

Nevertheless, the geography of aid still has a strong political flavour: Asia, for instance, receives less aid per capita than the average for all recipient countries, yet within this region Western strategic involvement has led in the past to South Korea and South Vietnam receiving above-average amounts of aid. Similarly, Turkey and Yugoslavia received above-average aid because of the special interest of the West in limiting the influence of the former Soviet bloc. Bilateral aid from several countries also reflects localized political aspirations and colonial ties. Thus much British and French aid is directed towards former African colonies, while Japanese aid is disbursed largely within Asia, and aid from the OPEC countries has been directed mainly towards the 'front-line' Arab countries (Hammond, 1994).

International *détente*, together with the balance of payments difficulties of several developed countries, have ensured that levels of aid have diminished. Thus, whereas official development assistance from OECD countries amounted to nearly 0.5 per cent of their total GNP in 1965, it had fallen to 0.33 per cent in 1993. The most striking decreases were those of the United States and the United Kingdom, whose overseas development assistance as a percentage of their GNP fell from 0.58 to 0.15 and from 0.47 to 0.31 respectively. The Japanese, meanwhile, have never exceeded one-third of 1 per cent of their GNP in aid. It is true that some countries – Denmark, the Netherlands, Finland, and Norway, in particular – have steadily increased their aid-giving; while France has maintained a relatively high level of donations.

Equally, the *impact* of aid on some countries is clearly significant. Sierra Leone, for example, received aid amounting to 164 per cent of its GNP in 1993; and eight other countries – Burundi, The Gambia, Guinea-Bisseau, Malawi, Mauritania, Mozambique, Tanzania, and Zambia – received aid equivalent to more than 25 per cent of their GNP. These figures, however, say as much about these countries' GNP as anything else. They are also exceptions. In general, the poorest countries are by no means the biggest recipients of aid; and the amount of aid received per capita is generally very low. In 1993, low-income developing economies received an average of just US$9.20 per capita in overseas development assistance, while lower-middle-income developing economies received US$17.80 per capita and upper-middle-income developing countries received US$10.20 per capita.

At these levels, aid cannot seriously be regarded as a catalyst for development or as an instrument for redressing North–South inequalities. On the other hand, because most of this aid is 'tied' in some way to donor countries' exports or to specific military, educational or cultural projects, it is argued by many that, insignificant as it is in relative terms, it *is* sufficient to reinforce the initial advantages of the 'donors'. Buchanan (1972), for example, wrote forcefully of military assistance that 'mercenarized' LDCs and of educational assistance that 'stole the souls' of their élite. Others have shown how 'cheap' loans have compounded the debt trap (see, for example, Payer, 1974). Paul Streeten put the whole argument very graphically:

> The Kings of Siam are said to have ruined obnoxious countries by presenting them with white elephants that had to be maintained at vast expense. In the modern setting this can be achieved best by tying a high-interest loan, called 'aid', to projects and to donors' exports and to confine it to the import content (or better still, some part of it) of the project. But even untied aid on soft terms can be used to promote exports of a white elephantine nature, because capital grants do not cover the subsequent recurrent expenditure which the elephant inflicts on its owners. Receiving aid is not just like receiving an elephant but like making love to an elephant. There is no pleasure in it, you run the risk of being crushed, and it takes years before you see the results.

(Streeten, 1968, p. 154)

Tourism and economic development

There are many parts of the world that do not have much of a primary base, that are not currently an important part of manufacturing commodity chains, and that are not closely tied in to the global financial network. For these areas, tourism can offer the otherwise unlikely prospect of economic development. Tourism has, in fact, become enormously important. By the year 2000 the largest single item in world trade will be international tourism; it is already the world's largest non-agricultural employer. One in every 15 workers, world-wide, is occupied in transporting, feeding, housing, guiding, or amusing tourists; and the global stock of lodging, restaurant, and transportation facilities is estimated to be worth about $3 trillion.

The globalization of the world economy has been paralleled by a globalization of the tourist industry. In aggregate, there were over 650 million international tourist trips in 1995, compared with just 147 million in 1970. Almost 70 per cent of these trips are generated by just 20 of the more affluent countries of the world. What is most striking, though, is not so much the growth in the number of international tourists as the increased range of international tourism. Thanks largely to cheap long-distance flights, a significant proportion of tourism is now trans-continental and trans-oceanic. Visits to peripheral countries in Africa, Asia, and Latin America now account for one-eighth of the industry. This, of course, has made tourism a central component of economic development in countries with sufficiently exotic wildlife (Kenya, for example), scenery (Nepal), beaches (the Seychelle Islands), shopping (Singapore and Hong Kong), culture (China, India, and Indonesia), or sex (Thailand).

But, although tourism can provide a basis for economic development in LDCs, it is often a mixed blessing. Tourism certainly creates jobs, but they are often seasonal jobs. Dependence on tourism also makes for a high degree of economic vulnerability. Tourism, like other high-end aspects of consumption, depends very much on matters of style and fashion. As a result, once-thriving tourist destinations can suddenly find themselves struggling for customers. Some places are sought out by tourists because of their remoteness and their 'natural', undeveloped qualities, and it is these that are most vulnerable to shifts of style and fashion. Nepal is a recent example of this phenomenon: it is now too 'obvious' as a destination and consequently the tourist industry in Nepal is having to work hard to continue to attract sufficient numbers of tourists. Bhutan, Bolivia, Estonia and Vietnam have been 'discovered', and are coping with their first real growth in tourism. Tourism in these more exotic tourist destinations is also vulnerable in other ways: to political disturbances, natural disasters, outbreaks of disease or food poisoning, and atypical weather. Fiji's military coup in 1987, for example, resulted in a devastating 70 per cent drop in tourism; while an outbreak of the plague in Surat in India in 1994 brought about a virtual halt to the flow of tourists to the whole country.

Although tourism is a multi-billion dollar industry, the financial returns for tourist areas are often not as high as might be expected. The greater part of the price of a package vacation, for example, stays with the organizing company and the airline. Typically, only 40 per cent is captured by the tourist region itself. If the package involves a foreign-owned hotel, this may fall below 25 per cent. Moreover, the costs and benefits of tourism are not only economic. On the positive side, tourism can help sustain indigenous lifestyles, regional cultures, arts and crafts, and provide incentives for wildlife preservation, environmental protection, and the conservation of historic buildings and sites. On the negative side, tourism can adulterate and debase indigenous cultures, and bring unsightly development, pollution and environmental degradation. Tourism can also involve exploitative relations that debase traditional lifestyles and regional cultural heritages as they become packaged for outsider consumption. Traditional ceremonies which formerly had cultural significance for the performers are now enacted only to be watched and photographed. In the process, indigenous cultures are edited, beautified, and altered to suit outsiders' tastes and expectations.

Such problems, coupled with the economic vulnerability of tourism, have led to the idea of 'alternative' tourism as the ideal strategy for economic development in peripheral regions. Alternative tourism emphasizes self-determination, authenticity, social harmony, preservation of the existing environment, small-scale development, and greater use of local techniques, materials, and architectural styles. Costa Rica, for example, has won high praise from environmentalists for protecting one-quarter of its territory in biosphere and wildlife preserves despite being a poor country. The payoff for Costa Rica is the escalating number of tourists who come to visit its active volcanoes, palm-lined beaches, cloud forests, and tropical parks that offer glimpses of toucan and coati. In 1995, Costa Rica received over 800 000 tourists. In that year, tourism exceeded banana exports as the country's main source of foreign exchange.

INTERPRETATIONS OF INTERNATIONAL INEQUALITY: UNDERDEVELOPMENT AND DEPENDENCY

By the 1960s it had became clear that international spatial inequalities were becoming more rather than less pronounced. In this context, a virtual avalanche of critical writings appeared, claiming that the prosperity of the developed countries in the world economy (the USA, Europe, and Japan, in particular) were based on *under*development and squalor in LDCs. The latter could not 'follow' the previous historical experience of developed countries because their underdevelopment was a structural requirement for development elsewhere. The development of Europe and North America, it was argued, *depended* on the systematic underdevelopment of LDCs. By means of unequal trade, exploitation of labour, and profit extraction, the underdeveloped countries were becoming *increasingly* rather than decreasingly impoverished.

The writing of André Gunder Frank exemplifies the explanations of international economic change that arose from this critique. Frank rejected the idea that underdevelopment is an original condition, equivalent to 'traditionalism' or 'backwardness'. To the contrary, it is a condition *created* by integration into the worldwide system of exchange that originated in the sixteenth century: the 'world capitalist system'. The concentration of poverty and the lack of development are not, he argued, a consequence of geographical isolation or the failure of western technology, capital, and values to spread. Rather, it stems directly from the nature of spatial relationships within the world capitalist system. The system is described by Frank (1967, pp. 146–7) as follows:

> As a photograph of the world taken at a point of time, this model of a world metropolis (today the United States) and its governing class, and its national and international satellites and their leaders – national satellites like the southern states of the United States, and international satellites like São Paulo. Since São Paulo is a national metropolis in its own right, the model consists further of its satellites, the provincial metropolises, like Recife or Belo Horizonte, and their regional and local satellites in turn. That is, taking a photograph of a slice of the world we get a whole chain of metropolises and satellites, which runs from the world metropolis down to the hacienda or rural merchant who are satellites of the local commercial metropolitan centre but who in their turn have peasants as their satellites. If we take a photograph of the world as a whole, we get a whole series of such constellations of metropolises and satellites.

This unequal structure to the world economy has been in place since Europeans first ventured out into the world in the sixteenth century. Although the form of the monopoly power of metropolis over satellites has changed (for example, with the switch from merchant to industrial capitalism in the nineteenth century and following political independence for former colonies), the system of 'surplus expropriation' (the transfer of wealth from satellites to metropolis) has continued to fuel growth in some places at the expense of others.

Frank's approach is an example of what can be called 'dependency theory'. This has been a very influential approach in explaining global patterns of development and underdevelopment. It states, essentially, that development and underdevelopment are reverse sides of the same process. *Development somewhere requires*

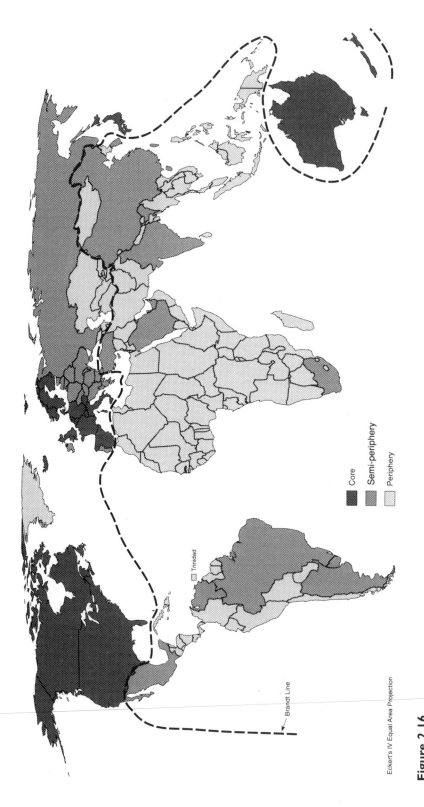

Eckert's IV Equal Area Projection

Core

Semi-periphery

Periphery

Trinidad

Brandt Line

Figure 2.16
The world-system: core, periphery, and semi-periphery

underdevelopment somewhere else. Independent development is impossible. Recent analyses have put more emphasis on the nature of the political and social structures created by external dependence and how these, in conjunction with the direct effects of external dependence, limit 'real' development (e.g. Slater 1992). It is plain that concepts of underdevelopment and dependency are based (without stating it) on the notion of perpetual spatial disequilibrium between metropolis and satellites. However, as we suggested in Chapter 1 and will argue in more detail in the next chapter, the logic and demands of the world economy are in a continual state of reconstitution. The rise of the US from satellite to metropolis, the challenge and the collapse of the Soviet Union, the appearance of the NICs, and the increasing importance of transnational firms are all contradictory to this notion of permanent spatial disequilibrium.

Immanuel Wallerstein's world-system theory (1984) takes this kind of disequilibrium into account. According to this perspective, the entire world economy is to be seen as an evolving market system in which an economic hierarchy of states – a *core*, a *semi-periphery* and a *periphery* – is a product of the long-wave economic rhythms that dominate the dynamics of the system. Because of the dynamics of these economic long-waves, the composition of this hierarchy is variable: countries can move from periphery to semiperiphery, core to periphery, and so on. The labels 'core' and 'periphery' are used by Wallerstein to refer to the dominant *processes* operating at particular levels in the hierarchy. Core processes are characterized by economic relations that incorporate relatively high wages, advanced technology and a diversified production mix, whereas periphery processes involve low wages, more rudimentary technology and a simple production mix. The label 'semi-periphery' refers to places where there is, at present, a mix of both sets of processes. In essence, semi-peripheral countries are seen as exploiting peripheral countries while being exploited by core countries.

Fig. 2.16 represents an attempt to capture the current composition of the three categories, based on countries' total GDP and GDP per capita. Countries with high scores on these two criteria are likely to have politically strong states, large internal markets, and predominantly high-wage, capital-intensive production – all, theoretically, defining characteristics of core status (Terlouw, 1989). Conversely, countries with a low national economic output and a low GDP per capita are likely to have weak states, and predominantly low-wage, labour-intensive production. These are rather sweeping assumptions; and the allocation of individual countries to particular categories is inevitably somewhat arbitrary. What is important in the present context is to recognize that the division of the world into a core, a semi-periphery and a periphery in this way represents more than an alternative classification of countries: it is a reflection of a particular conception of the dynamics of the world economy.

Summary

Brief reviews of the major aspects of international economic differentiation have been sufficient to have illuminated both the dominance of core–periphery patterns and the extent of the gradient between core and periphery. Other important points to have emerged include:

- The existence of a distinctive group of semi-peripheral countries with intermediate levels of living. This group includes two very different types: resource-exporting countries and newly industrializing countries.
- The speed and intensity of changes in patterns of economic activity and development associated with the globalization of economic activity under the influence of the strategies of transnational corporations.
- The degree of dependency of peripheral countries created by inequalities in resources, trading relationships, access to financial resources, and control of economic activity.

Key Sources and Suggested Reading

Barnet, R. J. and Cavanaugh, J. 1994. *Global Dreams: Imperial Corporations and the New World Order.* New York: Simon and Schuster.

Berry, B. J. L., Conkling, E. C. and Ray, D. M. 1976. *The Geography of Economic Systems.* Englewood Cliffs, NJ: Prentice Hall.

Blaikie, P. 1989. The use of natural resources in developing and developed countries. In R. J. Johnston and P. J. Taylor (eds), *A World in Crisis?* 2nd edn. Oxford: Blackwell, 125–50.

Bradley, P. N. and Carter, S. E. 1989. Food production and distribution – and hunger. In R. J. Johnston and P. J. Taylor (eds), *A World in Crisis?* 2nd edn. Oxford: Blackwell, 101–24.

Buchanan, K. 1972. *The Geography of Empire.* London: Spokesman Books.

Christopherson, S. 1995. Changing women's status in a global economy. In R. J. Johnston, P. J. Taylor and M. J. Watts (eds), *Geographies of Global Change.* Oxford: Blackwell, 191–205.

Corbridge, S. 1992. *Debt and Development.* Cambridge, MA: Blackwell.

Dicken, P. 1992. *Global Shift.* 2nd edn. London: Harper & Row. (1st edn. 1986).

Frank, A. G. 1967. *Capitalism and Underdevelopment in Latin America.* New York: Monthly Review Press.

Friedmann, J. 1956. Locational aspects of economic development. *Land Economics,* **32**, 213–27.

Grigg, D. 1985. *The World Food Problem.* Oxford: Blackwell.

Hammond, R. 1994. A geography of overseas aid, *Geography,* **79**, 210–21.

Hauchler, I. and Kennedy, P. M. 1994. *Global Trends.* New York: Continuum.

Hicks, J. R. 1959. *Essays in World Economics.* Oxford: Oxford University Press.

Hirschman, A. O. 1958. *The Strategy of Economic Development.* New Haven: Yale University Press.

Holloway, S. R. and Pandit, K. 1992. The disparity between the level of economic development and human welfare. *Professional Geographer,* **44**, 57–71.

Independent Commission on International Development Issues 1980. *The Brandt Report.* Oxford: Oxford University Press.

International Monetary Fund 1991. *World Imports and Exports.* Washington, DC: International Monetary Fund.

Johnston, R. J. 1984. The world is our oyster. *Transactions, Institute of British Geographers,* **9**, 443–59.

Johnston, R. J., Taylor, P. J. and Watts, M. J. (eds) 1995. *Geographies of Global Change.* Oxford: Blackwell.

Krugman, P. 1990. *The Age of Diminished Expectations: US Economic Policy in the 1990s.* Cambridge, MA: MIT Press.

Krugman, P. 1991. *Geography and Trade.* Cambridge, MA: MIT Press.

McGranahan, D. V. *et al.* 1970. *Content and Measurement of Socio-Economic Development.* Geneva: UNRISD.

Meadows, D. H. *et al.* 1972. *The Limits to Growth.* London: Earth Island.

Meier, G. M. and Baldwin, R. E. 1957. *Economic Development.* New York: Wiley.

Morris, M. D. 1980. *Measuring the Condition of the World's Poor.* New York: Pergamon.

Myrdal, G. 1957. *Economic Theory and Underdeveloped Regions.* London: Duckworth.

Payer, C. 1974. *The Debt Trap.* Harmondsworth: Penguin.

Roberts, S. 1994. Fictitious capital, fictitious spaces: the geography of offshore financial flows. In S. Corbridge, R. Martin and N. Thrift (eds), *Money, Power, and Space.* Oxford: Blackwell, 91–115.

Slater, D. 1992. On the borders of social theory: learning from other regions. *Society and Space,* **10**, 307–27.

Streeten, P. 1968. A poor nation's guide to getting aid. *New Society,* **18**, 154–6.

Tata, R. J. and Schultz, R. R. 1988. World variation in human welfare: a new index of development status, *Annals, Association of American Geographers,* **78**, 580–93.

Terlouw, C. P. 1989. World-System Theory and Regional Geography. *Tijdschrift voor Economische en Sociale Geografie,* **80**, 206–21.

Thrift, N. 1989. The geography of international economic disorder. In R. J. Johnston and P. J. Taylor (eds), *A World in Crisis?* 2nd edn. Oxford: Blackwell, 16–78.

United Nations 1984. *Land, Food, and People.* Rome: FAO.

United Nations 1991. *World Investment Report.* United Nations Centre on Transnational Corporations, New York.

Vance, J. Jr. 1970. *The Merchant's World: The Geography of Wholesaling.* Englewood Cliffs, NJ: Prentice Hall.

Wallerstein, I. 1984. *The Politics of the World-Economy.* Cambridge: Cambridge University Press.

World Bank 1986. *Poverty and Hunger.* Washington, DC: World Bank.

World Bank 1995. *World Development Report, 1995.* Washington, DC: World Bank.

World Bank 1996. *From Plan to Market: World Development Report 1996.* New York: Oxford University Press.

World Commission on Environment and Development 1987. *Our Common Future.* New York: Oxford University Press.

World Resources Institute 1986. *World Resources 1986.* New York: Basic Books.

World Resources Institute 1996. *World Resources 1996–97.* New York: Oxford University Press.

CHAPTER 3

THE GEOGRAPHICAL DYNAMICS OF THE WORLD ECONOMY

In the years immediately after the Second World War economic growth was viewed by most commentators as a constant even though its rate of increase might vary from time to time and from place to place. The memory of the Depression that had afflicted the world economy in the 1930s faded very fast. Since the late 1960s, however, confidence in the permanence and inevitability of economic growth has been shaken. Perhaps the clearest indication of the onset of a crisis in the expectation of growth came in the aftermath of the dramatic increase in world oil prices in 1973. National economies, especially that of the United States, which had appeared to be largely self-contained, autonomous, and invulnerable to 'external' pressures, suddenly appeared as anything but. This led some economic geographers, economists and historians to advocate historical–global perspectives on economic growth and change. Cyclical theories of economic development, long out of style among intellectuals, gained increasing attention (see Chapter 1). More particularly, questions of location and spatial structure were increasingly related to questions concerning the dynamics of an evolving world economy. In particular, changing spatial patterns of economic activities were traced to shifts in the workings of the world economy.

A basic issue involved the 'newness' of this world economy. From one point of view the 1960s marked, as we saw in Chapter 2, the beginning of a new, more interdependent world economy. In particular, faced with dramatic drops in their rates of profit from established capital–labour relations and associated patterns of location (and other pressures such as increasing environmental regulations) many firms operating in Western Europe and the United States, and some elsewhere, began to reorganize their operations and thus improve their profitability by subcontracting some of their activities and internationalizing their production facilities (see Chapters 6, 7 and 10). But from another viewpoint there was already a world economy in existence that was simply undergoing a process of 'globalization'. Beginning in the sixteenth century, but undergoing its greatest expansion and intensification in the nineteenth and twentieth centuries, a world economy had evolved out of more localized economic systems. As it has become progressively more integrated, covering ever wider geographical areas and more and more economic activities (resource extraction, capital investment, trade in manufactures, services, etc.), it has undergone shifts in its mode of operation (see Chapters 4, 5 and 8). The changes beginning in the late 1960s are the most recent manifestation of this evolutionary process.

The purpose of this chapter is, first, to sketch the historical development of the modern world economy; second, to pinpoint the geographical effects of state regulatory and macroeconomic actions; and, third, to identify the main causes and consequences of the current geographical reorganization of the world economy. The intention is to provide both a framework for understanding economic landscapes, such as those described most generally in Chapter 2 and more specifically in Parts 2 and 3, to offer an outline of the historical context for the emerging trends in world economic geography examined in Part 4, and give an overview of an important theoretical trend in the field of economic geography: geopolitical economy, the impact of states on the working of the world economy and the geography of economic activities.

THE HISTORY OF THE WORLD ECONOMY

Several theories have been proposed to account for the emergence and structure of a world economy. The most important ones are (1) long-wave theories, which focus on cycles of technological innovation and their economic effects; (2) world-system theories, which argue for the progressive incorporation of the world into a European world economy from the sixteenth century onwards with a global periphery providing the basis for the growth of the economy of the core (originally Western Europe); and (3) regulationist theories, which see the world economy as the outcome of interaction between national economies and their internal regulation. Each of these theories has an emphasis that the others lack. A thorough account of the world economy needs to incorporate each one's major insight: (1) the historical evolution of the world economy; (2) the geographical structure of the world economy; and (3) the role of national economic structures in shaping the evolution of the world economy. Of course, in the act of combining elements of each of these theories we end up with a perspective that is different in its entirety from any one of them.

We start from some basic distinctions provided by the world-systems theory of Wallerstein. In Wallerstein's (1979, p.17) view, at one time all societies were **minisystems**: 'A minisystem is an entity that has within it a complete division of labour, and a single cultural framework.' Such minisystems would include very simple hunting and gathering and some agricultural societies. But they no longer exist. As soon as they became tied to empires or the world economy they ceased to be separate systems (see this chapter pp. 77–80 and Chapter 4).

Empires and the world economy (world-economy for Wallerstein to symbolize its integrated nature) are examples of **world-systems**: units with a single division of labour but multiple cultural systems. In one case, that of a unit with a common political system, there is a world-empire, in the other, where there is no political integration, there is a world-economy.

minisystems
Local societies with a simple division of labour within a single cultural framework, such as hunter–gatherer and some agricultural societies.

world-systems
Any spatially-extensive economic system that has a single division of labour but multiple cultural systems.

Until the advent of capitalism in Europe, world-economies were unstable and tended towards disintegration or conquest by one regional group and, hence, transformation into world-empires. Examples of such world-empires emerging from incipient world-economies are all of the so-called great civilizations of ancient times such as China, Egypt, Rome. World-empires, however, undermined the growth of their territories by using too much of the economic surpluses they generated to expand their bureaucracies and military establishments.

> Around AD 1500 a new type of world economy, the modern capitalist one, began to take shape: In a capitalist world-economy, political energy is used to secure monopoly rights (or as near to them as can be achieved). The state becomes less the central economic enterprise than the means of ensuring certain terms of trade and other economic transactions. In this way, the operation of the market (not the free operation but nonetheless its operation) creates incentives to increased productivity and all the consequent accompaniment of modern economic development. The world-economy is the arena in which these processes occur.
>
> (Wallerstein, 1974, p. 16)

How and why Europe alone embarked on the path of capitalist development through incorporating and subjugating the rest of the world is beyond the scope of the present study, although some suggestions are provided in Chapter 4. Suffice it to say here that a set of practices and beliefs evolved in Europe that emphasized the centrality to life of capital accumulation and production for the market. The cultural system of exchange and accumulation that emerged in late medieval Europe allied to the political fragmentation of the region into territorial states (stimulating extra-European competitive expansion) was what made the peculiarly modern world economy possible.

The causes of capitalism's success as a means of organizing the world economy are also complex, but two seem fundamental. First, new transportation technologies, first, ocean-going sailing ships and later steamships and railways, allowed far-flung resource areas to be connected with markets, and European military technology provided the means to enforce favourable terms of trade. Second, released by the growth of the new territorial states (such as England and Holland) from the burden of maintaining relations of tribute within their zones of operation (they no longer had to buy off local despots or warlords once the 'rule of law' was established and private property rights were legally protected), merchant capitalists could attend single-mindedly to the geographical expansion of their interests. Of special importance were Dutch and English capitalists, who were able in the sixteenth century to overcome the Spanish–Habsburg attempt to turn the emerging world economy into a world-empire.

The single world market

Six basic features of the world economy can serve to organize our account of it, drawing from an array of different sources. First, the modern world economy consists of a single world market. Within this, production is for exchange rather than use: producers exchange what they produce for the best price they can get. As the price of a product is not fixed but set in a geographically extensive market, there is competition between producers, at least for periods of time. More efficient producers can undercut other producers to increase their share of total production and achieve monopoly. Until this century, control was largely achieved through territorial division among national enterprises that aimed for localized or regional rather global monopoly. Progressively, and particularly since the 1970s, the world as a whole figures in decisions concerning trade and investment.

There is controversy among economic historians over whether the world economy can properly be labelled 'capitalist' before the late eighteenth century and whether extra-European expansion was of much account before the nineteenth century. Certainly, price-setting markets in the modern sense took many years to become established. They did not spring up overnight in the sixteenth century. In particular, labour did not become a commodity to be bought and sold like any other (**wage labour**) until the late eighteenth century (Brenner, 1977). But it is probably academic pedantry of the worst sort to restrict the term capitalist to the last two centuries alone. Most commentators are agreed that between 1500 and the late eighteenth century there was a vast expansion of trade within Europe and between Europe and other parts of the world based increasingly on market exchange. However, the late eighteenth century does mark an important qualitative change in the nature of the world economy. European industrialization produced both a tremendous intensification of world trade and increasingly complex markets for raw materials and manufactured goods. Extra-European investment and trade also became much more important for the growth of Europe after this time. Previously, most economic growth in Europe had been generated internally (O'Brien, 1982). At the same time labour markets have also become more integrated. Major waves of international migration in the nineteenth and late twentieth centuries testify to the periodic increase in geographical scope of the pools of labour from which firms and regions can draw on to fuel their growth (see, e.g. Harris, 1996). This is another indicator, therefore, of the emergence of a single world market extending beyond the simple case of international trade.

wage labour
When people work in exchange for monetary payment rather than bartered goods, military protection or as a result of enslavement.

The state system

Second, the modern world economy has always had a territorial division between political states. This division both pre-dates and grew along with the geographical expansion of the modern world economy. In the present context its most important feature is that it has ensured that the world economy has not as yet become a world-empire. At the same time it provides through **tariffs**, trade quotas and financial incentives for the protection and stimulus of developing or **infant industries** (by restricting access to 'domestic markets' of foreign-made goods and encouraging development of domestic production) and a way for groups of capitalists to protect their interests collectively and distort markets to their advantage. The result of this process is a competitive state system in which each state attempts to the best of its ability to insulate itself from the rigours of the world market while trying to turn the world market to its advantage.

Even in cases such as Britain and the United States, where government economic intervention has long been subject to political dispute and opposition, the states and

tariffs
The schedules of duties or taxes imposed by a government on exported or, more typically, imported goods and services.

infant industries
Industries at an early stage of development that are protected by tariffs and other trade barriers until they can survive foreign competition without protection.

the world economy grew together. Elsewhere, for example in France, the connection has often been tighter. A strongly coercive political atmosphere and government investment especially in physical and social infrastructure (e.g. roads, railways, schools, health-care facilities), have usually lain behind successful attempts at overcoming initial disadvantages within the world market. The 'Prussian road to capitalism', as Karl Marx termed it, has been an important characteristic of all 'late-developing' states. The Netherlands, Britain and the United States are the exceptions. These were perhaps the only important cases where economic growth was largely in 'private' hands from the start. Even in these countries the state very quickly became involved in various allocation and stabilization activities that have progressively grown in scope during the last 100 years. Mobilization of national resources, in other words, does not usually arise naturally out of the free play of market forces but through the stimulus and direction given by government intervention.

competitive advantage
The advantage acquired in economic competition by some locations because of the benefits that accrue from an early start in production of a particular good and the continuing defence of that historic base through superior organization and adaptability.

Obtaining **competitive advantage** does not always require conventional protectionist measures such as tariffs on competing products. However, as Ettlinger (1991, p. 401) argues in an interesting comparative study of the growth of high-tech industries in Japan and California, 'import substitution and the creation of new industrial sectors are possible with negligible protection, provided that other institutions are in place to support and nurture a small and emerging small firm sector'. Both formal or governmental and informal or communal institutions are needed to provide the financial resources and cooperative economic environment in which new economic activities can flourish. Numerous historical examples attest to the importance of the mediating role of the state and other institutions, from the stimulus to US economic development from canal building and high industrial tariffs in the early nineteenth century and the high spending on military goods since the Second World War through Japan's disciplined conquest of market share in a wide range of industries to Taiwan's highly organized entry into the global microelectronics industry in the 1980s.

States have also become the most important means of defining political identity. People identify with the states of which they are subjects or citizens. As a result of pressure from popular groups, states also supply a range of goods and services that are either under-supplied or would not be made available to all through private provision. Preparation for war with other states and international economic competition have also expanded the scope of governmental activities. In recent years the seemingly inexorable growth of the state has been subject to a number of checks as governments have found themselves overextended relative to their resources. Ideological opposition to 'big government' has also increased, especially in the English-speaking countries where criticism of the state's power in relation to people's private lives has been strong for many years.

States have also been important in creating alternative adaptations to the world economy (see the later section on Alternative Adaptations). The most important 'experiment' in providing a consciously designed departure from the guiding principles of the world economy was the 'model' of economic development established in the Soviet Union in the aftermath of the Russian Revolution of 1917. As institutionalized under the dictatorship of Stalin this model involved the creation of a national economy commanded from the political capital of Moscow and designed to produce goods that would serve the 'national interest' in competition with foreign enemies whose economies remained part of the capitalist world economy. After the Second World War this model was imposed upon those countries in East-Central Europe liberated from German occupation by Soviet forces. Some other countries such as China, North Korea, Vietnam, and Cuba adopted the model or variants of it after their own revolutions. To a degree there was an organized interdependence between these countries through military alliances such as the Warsaw Pact in Eastern Europe and an economic arrangement such as the Council for Mutual Economic Assistance (CMEA). An equivalent of Fordist-type production became prevalent but it never extended to the realm of consumption. Trade between socialist countries, largely through barter (exchange of goods without monetary mediation), created a degree of specialization in economic activities between regions and countries. In the 1960s trade with the world economy began to expand, with important exports of weapons and oil and natural gas. In return 'hard currencies', such as US dollars and Swiss francs, foodstuffs, and raw materials were imported.

The 'crisis of the Soviet system' in the 1980s can be seen as emanating in part from a combined failure to deliver consumer goods and shift the organization of production from a focus on **scale economies** to **economies of scope** while at the same time the population became increasingly well informed about the consumer societies of their foreign enemies. The cynical manipulation of power for personal gain by ruling élites and the drain on national resources from the arms race with the United States also undoubtedly played some role in undermining the Soviet model. Surprising even the most astute observers, the Soviet experiment unravelled rapidly between 1989 and 1991. The former Soviet Union and the countries of Eastern Europe are now 'ordinary countries' again, struggling to find their place in a world economy that was abandoned for what turned out to be a failed attempt at 'socialism in several countries' (Deudney and Ikenberry, 1991/92; Hewett and Gaddy, 1992; also see Bradshaw, 1991). (A later section – pp. 84–97 – deals in more detail with the impact of the contemporary state system upon the geography of the world economy.)

scale economics
Cost advantages from large-scale production.

economies of scope
Cost advantages from large-scale flexible organization.

The three geographical tiers

Third, the modern world economy has established a basic three-tiered geography as it has expanded to cover the globe. This geography is defined by the global division of labour at any particular time. The early world economy consisted of Europe and those parts of South and Central America under Spanish–Portuguese control. The rest of the world was an external arena, essentially outside the workings of the world economy. Later, by means of plunder, European settlement and the reorientation of local economies to the world economy, the rest of the world became transformed into a periphery like the Latin American colonies. The world economy thus came to consist of a core (Western Europe at first, joined later by the United States and Japan) and a periphery. The core is defined by processes that produce control over the world economy, advanced technology, diversified production and relatively high average incomes. The periphery is defined in terms of processes that lead to dependence, primitive technology, undiversified production, and relatively low average incomes. The higher development of the core is fuelled in part by the lower development of the periphery. Uneven development at a global scale, therefore, is not a recent phenomenon or a mere by-product of the world economy; it is one of the world economy's basic features.

It would be a mistake to think, as we tend to do, that the world outside of Europe and a few other blessed spots was uniformly 'backward'. Indeed, Europe's industrialization was not based so much on acquiring raw materials from the rest of the world and carving out markets for finished products there, but on forcing industries elsewhere, overwhelmingly organized on a small-scale craft basis, out of business. It was about substituting European (and, later, American) mass-produced goods for locally produced craft goods. Before the European Industrial Revolution, Asia and other parts of what we have conventionally called the 'Third World' contained a far larger share of world manufacturing output than did Europe. Bairoch's (1982) calculations, though challenged somewhat by Maddison (1983), reveal a picture not simply of higher rates of industrialization in core countries but also of *de*industrialization in the periphery as cheaper mass-produced European goods forced traditional producers out of business (Table 3.1).

However, even as it expanded to incorporate ever-greater parts of the world, the system was not without mobility. There has been movement between the two categories of core and periphery, as attested to by the 'rise' of the United States and Japan, and vice versa, decline, as in the case of Spain and Portugal. A third geographical tier, the semi-periphery, applies to the processes operating in certain parts of the world to provide movement between periphery and core (and vice versa). This is a zone in which a mix of core and periphery processes are at work. Essentially, political conditions determine whether core processes come to dominate and allow for 'upward mobility' within the world economy (or vice versa and 'downward mobility'). The application of protectionist and other measures can initiate economic mobility (mentioned under the previous point) if undertaken in conjunction with favourable global economic conditions. This was the case for the United States in the aftermath of the Civil War (1870–1900) and Japan somewhat later.

There are a number of different ways of classifying states into the three tiers of the contemporary world economy. The two most common are based on relations between states (e.g. trade flows, military interventions, diplomatic exchanges, etc.) and the individual characteristics of states (e.g. size of GNP, part of GNP in world trade, etc.). Irrespective of method, the United States and most of Europe are always classified as core, whereas most of Africa is always classified as periphery. The major differences arise in identifying the semi-periphery and its stability over time. This suggests that 'the existence of a distinct semi-peripheral group of states is . . . very doubtful' (Terlouw, 1992, p. 45). This category is in a state of flux and cannot be given a fixed position within the spatial structure of the world economy. Any allocation to it, therefore, is tentative and subject to reversal.

Table 3.1: World manufacturing 1750–1900. (A) relative shares of world manufacturing output; (B) per capita levels of industrialization (Britain in 1900 = 100)

A	1750	1800	1830	1860	1880	1900
Europe	23.2	28.1	34.2	53.2	61.3	62.0
Britain	1.9	4.3	9.5	19.9	22.9	18.5
Habsburg Empire	2.9	3.2	3.2	4.2	4.4	4.7
France	4.0	4.2	5.2	7.9	7.8	6.8
Germany	2.9	3.5	3.5	4.9	8.5	13.2
Italy	2.4	2.5	2.3	2.5	2.5	2.5
Russia	5.0	5.6	5.6	7.0	7.6	8.8
United States	0.1	0.8	2.4	7.2	14.7	23.6
Japan	3.8	3.5	2.8	2.6	2.4	2.4
Rest of World	73.0	67.7	60.5	36.6	20.9	11.0
China	32.8	33.3	29.8	19.7	12.5	6.2
India/Pakistan	24.5	19.7	17.6	8.6	2.8	1.7

B	1750	1800	1830	1860	1880	1900
Europe	8	88	11	16	24	35
Britain	10	16	25	64	87	(100)
Habsburg Empire	7	7	8	11	15	23
France	9	9	12	20	28	39
Germany	8	8	9	15	25	52
Italy	8	8	8	10	12	17
Russia	6	6	7	8	10	15
United States	4	9	14	21	38	69
Japan	7	7	7	7	9	12
Rest of World	7	6	6	4	3	2
China	8	6	6	5	5	3
India/Pakistan	7	6	6	3	2	1

Note: Territories of Germany, Italy and India/Pakistan are always those of 1900, others are at dates shown.
Source: Bairoch (1982), pp. 294, 296.

Temporal patterns and hegemony

Fourth, the modern world economy has followed a number of cyclical patterns of growth and stagnation. The nature and causes of these temporal patterns are the subject of considerable controversy (Tylecote, 1992; Berry, 1991). Some of the main cycles that have been identified in the world economy are reviewed in Chapter 1. However, there is evidence from price and some growth data that since the late eighteenth century the world economy has gone through four major cycles of growth (A) and stagnation (B) (a diagram showing one estimate of the four cycles and the specifics of American economic experience as measured by the US wholesale price index is provided in Fig. 3.1). It is important to note that the evidence about

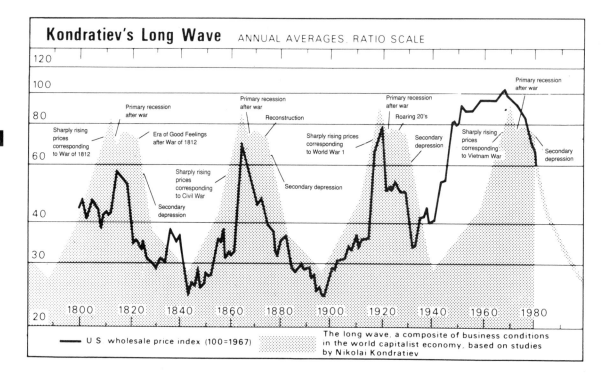

Kondratiev's Long Wave ANNUAL AVERAGES, RATIO SCALE

U S wholesale price index (100=1967)

The long wave, a composite of business conditions in the world capitalist economy, based on studies by Nikolai Kondratiev

Figure 3.1
Kondratiev's long-wave cycle and American experience

Source: Agnew (1987), p. 22

trends in the world economy is generally better concerning the more recent cycles; the nineteenth-century ones are both more difficult to discern and the data available are less reliable. When four cycles are identified they are usually presented in terms of the following approximate dates:

I: 1780–90 – A – 1810–17 – B – 1844–51
II: 1844–51 – A – 1870–75 – B – 1890–96
III: 1890–96 – A – 1914–20 – B – 1940–45
IV: 1940–45 – A – 1967–73 – B – ?

A widely accepted, if hardly definitive, explanation for these cyclical trends (often called Kondratiev long-wave cycles after the Russian economist who first identified them) lies in the periodic growth and exhaustion of technology and organizational systems under industrial capitalism. In one account, the B phases are ones in which new technological systems are not yet matched by corresponding socio-institutional frameworks that successfully harness the technologies for capital accumulation. After a period of crisis this eventually happens but by then a new technological system is under way and the mismatch occurs again (Perez, 1983). As we saw in Chapter 1, five technological systems and their times of crystallization or initiation can be identified:

I: Water style (around 1785 in England)
II: Steam transport style (late 1820s)
III: Steel and electricity style (late 1870s)
IV: Fordist style (around 1915 in the United States)
V: Microelectronics and biotechnology style (late 1970s–).

A different way of putting the process of technological innovation and its institutionalization would be to say that once dominant technologies emerge they become progressively more 'locked in'. These are all technologies that seem to produce increasing returns to adoption. Small initial advantages in use of the critical new technologies and subsequent refinements in them bring much larger or increasing returns to those firms (and places) that have them (Arthur, 1989). At a later date, however, this can become a disadvantage. Locked in to older technologies, firms (and places) find greater difficulty in adopting and adapting to new technological systems. Small initial differences, therefore, can be seen as producing new technological systems that then get locked in and produce competitive advantage for those who have early access to them. After a certain point, however, diminishing returns set in and the ground is set for a new round of technological innovation.

The so-called knowledge-based industries that are particularly important today are subject to increasing returns because the businesses involved share three features. They have high fixed costs, particularly in research and development, but very low variable costs. For instance, the cost of writing computer software is the same without regard to the number of copies that are sold. As a result, the higher the sales, the greater the profit margin. Second, once in place with a large number of users the software can become a standard for all users (think of Windows 95) because others will want to produce new software that is compatible with the standard. This gives the 'market leader' a distinct and continuing advantage. Third, customers get locked in to established standards, largely because they must learn how to use them. Customer loyalty guarantees future sales when the software is upgraded. It is hard to say when these competitive advantages will be exhausted. In the past they always have been, eventually. One difference from the past is that today technologies spread quickly between countries with similar communication infrastructures and support networks so that the 'national appropriation' of technology is less and less feasible. So, even if some firms can capture competitive advantages, it is much harder for countries to do so.

Secondary bursts of growth and stagnation in the world economy conform to cycles within the long-wave cycles (called Kuznets cycles) that relate to the consequences for aggregate economic growth of overinvestment in new technologies, overshoot in supply and collapse in demand. These 11–15-year acceleration and 11–15-year deceleration phases create the familiar business cycle of expansion/recession that attracts so much public debate (see Chapter 1).

In institutional terms some of the key features of the organizational evolution of capitalism correlate with the long-wave cycles. Competitive capitalism prevailed (before and) during the two Kondratiev cycles from the late eighteenth to the late nineteenth century. Small industrial firms operating in competitive national markets with foreign trade under rigid government control displaced more free-wheeling merchant capitalism. Organized capitalism developed through the next two cycles. In this period average firm size grew significantly and the new firms began to operate globally while maintaining strong regional bases within their home countries. Declining rates of profit under existing locational arrangements provided a powerful stimulus to global operations, particularly the establishment of branch plants free of the labour market constraints and environmental regulations increasingly prevalent in the USA and some European countries. The period since the early

regime of accumulation
A particular form of organizing economic production, income distribution, consumption, and public goods and services.

1970s marks the emergence of a disorganized capitalism: national markets are less regulated by nation-based firms and governments in general have less regulatory control over their national economies, average plant size has decreased, production has spread globally and service industries have grown faster than manufacturing ones (see Chapter 1 and Chapters 4–8 and 10 for more detailed descriptions). Each of these represents a distinctive **regime of accumulation** – an inter-related complex of production, consumption, and income distribution based on the organization of capitalist firms. Regimes of accumulation evolve in response to the opportunities and constraints created by new production, transportation and communications technologies. Associated with each is a specific mode of regulation or set of economic and political arrangements that emerges to provide appropriate management for the regime of accumulation (e.g. monetary and wage regulation, particular government–business relationships, trading regulation, etc.).

The shift from organized to disorganized capitalism – as with all transitions – has been more an erosion than a dramatic break. Beginning in the 1960s there was a slow-down in productivity in most industrialized economies and a squeeze on profits that led firms to adopt investment strategies involving the increased use of new technologies and locational adjustments (the establishment of branch plants in low-wage locations, etc.). In a detailed examination of the 'profit squeeze' hypo-thesis one group of economists argue that 'The writing was already on the wall for the Fordist system of production in the 1970s' (Glyn *et al.*, 1990, p. 88). As a consequence a 'Golden Age' of cooperative or organized capitalism within the main industrialized national economies gave way to a much more volatile and globalized system of production. In other words, national economies became less and less integral entities but were fragmented under the growing pressure of globalization. From a slightly different point of view, the economic growth through spiralling wages and consumption in the developed world that had produced the affluence of the postwar years came to an end. The Fordist regime of accumulation based on domestic production/consumption management within the states of the core was 'exhausted' (Lipietz, 1986).

Certain cultural-political correlates can be matched closely to specific long-wave cycles. In particular, competition between states with different cultural-political characteristics results in the geographical redistribution of cyclical effects, especially technological and organizational advantages, from some states and to others. States can differentially exploit or suffer from cyclical effects depending on their pro-ductive efficiency, commercial supremacy, ability to restrict competition from rivals and facility in responding to crises. According to Wallerstein *et al.* (1979) the four Kondratiev cycles, when put in a political context, can be described as two 'paired Kondratievs' (see Table 3.2). The first pair covering the nineteenth century saw the rise and fall of British economic and political dominance and the second pair suggest a similar trend for the United States in the twentieth century.

There is a danger of seeing the four cycles and their political correlates as cases of 'history repeating itself'. The world economy has evolved considerably in its workings over the past 200 years. The 'water style' was uniquely associated with British economic development. Later technological systems have had external effects that have transcended national boundaries. Until the late nineteenth century the operations and markets of industrial firms were largely confined within specific

Table 3.2: Wallerstein's model of hegemony and rivalry applied to Britain and the United States

	Great Britain		USA	
	1790–98		1890–96	
A$_1$ Ascending hegemony		Rivalry with France (Napoleonic Wars). Productive efficiency: industrial revolution		Rivalry with Germany. Productive efficiency: mass production techniques
	1815–25		1913–20	
B$_1$ Hegemonic victory		Commercial victory in Latin America and control of India: workshop of the world		Commercial victory in the final collapse of British free trade system and decisive military defeat of Germany
	1844–51		1940–45	
A$_2$ Hegemonic maturity		Era of free trade: London becomes financial centre of the world economy		Liberal economic system of Bretton Woods based upon the US dollar: New York new financial centre of the world
	1870–75		1967–73	
B$_2$ Declining hegemony		Classical age of imperialism as European powers and USA rival Britain. 'New' industrial revolution emerging outside Britain		Reversal to protectionist practices to counteract Japan and European rivals

Source: Agnew (1987), Table 1, p. 6.

75

countries. Since then, but especially since the 1960s, many industrial firms have become increasingly transnational in operation. If British and American political-economic dominance is accepted as a feature of at least part of the time-periods that they are assigned in Table 3.2, it is nevertheless clear that the nature of their dominance has been different (even though in each case it rested in part on their command over the contemporary technological systems). Differing cultural logics (or ideologies) guided practice.

For example, Britain followed a mixed strategy of formal and informal imperialism (empire-building and extensive investment outside its empire) whereas the United States has largely avoided formal imperialism. Instead it sponsored a set of international economic institutions (such as the International Monetary Fund (IMF), the International Bank for Reconstruction and Development (the World Bank), and the General Agreement on Tariffs and Trade (GATT)) and international and regional security institutions (such as the United Nations and the North Atlantic Treaty Organization (NATO)) to represent and administer the conception of a liberal international capitalism that dominated American political élites after the Second World War. American sponsorship of multilateralism, the opening up of trade and investment through agreements among many countries, has been a major feature of American influence over the world economy. The United States also faced a political-military challenge from a state representing a different image of world order and an alternative mode of political-economic organization based on state-direction: the Soviet Union. Britain never experienced such a threat to its mode of operation or global status. Under American influence the world economy has also become much more interdependent than it was previously. Large American-based multinational companies have been pioneers in extending production operations across political boundaries. The frontiers of American business have not been readily contained by the political boundaries of the United States (Agnew, 1992).

The term hegemony is often applied to instances of dominance such as those of Britain and the United States. This is a difficult and controversial concept but most definitions would accept the criterion that the leading state cannot be simply the strongest militarily but also must have the economic and cultural power to set and enforce the rules of international conduct that it prefers. American hegemony, therefore, does not only signify a shift in the identity of the hegemonic power from a previous one (as Table 3.2 implies) or a world without one (as can plausibly be argued was the case from 1890 until 1945 as Britain lost hegemony but the USA had not yet attained it), but also the institutions and practices the United States brought to the world by virtue of its dominant position. These have included mass production/consumption, limited state welfare policies, electoral democracy based on weak mass political parties, and government economic policies directed towards stimulating private economic activities. Ruggie (1983) calls these policies taken together 'embedded liberalism'. Such multilateral international institutions as the World Bank, the IMF (International Monetary Fund) and the GATT (General Agreement on Tariffs and Trade) were created to bring this vision to realization.

The emergence of Japan as a potential economic challenger to the United States puts the 'cultural logic' of American hegemony in clearer focus. The Japanese approach to industrial organization and national economic policy is distinctly different, based upon a particular mixture of state-oriented values and political collectivism. This threatens the legitimacy of the entire American approach to capital accumulation as it proves more effective at accumulating capital.

The capitalist world economy, therefore, not only persists, it changes as rising powers deploying different cultural logics achieve influence or hegemony. As yet, the 'transnational liberal order' (Gill, 1990) created under American auspices appears likely to outlast the hegemony of its creator. Some recent trends in productivity and technological innovation suggest that the US territorial economy now suffers rather than benefits from the globalized economic system that American efforts did so much to develop. Japanese firms (and some others) have successfully exploited the contemporary world economy from cultural contexts that diverge from many of its basic practices (Agnew, 1993).

Incorporation, subordination and resistance

Fifth, the world's population resists or adapts to incorporation into the world economy rather than invariably accepting or succumbing to it. A danger in elevating the concept 'world economy' to a key position is that local histories can be deprived of their integrity and specificity. All too often the world outside of Europe and North America is portrayed in drab, uniform colours and as reacting to 'core' influences in a uniform manner. Thoughtless use of such concepts as 'the less developed countries' and 'the Third World' can lead to an easy over-homogenization of a much more geographically variegated pattern. Different parts of the world have reacted distinctively to the expansion of the world economy.

Before examining the factors that have made for these differences it is necessary to say something in general about the nature of the 'independent systems' that have been incorporated into the world economy. Contrary to popular thinking, these were not mired in stasis or backwardness. With but few exceptions they have been open systems 'inextricably involved with other aggregates, near and far, in weblike, netlike connections' (Lesser, 1961, p. 42).

Two erroneous assumptions about these systems must be dispelled before a more nuanced picture is possible. One is that societies move in a linear sequence from 'traditional' to 'modern'. In fact, as Polanyi (1957) has shown, several different 'principles of economic integration' usually coexist in different societies with the balance changing in different historical circumstances. Only with the coming of the world economy does market exchange predominate over the other principles of reciprocity and redistribution to create a market society. But they continue to operate, even if in a subordinate status. Exchange, as a form of integration, refers to a system of price-setting markets in which the prices extend beyond a particular market-place. Reciprocity denotes the interaction between groupings on the basis of symmetry and mutuality, such as in kinship, neighbourhood and some religious groupings. Redistribution refers to centralized allocation of goods by virtue of custom, law or *ad hoc* central decision. Redistribution is present whenever there is collecting into and distribution from a centre as is the case in most societies from a hunting tribe to the modern welfare state. These forms of integration are not stages of development. Several subordinate forms can be present at any one time and any one can re-emerge as dominant after periods of eclipse. For example, reciprocity re-emerged among the inhabitants of the Amazon Basin after the (redistributive) empires of which they were part collapsed with the arrival of Europeans in South America (see, e.g. Roosevelt, 1992) and redistribution gained ground under organized capitalism in the core of the world economy in the form of the welfare state.

The other erroneous assumption is that societies without the dominance of market exchange exhibit a timeless immobility. Trade in staple crops and dynamic political change were in fact the rule rather than the exception prior to incorporation into the world economy. For example, contrary to the widespread image of sub-Saharan (black) Africa as a continent mired in stasis before the arrival of European traders, missionaries and soldiers, recent research suggests that its history was 'one of ceaseless flux among populations that, in comparison to other continents, are relatively recent inhabitants of their present habitat' (Kopytoff, 1987, p. 7). Modern ethnography shows that rather than a simple evolution of more complex from more simple independent systems

what we see are building-blocks of different sizes, and chips off the blocks, and the moving kaleidoscope of their grouping and regrouping . . . It was a [political] ecology that made for the fact that 'states and stateless societies have existed side by side for over nearly two millenia'.

(Curtin *et al.*, 1978, p. 72, in Kopytoff, 1987, p. 78)

However, it would be equally erroneous to suppose that the 'independent systems' encountered and incorporated by the world economy were all the same. A key variable has been the ability of local systems of political–cultural organization to resist and/or adapt to outside influences. Three basic types of independent system can be identified in terms of their cultural–political organization and geographical attributes: the Family-level Group, the Local Group, and the Regional Polity (Johnson and Earle, 1987). Reciprocity dominates in the first of these, whereas redistribution and elements of market exchange predominate in the latter two. The three basic types can be divided into more empirically meaningful sub-types (Table 3.3).

In the family-level society under the dominance of reciprocity there is a spatial diffuseness and minimal degree of geographical organization and integration. With the threat of outside force or famine, family groups can coalesce into larger

Table 3.3: Types of independent system, with examples

Type	Example	Size of community	Size of polity
Family Group			
Camp	Shoshone. North America	30	30
(without domestication)	!Kung San. Africa	20	20
Hamlet	Machiguenga. South America	25	25
(with domestication)	Nganasan, Siberia	30	30
Local Group			
Acephalous local group	Yanomano, South America	150–250	150–500
	Eskimos, North America	150–300	150–300
	Turkana, Africa	20–25	100–200
Big Man collectivity	Indians of Northwest and North America	500–800	500–800
	Central Enga, New Guinea	350	350
	Kirghiz, Asia	20–35	1800
Regional Polity			
Chiefdom	Trobriand Islanders, Pacific	200–400	1000
	Basseri, Iran	200–500	16 000
	Hawaii Islanders, Pacific	300–400	100 000
Early state	Inka, South America	±400	14 000 000
Nation–state	Northeast China	±300	600 000 000
	Central Java	±300	100 000 000
	Northeast Brazil	±300	80 000 000

Source: Johnson and Earle (1987), Table 1 (p. 23) and Table 10 (p. 314).

groupings. When faced by hostile neighbours or intruders the tendency is for such acephalous groups to form alliances under the tutelage of so-called Big Men. They represent the political unity of their groups but their political power is contingent and transitory depending on external threat and their ability to manage the flow of internal debts and credits. The local group, therefore, shows the beginnings of a switch to redistribution as the dominant principle of integration. This, however, is especially marked in the case of regional politics. Through the centralization of power in the hands of chiefs and bureaucratic hierarchies, the political-economic landscape acquires a characteristic nodality. Space is ordered in core–periphery terms. Territory is accorded primacy over kinship.

The state differs from the chiefdom in terms of a larger scale, a larger population and much more rigid social stratification. Integration tends to be beyond the personal control of an hereditary élite. It requires a state bureaucracy, a state religion, a judicial apparatus, a military establishment and a police force. With the nation state, trade and markets are defined as national attributes. The state provides military protection, risk management and investment in large-scale technologies. But most economic transactions are undertaken through market exchange. With the addition of competitive, price-setting markets more and more economic transactions are drawn outside the control of the state. Price-setting markets appear to be intrinsically expansive. This is what turned the expansiveness of the European states in the sixteenth century into a world economy rather than a world-empire. Even then, however, the state can and does, through its political–military monopoly over a rigidly defined territory, regulate and protect 'its' economy from external competition.

Skocpol (1976) suggests that the key to whether an indigenous 'agrarian order' (she excludes less complex groups) could preserve itself or resist incorporation lay in the degree of state autonomy from the short-run interests of the local economic élite. However, self-preservation often involved adapting or organizing a response to the world economy rather than long-run resistance to it. Over the long haul success has been variable: 'The Ottoman empire disintegrated while Japan modernized successfully; Africa was partitioned while Latin American governments were more effectively expanding and consolidating their rule' (Smith, 1979, p. 276).

The potential for resistance or adaptation, therefore, would vary depending on the internal strength of the 'independent system' and its compatibility with the world economy. But the nature of the process of incorporation has also been of fundamental importance. Hall (1986) provides a very useful typology of world-economy impacts. Along a continuum of incorporation he distinguishes a 'weak' pole of areas external to the world economy (external arenas) and a strong pole of fully fledged dependent peripheries. In between are areas where contact has been slight (contact peripheries) and an intermediate category of marginal peripheries (Table 3.4).

The processes involved as an area shifts from the status of an external arena to a dependent periphery are also indicated in Table 3.4. Market articulation refers to the nature of the capital and product flows between the expanding world economy and an area undergoing absorption. At the weak pole of the continuum are areas only slightly connected to the world economy, with the primary flow of influence from the core to the periphery. By way of example, the North American fur industry

Table 3.4: The continuum of geographical incorporation into the world economy

The continuum of incorporation	None	Weak	Moderate	Strong
Type of periphery	Exernal arena	Contact periphery	Marginal periphery	Dependent periphery
Market articulation	None	Weak	Moderate	Strong
Impact of core on periphery	None	Strong	Stronger	Strongest
Impact of periphery on core	None	Low	Moderate	Significant

Source: Hall (1986), p. 392.

in the eighteenth century was not vital to European economic development, but the fur trade produced massive social and economic change among the native societies which became tied into it. At the strong pole of the continuum, the exchange involved is important to core development. This was the case, for example, with cotton from the US South flowing to England in the early nineteenth century. Though influence does flow in both directions, across the entire continuum net product and capital flows favour the core.

However, it is clear from this approach that the 'motor' of incorporation is not simply profitable trade and capital accumulation in the 'interest' of the core. Contact and marginal peripheries do not contribute much if anything to capital accumulation in the core or for any class in it. Motivations such as religious zeal, pre-emptive colonization (getting there before a rival), state expansion for its own sake, degree of collusion between colonizers and colonized, and entrepreneurial exploration (individuals on the make) must be seen as important elements in the expansion of the world economy.

Furthermore, movement along the continuum is contingent rather than necessary, both in terms of pace and eventual degree of dependence. A number of conditions seem of particular importance. First, geographical integration into the world economy is more or less complete when early contact leads to the complete subjugation or liquidation of indigenous groups. Later, settlement of Europeans created conditions for the dominance of market exchange as, for example, in the USA and Australia. The emergence of independent settler political forces also favoured the pursuit of capitalist economic development and full integration into the world economy.

Second, the pace of movement towards strong incorporation depends on the strength of the state engaged in expansion and the nature of the world economy at the time. For example, in the sixteenth century Spanish expansion led less immediately to effective incorporation and the spread of market exchange than did British expansion in the nineteenth century. Plunder and religious zeal were more important than 'bringing to market'. In the nineteenth century, market exchange was effectively internationalized under British hegemony as production for the market everywhere replaced the mere exchange of commodities (Table 3.5). The British national economy had become the 'locomotive' of the world economy. But as its markets in Europe became more competitive it was pushed into a widening of its markets elsewhere. The British Empire was an important part of this expansion; although, as Table 3.5 shows, Latin America and China, regions outside of formal British control, were also significant. The internationalization of the British economy in the nineteenth century was a crucial element in the quickening pace and increased strength of incorporation world-wide (see Chapter 8).

Table 3.5: The geographical development of the world economy in the nineteenth century

Stage: Factor intensity:	Developed		Developing	Underdeveloped	
	Capital	Labour	Land	Land	Labour
1800	Britain	Europe		USA	India
1840	Britain	Europe	USA	Latin America Australia Canada	India China
1870	Britain Europe		USA	Australia Latin America Canada Africa	China
1900	Britain Europe USA		Australia Canada Argentina Mexico South Africa	Latin America Africa	India China

Source: Hansson (1952), pp. 49–82.

As a result of the nature of specific independent systems and the process of incorporation, the responses to European expansion and the penetration have been diverse. In Africa, for example, 'Some societies accepted colonial rule; others resisted. Some chose to cooperate with the new rulers in order to manipulate them to their purposes; others tried to opt out of the imperial system by force or by stealth . . .' Even after long exposure of equivalent intensity, responses could still differ. In Africa again, 'After a period of initial revulsion, the Kikuyu [in the British colony of Kenya] took easily to education and wage employment; the Kamba and the Masai, on the other hand, for long ignored schools' (Gann and Duignan, 1978, pp. 361–2).

Resistance and adaptation have continued. At one extreme, as with Japan, lies the conjunction of adoption and adaptation. At the other, as with Iran since 1979, lies overt rejection. In the writings and pronouncements of the late Iranian leader, the Ayatollah Khomeini, 'the West', through its association with the corruption and decadence of market exchange, is viewed as the enemy of the legitimate system of

Islamic rule. This example illustrates a more general point: that European-style states and their associated political ideas are often incompatible with local subordinate cultural logics. In the face of totalizing cultures such as Shi'ite Islam, the principles of structural differentiation and autonomy on which distinctions such as 'the economic', 'the religious' and 'the political' rest, and from which Western views of development derive – essentially the 'freeing' and nurturing of the economy – fail to take root and to gain popular legitimacy.

Alternative adaptations

Sixth, and finally, every part of the world has had its own particular relationship to the evolution of the world economy. In the case of the United States, for example, the existence from an early period of two contrasting and incompatible modes of socio-economic organization within one territorial state, a plantation agriculture based on African slave labour in the South and a 'classic' capitalist or free enterprise economy in the North, was peculiarly American. Its heritage in terms of the relative regional underdevelopment of the South and racially polarized politics wherever there are concentrations of African-Americans continues to this day. Likewise, the apartheid system of racial categorization and control in South Africa was a peculiarly South African response to the history of European settlement and economic exploitation in southern Africa. The nature of state–economy relations differs importantly between such nominally 'capitalist' states as Britain, Italy, France and Germany. This is the result of both the historical development of connections with the world economy and the extent of organized attempts to create international competitive advantage. Italian economic development since the Second World War, for example, has involved a combination of government investment in heavy industry and small-scale private investment in the production of consumer goods for export that is quite different from the decline of manufacturing industry that has resulted from government favouring of financial industry in the British case. The whole process of development and underdevelopment, therefore, is mediated geographically through the actions of state-level regulation. China and Japan offer fascinating examples of this process.

At the time of initial contact with Europeans, China had a long history of political unity as an independent system. China was thought by its élites to be the centre of all civilization and the Chinese emperor had the *de jure* right to rule over all human affairs. The geographical setting of China both inspired and reinforced the sense of China's uniqueness. Hemmed in by mountains to the west and south, by deserts to the north and ocean to the east, China appeared as a centre in relation to a ring of surrounding 'barbarians'. Tribute flowed from this periphery into the centre. This changed with the increasing pressure of European and American involvement in Chinese affairs. A series of humiliating military defeats from the Opium War (1839–42) to the war of 1856–57 forced China into acknowledging its equal status with foreign states.

The overthrow of the imperial government in 1911 was the outcome of the decline initiated by the earlier defeats. A rising generation of intellectuals believed that the 'dead hand' of Confucian thought (which celebrates the 'wholeness' of culture and the decisive role of ideas in human affairs) had retarded Chinese responses to the West. This generation nevertheless accepted the traditional idea that ideology is the foundation for economic and political change. Indeed, the history of China since that time has followed the 'totalistic' nature of those rejecting the substance of Confucian thought without discarding the form. From 1949 until the mid-1970s, political thought and the policies of the Chinese government were dominated by Mao Tse-Dong. Mao's repeated stress on the importance of remoulding the image of the world in order to change it clearly reflects the traditional influence. In remaking Marxism–Leninism in a Chinese image, therefore, Mao gave a primacy to consciousness and will rather than to the materialistic factors stressed by others in this Western political tradition. Since Mao's death in 1976 China has 'opened up' to overt dealing with the world economy. There is now large-scale foreign investment in labour-intensive manufacturing in coastal China. But the ambivalence about trading with the 'barbarians' remains, even if China cannot without great cost continue to challenge the 'new world' it came into contact and conflict with in the nineteenth century.

Japan, even more than China, has a population that is highly homogeneous ethnically. But nineteenth-century Japan was less unified politically than China. It consisted of a number of semi-autonomous fiefdoms rather than a single politically unified nation. Although there was a common core of cultural attributes, up until the 1930s there were many mutually unintelligible dialects and much regional variation in social and economic life. The modern homogeneity of Japan has been created and is maintained through national social policy (see pp. 171–7).

However, the amazing economic mobilization of the national population characteristic of modern Japan is not based on coercion. Rather, it exploits the absence of a highly stratified caste or class system in the sense that these developed, for example, in the Indian states or in Europe (Nakane, 1970, pp. 105–6). The modern Japanese state, emerging after a *coup d'état* in 1868 (the Meiji Restoration), has faced minimal alternative socio-cultural foci to that of nationality in mobilizing the population around its objectives. Before the mid-nineteenth century Japan escaped from European penetration because it was perceived as poor in markets and resources. But when faced by a dramatic increase in foreign interest it was the ability of the state to mobilize people and resources that enabled Japan to join the world economy in the nineteenth century on its terms rather than on those of a colonial power.

Since the Second World War and the defeat of Japan's effort to turn its successful industrialization into a territorial empire, the institutions developed after the Meiji Restoration – central government planning, close government–business cooperation, strong links between banks and industry, etc. – have turned out to be at least as well adapted to the contemporary world economy, with its general trend towards reductions in barriers to trade and investment, as those of either Europe or the United States. The geopolitical relationship between the USA and Japan provided perhaps the most important prerequisite for this development. In return for providing an offshore bastion in the containment of the Soviet Union during the early Cold War, Japan was allowed to re-enter the world economy on terms that were favourable to its economic growth. For example, high Japanese tariffs were tolerated at the same time that the USA was sponsoring a world economic order based upon tariff reduction or removal. Throughout the period of USA occupation of Japan (1945–52) the USA authorities not only discouraged direct investment in Japan they also forbade US firms with previous investments from surveying the condition of their holdings (Mason, 1992).

STATES AND THE WORLD ECONOMY

The emergence of the world economy in Europe coincided with and was dependent on the territorial consolidation of Europe's territorial states. From 500 political entities in Europe in 1500, there were not more than 25 territorial states in that region in 1900 (Tilly, 1975). The modern territorial state has ever since been the political form of the modern world economy. Thus, the expansion of the world economy has been accompanied by a parallel expansion of the inter-state system as the sole form of political, military and administrative organization. As colonies achieved independence from the European empires they expanded the list of the world's states.

The state is not a standard political-organizational entity in all countries. It includes very different institutions (different representative assemblies, bureaucracies, police forces, militaries, etc.), products of particular histories, that adapt in phase to shifts in national development and position within the world economy. The relative power of states can be thought of in terms of three dimensions: (1) relative to one another (the hierarchy of states), (2) relative to their inhabitants, and (3) relative to the 'globalizing' world economy (Harris, 1986, pp. 145–69). The first has shown considerable variation historically, e.g. in the rise and decline of Britain, in the rise and decline (?) of the USA, and in the rise of Japan and the NICS. The second seems to grow incessantly, particularly in the more developed economies, with respect to regulation, taxation and surveillance. However, the third, the power of states relative to other actors in the world economy, appears to be declining in general, and to decline more as countries develop economically. As Harris (1986, p. 148) puts it: 'the more industrialized the country, in contemporary conditions, the more the condition of continued growth of the wealth and power of the state depends upon permitting the integration of the local with the world economies'. What is the truth to this? It is this third dimension that attracts our attention before we proceed to the first and the second.

States and the globalizing world economy

The latest phase in the evolution of the world economy, labelled earlier by the term disorganized capitalism, has involved the erosion of national economies as the basic building blocks of the world economy. A transnational element has been in the ascendence in the form of the growth of an 'immediately global' market which is supplied by firms that organize their production and distribution without much reference to national boundaries. The 'global shift' in production has given rise to an explosion of foreign direct investment and to the emergence of trade within firms as the most rapidly expanding component of total world trade (Grant et al., 1993).

Perhaps 30–40 per cent of US imports, for example, consist of parts and finished goods from the foreign subsidiaries of US headquartered firms. Transnational corporations, therefore, are major engines in the growth of world trade. Chapters 7, 8 and 10 are taken up with documenting the impact of the global shift on the geography of the world economy. However, one point that must be made here is that rather than marking the demise of national boundaries, the globalization of production has occurred in large part because of them. The spatial strategies of multinational corporations are designed to exploit national differences in labour forces, market conditions, regulatory environments and macroeconomic (fiscal and monetary) conditions. Without the variation institutionalized by the territorial states into which the world is divided, the attraction to firms of shifting investments and moving production would be reduced. The challenge to states is to compete in this more integrated and volatile economic environment as effectively as possible (Picciotto, 1991; Dicken, 1992b).

There is some danger of exaggerating the extent to which the globalization of production has involved the rise to power of global firms whose worlds 'know no boundaries' (Pauly and Reich 1997; Dunning 1997). Indeed, most so-called global companies are still strongly attached to their home countries (see the later Box: Does a home base still matter?). For example, in 1991 only 2 per cent of the board members of big US companies were foreigners. In Japanese companies, 'foreign directors are as rare as British sumo wrestlers' (*The Economist*, 1993, p. 69). This bias is manifested in the activities of even the most 'global' firms. When total sales decline, home markets tend to be protected at the expense of foreign ones. When firms expand abroad they continue to rely heavily on suppliers from their home countries. Antitrust laws and nationalism still make it hard for foreign firms to expand through takeovers. The difficulties encountered by British Airways and other European airlines in trying to join forces with various US airlines are a case in point.

Many states and trading blocs (such as the European Community) have industrial and trading policies that serve to protect and enhance their national economies by encouraging domestic and discouraging foreign investment (e.g. Airbus Industrie, the producer of the European Airbus is a consortium of aerospace firms from different European countries collaborating to take on the big American producers such as Boeing). The United States and Britain are unique in the extent to which they have allowed their economic frontiers to expand beyond their national boundaries. Perhaps because they have been hegemonic powers that have at one time or another been in positions of global dominance they have forgotten the lessons of their own economic development (see, e.g. Bourgin, 1989). In neither case did they rely on 'open borders'. Both had policies which protected infant industries and provided public investment in infrastructure (roads, railways, schools, hospitals, etc.) and research and development (as in the federal support for the land-grant universities in the USA and government support of universities in general in both countries). Both are now faced with problems of deindustrialization that are more severe than those of many other countries, at least in part because of their neglect of the connection between government action and economic growth (see, e.g. Martin and Rowthorn, 1986; Cohen and Zysman, 1987; Heim, 1991).

The international financial system

Of even greater alleged importance in reducing the power of states relative to the world economy than the 'global shift' in production is the international financial system, which provides the structural framework for trade and investment decisions and is much more integrated in the 1990s than it was even 10 years ago. All forms of capital have become more mobile, in time and space. Corporate and financial market decisions are now made within much tighter time schedules than was only recently the case. Currency exchange rates now can change several times a day compared to less than annually before 1970. Decisions about employing and shedding labour are now made weekly and monthly compared to quarterly before 1970 (Fig. 3.2). Commodity, exchange and stock markets in world cities work around the clock shifting investments and setting prices that have different effects in different places because of their particular mixes of economic activities. Banking has become a global industry with the reduction of national and local barriers to operations and capital movements (e.g. Smith and Walter 1996).

Flows of capital rather than trade now drive the world economy. Foreign direct investment (FDI) among industrialized countries has expanded at a much faster rate in the 1980s than has world trade (Agnew, 1992). FDI in so-called less developed countries (LDCs) also shows signs of increase, although it is unevenly distributed. The economies of East Asia now take the largest share. Between 1986 and 1990 the LDCs received, on average, US$26 billion of FDI a year. Of that, East and Southeast Asia's share was US$14 billion, Latin America's US$9 billion and Africa's US$3 billion. To put this in perspective, during the same period annual flows of FDI among the industrialized countries averaged around US$126 billion (Crook, 1992, p. 17). In 1995, the United States, Britain, France and Australia were the biggest recipients of FDI. The USA received US$60 billion of FDI, twice as much as second-place Britain. Outside of North America and Europe, only Asia was increasingly attractive to outside businesses. Much of the world's capital, therefore, continues to circulate largely between the already rich economies with smaller flows to selective destinations (particularly China in the early 1990s) elsewhere (UNCTAD, 1996).

Government policies towards FDI have liberalized world-wide, partly as a result of privatization of state-owned industries but also because of pressure from multilateral institutions such as the IMF and the World Bank which insist on increased openness to FDI in return for favourable loan conditions. In some LDCs the IMF and the World Bank have cooperated to intervene in domestic politics on behalf of programmes based on fiscal and monetary restraint combined with deregulation and liberalization of national markets (Crook, 1991; Rodriguez and Griffith-Jones, 1992). Even the most economically powerful countries, in terms of the importance of their currencies in world investment flows and the absolute size of their national economies, such as the United States and Japan, are now less autonomous and more vulnerable to both foreign ownership of economic assets and shocks emanating from world financial markets. One commentator has gone so far as to proclaim in the sub-title of his book that global financial integration signifies the 'End of Geography' (O'Brien, 1992). By this he means the declining ability of states, even the most powerful ones, to regulate the financial sectors within their

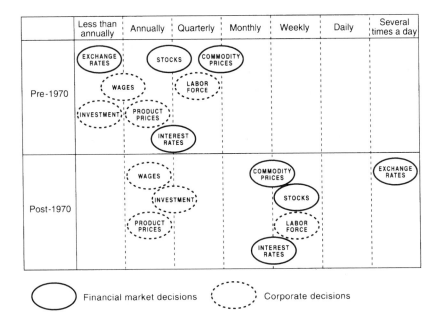

87

Figure 3.2
The increasing pace of
the world economy
Source: The Economist (24 Sept.
1983), p. 11

economies. This may well be true, but there are still very important institutional and cultural barriers to the free movement of capital. For example, knowledge about financial opportunities is still bound up with knowing the local investment 'scene' which largely depends on having access to local social networks that rely on relations of trust built up through social interaction and common social bonds (Leyshon and Thrift, 1997).

Financial markets are undoubtedly more open than only 30 years ago, however. This opening up dates back to the collapse of the Bretton Woods system of pegged exchange rates that prevailed between 1945 and 1971 (see Chapter 2). Under that system, free trade and fettered finance ultimately proved incompatible; once the former became more significant in the 1960s the latter proved impossible to sustain (Nau, 1990). Confidence in the stability of currencies began to wane as governments (such as that in Britain in 1967) proved incapable of controlling international capital movements. The new, more flexible exchange rates after 1971 have proved relatively advantageous for large, relatively closed economies such as the United States and Japan. The costs of floating, however, have become increasingly high for smaller, open economies. Volatile exchange rates disrupt domestic policy-making and reduce the ability of firms to make calculations about long-term rates of return on investment. The desire to seek cooperative arrangements that tie currencies together (such as in Europe today with the proposed 'Euro' currency) is explicable, therefore, in terms of the costs that the present system imposes on countries (and currencies) other than the most powerful. In the long term this may also encourage the development of monetary blocs organized around the world's three major currencies – the US Dollar, the German Mark, and the Japanese Yen (on the history of the international monetary system, see Eichengreen, 1996).

Collaboration and independence

Increased financial integration does not mean, however, that the world economy now runs rampant over national economies. Indeed, O'Brien, the prophet of the 'End of Geography', suggests all kinds of barriers to the homogenization that his sub-title announces so dramatically. Individual governments (or state institutions such as central banks) can still influence international financial markets through industrial policies, budget deficits, and by manipulating interest rates. But no one government can now control the system or completely control its own national economy. Even collaboration between governments, as in the Group of 7 meetings of leaders from the main industrialized countries, cannot always produce predictable effects on currencies, investment, and trade flows. But collaboration, both formal and informal and largely under American sponsorship, has become an important feature of the world economy. Indeed, in the 1980s the USA was able to win coordination of monetary and trade policies on its own terms 'even though American budget deficits have been the single most important source of international economic imbalance since the early 1980s' (Webb, 1991, pp. 341–2). Negotiated mutual adjustment among the major industrialized countries has definitely become more intensive. The willingness of governments from the 'strongest' national economies (such as Japan and Germany) to pay a price for American leadership has persisted because increased international economic integration gives all countries a stake in maintaining stability. The critical question concerns when that price might become too high (Friedman, 1989; Corbridge and Agnew, 1991).

Two particular features of the present world economy are important for what they say about the novelty and the fragility of the present situation. The world economy at the turn of the last century was in fact more integrated in several ways than it is today. Take, for example, world trade. In 1913 trade was a larger share of British national income than it was before or has been since. Trade has never been that important to the USA and in the late 1980s it was only beginning to approach turn-of-the-century levels (Fig. 3.3). In each case economic nationalism and associated increases in protectionist measures brought dramatic reductions in trade. This may happen again if regional trading blocs such as the European Community constrain the total growth of world trade (Krugman, 1990, p. 195). Given the early growth of trade, the most important prerequisites for a highly integrated world economy existed as early as the 1880s; steamships, the telegraph and railways were more important historically in laying its foundations than more recent innovations such as containerization, the jet airplane and the fax. So the present trend towards a global trading system may be as subject to reversal as the last one. The major differences this time around are the extent of trade within the production process (exchange of components, often within firms and industries) compared to the simple exchange of primary products and finished goods that prevailed in the past and the increasing openness of most countries to trade and investment. Participating in the global economy is now an important part of doing business anywhere.

Second, over the period 1880–1913 national investment and national saving (both as a per cent of GNP) were much less correlated for a range of industrialized countries than a comparable correlation over the period 1965–86 (Fig. 3.4). This suggests that investment in different national economies in recent years has

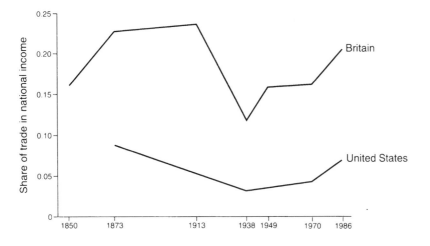

Figure 3.3
Britain and the United States: changing share of trade in national income

Source: Krugman (1990), p. 195

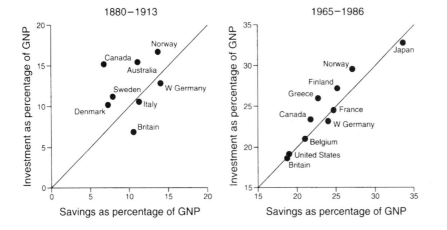

Figure 3.4
National savings–investment correlations

Source: Bayoumi (1989), p. 11

depended more on domestic saving and less on imported capital than was the case 90 years ago. From this point of view, capital is less mobile now than it once was, perhaps because of increased volatility in exchange rates between currencies in the absence of a gold standard (as prevailed during the period 1880–1913) or a semi-fixed monetary system such as the Bretton Woods agreement that prevailed in large parts of the world from 1944 until 1972 (Bayoumi, 1989). Whatever the precise cause of this particular historical difference, the evidence on both trade and capital markets suggests a continuing political division of the world economy whose trend towards increased integration over the recent past could easily be reversed.

Management and mediation

The crux of the contemporary world economy is the coexistence of national and global structures and attempts by states to manage the tensions between them. Transnational activities operate within states and under conditions imposed by them. Trade and tariff policies are still of vital importance in regulating trade and investment. For example, even as industrial countries' tariff levels have dropped steadily as a result of the GATT negotiations of the post-Second World War period, non-tariff barriers (quotas, voluntary trade restraints, etc.) have often increased, usually as a result of pressure on governments from domestic political coalitions (Laird and Yeats, 1990; World Bank, 1991). In a number of countries, especially the United States, there is a growing controversy among economists and politicians over the geographical distribution of the costs and benefits of open as opposed to more 'managed' trade. Some commentators argue that some countries, particularly Japan, do not trade 'fairly', they impose too many restrictions on imports while they benefit from the relative openness of other countries to their exports. Others argue that an open world economy is an end in itself. Everyone suffers in the end when there are restrictions on trade. Opinion in the United States seems to be tilting towards the advocates of managed trade or 'strategic' trade policy, particularly in those regions whose industries have been most affected by foreign competition (such as the Great Lakes region) or whose sectors would benefit most from government stimulus and protection (e.g. so-called high tech sectors) (see Destler, 1992; Tyson, 1992).

Largely as a result of national regulatory policy, national financial systems still differ profoundly in their connection with local industry and in their openness to foreign penetration. Financial systems are important to the geography of the world economy through the mediation they provide between political systems, on the one hand, and the productive system, on the other. The German and Japanese financial systems, for example, are intimately connected with their co-national industrial companies whereas the British system is not only independent of but often appears to be at cross-purposes with the needs of British manufacturing industry (Hayes and Hubbard, 1990). This difference has important consequences, particularly in terms of the geography of investment. In Germany and Japan the fortunes of finance and industry are closely tied. Each has a stake in the other. In Britain industry has had to compete with a wide range of alternative and often more attractive outlets for bank investment, inside and outside the country. At the same time financial markets in Britain have been more readily accessible to investment from outside, creating in London a truly global financial centre. The relations of industry and finance in Germany and Japan would have to change significantly for Frankfurt and Tokyo to become equivalently global financial centres. One of the difficulties involved in creating an EC monetary system is the fundamentally different nature of the financial systems in member countries such as Britain and Germany and the impossibility of overcoming institutional differences through the mere linking of currencies within an exchange-rate mechanism such as that in the European Monetary System (EMS) (Agnew, 1993).

National financial systems differ along three dimensions (Zysman, 1983, p. 69). The first is the process whereby savings are transformed into investments and allocated among competing uses (intermediation). The second dimension is the degree to which prices are set in financial markets, by dominant financial institutions or by government (marketization). The third is the amount of government intervention in the financial system (regulation). The combinations of the three dimensions differ significantly between countries and thus provide distinctive

Table 3.6: The institutions and ideological basis to the world's dominant capitalisms

A comparison of four systems

Characteristic	American Capitalism	Japanese Capitalism	European Social Market	British Capitalism
Basic principle				
Dominant factor of production	capital	labour	partnership	capital
'Public' tradition	medium	high	high	low
Centralization	low	medium	medium	high
Reliance on price-mediated markets	high	low	medium	high
Supply relations	arm's-length price-driven	close enduring	bureaucracy planned	arm's-length price-driven
Industrial groups	partial, defence etc.	very high	high	low
Extent privatized	high	high	medium	high
Financial system				
Market structure	anonymous securitized	personal committed	bureaucracy committed	uncommitted marketized
Banking system	advanced marketized regional	traditional regulated concentrated	traditional regulated regional	advanced marketized centralized
Stock market	v. important	unimportant	unimportant	v. important
Required returns	high	low	medium	high
Labour market				
Job security	low	high	high	low
Labour mobility	high	low	medium	medium
Labour/management	adversarial	cooperative	cooperative	adversarial
Pay differential	large	small	medium	large
Turnover	high	low	medium	medium
Skills	medium	high	high	poor
Union structure	sector-based	firm-based	industry-wide	craft
Strength	low	low	high	low
The firm				
Main goal	profits	market share stable jobs	market share fulfilment	profits
Role top manager	boss-king autocratic	consensus	consensus	boss-king hierarchy
Social overheads	low	low	high	medium, down
Welfare system				
Basic principle	liberal	corporatist	corporatist social democracy	mixed
Universal transfers	low	medium	high	medium, down
Means-testing	high	medium	low	medium, up
Degree education tiered by class	high	medium	medium	high
Private welfare	high	medium	low	medium, up
Government policies				
Role of government	limited adversarial	extensive cooperative	encompassing	strong adversarial
Openness to trade	quite open	least open	quite open	open
Industrial policy	little	high	high	non-existent
Top income tax	low	low	high	medium

Source: Hutton (1995, p. 282).

financial environments for internal economic development and affect the degree of external orientation of local firms. The USA and Britain are capital market financial systems in which the intermediation of financial institutions is relatively weak, where prices are set in markets and there is limited state intervention in financial markets. There is little if any bias towards local industrial firms or long-run national economic development. Germany typifies a second model of a financial institution-dominated credit-based financial system in which bank intermediation dominates, prices are set in markets but the state and financial institutions are closely related. Large banks provide the bulk of the credit to domestic industrial firms and are usually represented on the governing boards of these firms. Finally, there are government-dominated credit-based financial systems, exemplified by France and Japan. In this case, state entanglement with industry (nationalized industry in France, private industry in Japan) has fundamental limiting effects upon the autonomy of markets and financial institutions. Particularly in Japan the financial system is structured to give a long-run orientation to industrial development. For example, the government-run Postal Savings system in Japan channels domestic savings through commercial banks into industrial investments.

For political-institutional reasons, therefore, national economies cannot always be confined solely to creating the optimal conditions for the operation of global industries within them, because national politics reflects conflicts of group and regional interests over tariffs and trade, and national economies have had distinctive trajectories in the development of the financial systems that underpin investment decisions by local firms (see Table 3.6). At the same time, the global economy is developing explicitly to optimize conditions for private business activities, at whatever cost to this or that national economy, including a firm's 'own'. An increasingly integrated world financial system is one mechanism for this. The 'global shift' in production is another. In some LDCs international economic institutions such as the IMF and the World Bank have become so powerful in setting conditions under which loans will be granted that they have become the *de facto* governments of those countries. In a large number of African states the 'internationalization of the state' has gone so far that some commentators now refer to them as quasi-states (e.g. Jackson, 1990), unable to direct their own economies without massive external assistance and not offering much in the way of citizenship or economic benefits to their populations. Harris (1986), therefore, may have systematically exaggerated the weakness of the states in the DCs and their strength in many LDCs. But what, if anything, is the continuing significance of states in the contemporary context of expansive globalization?

States and the geography of the world economy

The continuing importance of states in the world economy, if in a dramatically changed context from the past (as sketched above and as we shall attempt to show later in this section), is manifested in a number of ways: first, in the organizing and mobilizing roles of the states in the NICS; second, in the continuing geopolitical rivalry of the developed, industrialized countries; third, in the macroeconomic policies pursued by national governments and central banks to stabilize and reorganize their economies; and, fourth, in the latitude and initiative of lower-level governments in attracting and keeping economic activities within their jurisdictions.

The NICs

It is remarkable that if the larger 'underdeveloped countries' (i.e., excluding Hong Kong and Singapore because of their peculiarity as ethnic Chinese city states) are examined in detail, the fastest growing and otherwise best-performing countries have all had national governments that have directly and actively intervened in their economies. In South Korea, for example, successive governments have played major parts in fostering economic growth. Adding government savings to deposits in nationalized banks, the South Korean government controlled two-thirds of South Korea's investment resources during the country's period of most rapid growth in the late 1970s (Sen, 1981). This power was used to guide investment in chosen directions through differential interest rates and easy credit terms. Korean export expansion, the main method of economic growth, was itself built on an economic base that was stringently protected from foreign imports. Economic growth was orchestrated by activist governments. 'In exchange for subsidies, the state . . . imposed performance standards on private firms' (Amsden, 1989, p. 8). Elsewhere subsidies have not always been tied so closely to performance and the result has been lower growth.

There is not a little difficulty, therefore, in reading the recent experience of South Korea, or some other NICs such as Taiwan, as an entirely 'market' phenomenon. Economists on the political right (e.g. Beenstock, 1983) and on the political left (e.g. Harris, 1986) err when they ignore or systematically devalue the importance of state action in organizing and mobilizing resources for economic growth. Of course, not all states have either the institutional foundations or the resources to 'mobilize' for economic growth. Their ethnic homogeneity, transport infrastructure inherited from Japanese colonialism, history of land reform and American investment during the Cold War to help 'contain' China and the Soviet Union, gave South Korea and Taiwan decisive advantages. In many other cases ethnic divisions and organized corruption have turned states into the enemies rather than the facilitators of development. Their action or inaction continues to afflict their populations. One thinks of such examples as Zaire and Somalia in Africa (only two examples from a much longer list), Argentina and Venezuela in Latin America and Sri Lanka and the Philippines in Asia (see Chapters 8 and 10).

States and geopolitical rivalry

States in the 'developed world' show few signs of disappearing either (see Chapters 6 and 7). Indeed, under the conditions of economic restructuring that have affected most industrialized economies over the past 20 years, there has been a deepening of rivalry between many countries over trade, monetary and foreign policy. A key 'geopolitical fact' has been the relative decline of the US economy combined with the US government's ability to slow this down by use of US military and, above all, monetary power. Successive US governments have used the devaluation and revaluation of the dollar, the main metric of world trade and investment, to insulate the US territorial economy from the negative impacts of increased foreign competition and import penetration upon the US economy. This has of necessity been at the expense of other currencies and other countries, especially Japan and the countries of Western Europe. The massive tax cuts and spending increases by the Reagan administration in the early 1980s temporarily stimulated the US economy but only through the promise of debt financing by foreign investors in US bonds and assets (Parboni, 1988; Corbridge and Agnew, 1991).

This is an important example of how the most powerful states can use fiscal and monetary policies to stabilize and reorganize their economies at the expense of other national economies. Within their own boundaries they can also encourage or slow down processes of restructuring emanating from changes in the world economy. In the USA in the 1950s and 1960s defence-spending and housing and transportation policies that stimulated suburban growth helped the development of Fordist firms oriented to national markets (e.g., in automobile production). In Japan during the same period the government Ministry of International Trade and Industry (MITI) used an industrial life-cycle model to guide investment in new industries as 'old' ones achieved 'maturity'. Other countries had less direct industrial policies, often operating through government-owned industries (such as electricity or steel) to stimulate new industries and stabilize production. In the USA in the 1980s defence-spending again and new financial services (stimulated through deregulation of banks and other financial institutions in the 1980s) became the focus of government attention. The large deficits in the federal government budget that began to accrue in the early part of the decade further stimulated the financial service industries through the need to attract and reward foreign investment. Regions and localities specializing in military production and finance were the beneficiaries. Increases in military spending were justified in terms of the security threat posed to the United States and its allies by another state with a different political-economic system: the (now) former Soviet Union. The demise of the Soviet Union promises both the reduction of military spending and the removal of the 'threat' upon which the political commitment to the United States of such countries as Germany and Japan was based. Without the Cold War and the associated militarization US governments will have to find a new approach to industrial policy and a new way of keeping allies (especially Germany and Japan) in line.

Macroeconomic policy-making

The possibility of international conflict is also increased because of important differences between countries in macroeconomic institutions and policy-making. In particular, with the rise of international financial integration and highly speculative financial markets (in world cities), has come a concomitant rise in the political power (if not economic effectiveness) of central banks (such as the Federal Reserve in the USA, the Bank of England in Britain, and the Bundesbank in Germany). This is because central banks in large financial markets have retained an ability to affect conditions in domestic and global markets through manipulating interest rates and the money supply. But these banks function differently in different countries and are subject to distinctive pressures emanating from the conjuncture of external influences and distinctive national institutional-policy environments. When these lead to conflicting policies there is the possibility of serious international conflict and disruption of international investment flows (both bank and portfolio – bank deposits, stocks, etc. – and direct investment – takeovers, new subsidiaries, etc.). For example, where a central bank is politically independent and connections between industry and finance are weak, as in the USA, the central bank will be a 'rentier' bank, following restrictive monetary policies that benefit banks and other financial interests. However, where a central bank is independent but industry and finance are closely linked, as in Germany, the central bank will try to benefit business as a whole, choosing specific policies depending on the relative influence of labour. The Bundesbank has given a very high priority to maintaining price stability through fighting inflation. At least since the mid-1970s this seems to have produced higher aggregate economic growth than would otherwise have been the case. But it has required a general popular fear of inflation (perhaps based on the collective memory of the drastic price inflations of the 1920s and mid-1940s) for acceptance of the tight monetary policies that have been pursued. Finally, where the central bank is part of the government, industry and finance are linked and labour is cooperative, as in Sweden, central bank policy can be expected to be expansionary and, potentially, inflationary (Epstein, 1992; *The Economist*, 1992).

Governments and peoples

The corporate-welfare state that developed under organized capitalism is under threat, however, in all of these countries (Johnston, 1992). The state is certainly still 'big' throughout the industrialized world and beyond (for example, in India and China). But the Reagan administration in the USA and the Thatcher governments in Britain in the 1980s popularized the view that government spending on social services (Mrs Thatcher's 'nanny state') was a drag on national investment and growth because of the taxes that were required to pay for them. As yet, this perspective has not spread much beyond the bounds of the English-speaking world. Certainly, national governments in the core of the world economy now find themselves in the fiscal crisis that is a 'normal' condition in the rest of the world with demands upon their resources increasing (e.g. ageing populations, increasing poverty) as their ability to meet them (e.g. loss of higher paying jobs and associated revenues) declines. However, governments in the industrialized world remain as the most important agents of social and political order within their territories. They differ only in the degree of 'bigness' as measured by laws passed, range and comprehensiveness of programmes, scope of government agencies, and number and initiative of employees (Rose, 1984). The overall intrusiveness of most governments in the lives of their citizens shows few signs of decline.

Lower-tier governments and economic development

Finally, with respect to the political stimulation of economic development, in some countries local levels of government are often able to pursue policies of their own with respect to attracting and keeping economic activities. As some manufacturing and service industries have become more 'footloose' following the technological and organizational changes of the recent past (less tied through agglomeration economies to specific locations), a variety of factors once marginal to a firm's locational calculus have assumed greater prominence. Some of these can be subsumed under the rubric of local 'business climate'. In the United States, for example, the northern states of the manufacturing belt (the region stretching from Illinois to New York where most US manufacturing industry was concentrated from 1880 to the 1960s) tend to have higher personal income tax rates and greater provision of public goods and services (public education, social services, etc.) than states in the South and West. These conditions provide for less favourable business climates than are found in the other regions.

Many states and localities throughout the country, however, have actively sought out businesses by offering tax breaks and subsidies. Many of the US states have offices in Europe and Japan designed to attract foreign investment to their states. But much of the competition for businesses seems to involve attracting established firms and branch plants from other states. This has undoubtedly contributed to the decentralization of manufacturing industries within the country. The net contribution to national economic welfare is less substantial than might appear. As a result, preparing local labour forces for higher-skilled jobs through education and training and thereby attracting 'higher-value' industries is now becoming a much more important local economic development strategy (Herzog and Schlottmann, 1991; see Chapter 12).

Conclusion

In a number of respects, therefore, territorial states and the political regulation of the economy that they provide are of fundamental importance in understanding the evolution of economic landscapes. But these states, far from being primordial and pre-existing 'essences' that are the same everywhere and stay the same over time, are intertwined in complex and contradictory ways with the world economy. States today must operate in a global economic environment in which they have become managers of internal–external transactions rather than, in coordination with national business, monopolists over discrete national territories.

Indeed, it has become a commonplace to observe that states must operate in a world in which they are less and less 'full societies' (e.g. Williams, 1983). They are at one and the same time too large and too small for a wide range of social and economic purposes. They are often too large territorially to create full social identities and real national interests. This can be seen in the proliferation of ethnic and cultural movements around the world. It can also be seen in the difficulties involved in achieving national consensus in many states on national institutions and policies. For example, the postwar consensus on the 'welfare state' has been under attack in Europe and the United States. But existing states are for many economic purposes also too small geographically. They are increasingly 'market sectors' in an intensely competitive, integrated, and unstable world economy. There is something of a contradiction, therefore, between the political claims and military pretensions of even large states such as the USA and the economic realities they now face (Agnew, 1992).

Two propositions follow from this perspective that inform the rest of this book:

- First, economic power is no longer a simple attribute of states that have more or less of it. The growth of world trade, the activity of transnational companies, the globalized world financial system, global production, and regional trading blocs such as the EC and the North American Free Trade Area point towards an emerging new global world order to which states must adjust.

- Second, state and society/economy are no longer mutually defining. Uneven economic development within and between states has produced redefinitions of economic interests and political identities from national to regional and local levels. The economic restructuring associated with globalization has tied local areas directly into global markets. Local areas are thus 'communities of fate' in a world in which there is less possibility of shielding from competition within large territorial units than used to be the case. When such places have different orientations to the world economy, i.e. different commodities in trade, different trading partners and different exposures to foreign competition, the possibilities for national consensus on trade policy are much reduced. The growing redundancy of national governments as supranational entities (such as the EU) increase in importance and the challenge to national regulation from global markets, have conspired to stimulate new and revive old political identities, especially when ethnic and cultural divisions are defined geographically. In this context, therefore, the recent flowering of nationalist and separatist movements around the world is not so surprising. New spaces for political regulation at this level may have to coexist with older ones on a larger scale. Institutions at both levels will have to cope with real if also reversible pressures for global interdependence.

'MARKET-ACCESS' AND THE REGIONAL MOTORS OF THE NEW WORLD ECONOMY

Wide acknowledgement that the world economy is currently undergoing a fundamental reorganization has not meant that there is agreement as to how and why this is happening. Agreement is confined only to the sense that the world economy has entered a phase of flexible production in which business operations around the world are increasingly taking the form of core firms (often transnational in scope) connected by formal and informal alliances to networks of other organizations, both firms, governments, and communities (also sometimes known as disorganized capitalism). The paradox of this trend, and hence why it has generated intense debate, is that while networking allows for an increased spanning of political boundaries by concentrated business organizations, it also opens up the possibility of more decentralized production to sites with competitive advantages. At the same time, networks take on different forms with different sectors and in different places. Some networks have large corporations at their centres with geographically dispersed sub-contractors and allied firms (for example, many car manufacturers), whereas others are clusters of firms in high-tech regions (such as California's Silicon Valley) or specialized industrial districts, such as those in Emilia-Romagna and Tuscany (parts of the so-called Third Italy – Central and Northeast Italy – where economic growth has been based on clusters of small firms specializing in the same industries, such as machine tools, shoes or woollen textiles). In all cases, however, sites are never isolated worlds unto themselves. They are connected together through social connections and the benefits that come from either spatial divisions of labour (splitting different activities between different locations) or **external economies** (the benefits that accrue from locating close to similar and complementary producers). The outcome is a world economy in which networks and flows bring together sites widely scattered around the world. Although, it is important to note, the vast majority of the tightest connections are found in and between Europe, North America, and East Asia. The globalization of the world economy is not yet world-wide in its geographical scope. The extent to which the 'old' system rested on mass production and an absence of networking can also be exaggerated (Cooke 1988). After all, the Italian industrial districts and Silicon Valley, just to name two examples, have historical origins that pre-date the crisis period of the late 1960s and early 1970s. The language of mass or Fordist production versus flexible production and organized versus disorganized capitalism exaggerates the degree of change. Writers usually have in mind 'tendencies' or trajectories more than total shifts. However, in the writing – and reading – this is sometimes forgotten.

One account of the source of this shift in the world economy from big, vertically integrated firms organized largely with reference to national economies to globe-spanning networks of production and finance emphasizes the declining rates of productivity and profits of major corporations in the years between 1965 and 1980. Profit rates, averaged across the seven largest national industrial economies and defined as net operating surplus divided by net capital stock at current prices, declined in these years in the manufacturing sector from 25 per cent to 12 per cent. Across all sectors, the average rate of profit fell from 17 to 11 per cent (Glyn *et al.* 1990, p. 53). There was considerable variability in rates of profit in the 1950s and 1960s, however, so the story of a long boom (or 'golden age') shared by all industrialized countries followed by a sudden collapse is open to question (Webber and Rigby, 1996). What appears to have happened is that the period from 1960 to the early 1970s was one of generally rising profit rates. Thereafter, but at different rates of decline and following different trajectories, rates of profit began to decline (see Fig. 3.5). These seem tied more to declining rates of productivity (efficiency in the use of equipment and resources) than to increasing labour costs. Although there has been a recovery of rates of profit in some economies (such as the USA) since the mid-1980s, this seems fuelled in part by suppressing wages and other labour benefits more than by returns to new technologies (such as computers) or new investment (Webber and Rigby, 1996, p. 325). It also reflects the results of the 'global turn' taken most aggressively by large (and other) American firms since the 1970s. Individual cases, such as General Motors or Ford, suggest as much. Each has come to depend increasingly on the profitability of its world-wide ventures to compensate for the loss of market share and profitability in the United States.

98

external economies
In this usage, refers to the specific benefits that accrue to producers from associating with similar producers in places that offer services that they need.

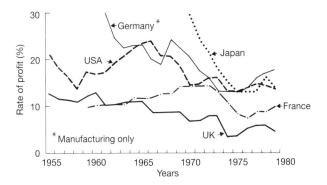

Figure 3.5
Twenty-five years of falling profits in major industrialized countries

99

Source: Edwards (1985), p. 155

Be that as it may, manufacturing industries in all of the major industrialized countries did experience a productivity and profitability crisis beginning in the 1970s. Out of this crisis came the urge to rationalize operations, to 'downsize', to divest relatively unprofitable activities, and to use relocation and diversification strategies to produce higher rates of return for investors. This massive shake-out had a number of consequences. They include an overall increase in the number of firms in many sectors, the rapid spread of new technologies, the compression of the 'shelf life' of commodities to keep up demand, and increased competition to deliver goods and services quickly. Of particular importance, new transportation and communications technologies (such as containerization and fax machines, respectively) made it possible for businesses to move physical assets (such as machines), components, finished products and services and financial capital ever more rapidly from place to place or from one use to another. In the background lay attempts by the governments of the most powerful states, particularly the United States, to open up the world economy to increased trade and investment across international boundaries. These reflected both the perceived interests of certain businesses coming from these countries in 'going global' to solve their problems and the ideological imperative (strongest in the United States) to build a 'free world' economy as an alternative to the closed-off and state-centred economies of the Soviet Union and its satellites (seen in the renewed emphasis by US governments on negotiating multi-lateral reductions in tariffs and quotas in the General Agreement on Tariffs and Trade (GATT) beginning in the 1960s) (Dunning, 1997; Hoekman and Kostecki, 1996). The exhaustion of the Fordist regime of accumulation also coincided with a number of general changes in the workings of the world economy, such as the collapse of the Bretton Woods system for fixing currency exchange rates in 1971, the oil-price increases forced on world consumers by the producer cartel, OPEC in 1973 and, again, in 1979 and the world debt crisis following the failure of borrowers (such as semi-peripheral countries including Mexico and Brazil) to pay back the loans made available to them from the 'petrodollars' recycled into the world economy by the oil producers.

The emerging character of the new world economy can be thought of both from the point of view of states and how they fit into the picture as firms reorganize and from the point of view of the firms and how they organize their networks geographically. The former is referred to by some commentators as an emerging 'market-access economy' in which states increasingly standardize the rules govern-ing trade and investment in order to situate themselves more advantageously within the evolving international division of labour. The latter can be seen as territorially based production systems held together through networks and alliances of firms, governments, and communities.

The 'market-access' regime

Globalization is partly about firms attempting to cash in on the comparative advantage enjoyed in production by other countries and localities and gain unim-peded access to their consumer markets. But it is also about governments wanting to attract capital and expertise from beyond their boundaries so as to increase employment, learn from foreign partners, and generally improve the global com-petitive position of 'their' firms. The combination of the two has given rise to a 'market-access' regime of world trade and investment (Cowhey and Aronson, 1993). This is eroding the free-trade regime that had increasingly predominated in trade between the main industrial capitalist countries in the post Second World War period. In its place is a regime in which acceptable rules governing trade and investment have spread from the relatively narrow realm of trade to cover a wide range of areas of firm organization and performance.

Six 'pillars' of this system can be identified (Table 3.7). The first is a move away from the dominance of the American model of industrial organization in inter-national negotiations towards a hybrid model in which there is less emphasis on keeping governments and industries 'at arm's length' and commitment to encourag-ing inter-firm collaboration and alliances across as well as within national bound-aries. In this new model, foreign firms are allowed to contest most segments of national markets, except in cases where clearly demarcated sectors are left for local firms. A second pillar involves the increased cooperation and acceptance of common rules concerning trade, investment, and money by national bureaucracies with an increasingly powerful role also played by supranational and international organizations (such as the European Commission for the EU and the World Trade Organization, respectively; see Chapter 11). Two consequences are the blurring of lines of regulation between 'issue areas' (such as trade and foreign direct invest-ment, which increasingly can substitute for one another) and the penetration of 'global norms' into the practices of national bureaucracies. The third pillar is the increasing trade in services beyond national boundaries and the concomitant increased importance of services (banking, insurance, transportation, legal, advert-ising, etc.) in the world economy. One reason for this is that high-tech products (computers, commercial aircraft, etc.) contain high levels of service inputs. Servicing the 'software' that such products require has led to an explosion in producer services. Another is that producers are demanding services that are of high quality and competitively priced. They can turn to foreign suppliers if appropriate ones are not available locally. Banking and telephone industries are two that have experi-enced a dramatic increase in internationalization as producers have turned to non-traditional (frequently foreign) suppliers. Fourth, international negotiations about trade and investment are now organized much more along sectoral and issue-

Table 3.7: The old and the new pillars of world trade

Old pillars of the free-trade regime	New pillars of the market-access regime
Structure	
1 US model of industrial organization	Hybrid model of industrial organization
2 Separate systems of governance	Internationalization of domestic policies
3 Goods traded and services produced and consumed domestically	Globalization of services; eroding boundaries between goods and services
4 Universal rules are the norm	Sector-specific codes are common
Rules	
5 Free movement of goods; investment conditional	Investment as integrated coequal with trade
6 National comparative advantage	Regional and global advantage

Source: Cowhey and Aronson (1993, p. 60).

specific lines than was the case in the past. One rule no longer fits all. But many of the new rules are essentially *ad hoc*, rather than formal. This has opened up the possibilities of bilateral and minilateral (more than two parties, but not everybody) negotiations but at the expense of the greater transparency that would come from a consistent multilateral focus.

The final two pillars concern the content of the rules of the market-access regime. One is equivalence today between trade and investment, due largely to the activities of transnational corporations in expanding the level of foreign direct investment to astronomical highs. Local content rules about how much of a finished product must be made locally (within a particular country) and worries about the competitive fairness of firm alliances, however, also led to new efforts by governments in industrialized countries to regulate the flows of foreign investment. 'Levelling the playing field', to use the American parlance, has meant pressure and counter-pressure between governments to ensure at least a degree of similarity in regulation (in, for example, cases of presumed monopoly or anti-trust violations). The final pillar involves the shift on the part of firms from a concern with national or home-base comparative advantage to a concern with establishing global or world-regional competitive advantages internal to firms and their networks. This reflects the overwhelming attractiveness of 'multinationality' to many businesses as a way of both diversifying assets, increasing market access, and enjoying the firm economies of scale that come from supplying larger markets. At the same time plant economies of scale (reductions in unit costs attributable to an increased volume of output) have tended to decease across a wide range of sectors, as noted first by Bain (1959). This means that large firms can enjoy firm economies of scale without having just a few large factories. They are not restricted by the lure of high average plant economies to one or few production locations. Production facilities can be located to take advantage of other benefits that come from operating in multiple locations, particularly those offered by foreign sites. Foreign direct investment is often seen as the result of three sets of factors, all of which became more important as established firms faced a profits' crisis beginning in the 1970s (Dunning, 1979):

1. the advantages that accrue to firms abroad because of their technology and market power relative to competitors (ownership advantages);
2. the need to ensure returns on research and development and other prior investments by controlling production and marketing rather than licensing to foreign firms (internalizing markets); and
3. favourable foreign location conditions that encourage foreign operations rather than export (market size and needs, production costs, trade barriers) (location advantages).

Business economists tend to emphasize the first two, whereas economic geographers tend to give more weight to the third. In particular, the geographers have tended to use the **product life-cycle** model to explain the trend towards increased relocation of certain production processes in foreign settings. In its original form this idea did not have any locational significance; it referred entirely to the tendency for products to move from being novelties to mass production to obsolescence. Raymond Vernon (1966) gave the product life-cycle a locational component by arguing that as production requirements change as a product 'ages' so do locational require-ments. In particular, mass production is more labour-intensive than the earlier phases of production. Hence, when a product reaches this phase in its life-cycle, it pays in terms of profitability to move production to where cheaper labour is available. Patterns of imports and exports adjust accordingly (Fig. 3.6). This model faces a number of criticisms when applied without attention to the specific production requirements of different sectors, such as the overemphasis on labour intensity as a feature of mass production, neglect of the importance of automation in much mass production, the implicit assumption that industries and their products 'mature' rather than adjust or fail to adjust to changing conditions of production, the lack of attention to the increased importance of customized production in many product categories today, and the neglect of the role of regulatory factors (such as tariffs and other import restrictions) in encouraging the movement of production

product life-cycle
Locational requirements for production change as products move from being novel and expensive to being standardized and cheaper. In particular, labour costs can become more important than adjacency to markets.

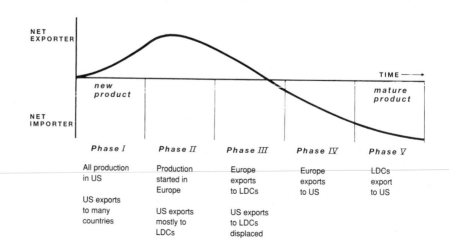

Figure 3.6
The product life-cycle model and possible effects upon US production and trade

Source: Vernon (1966)

facilities to countries other than the home one. (This last factor is one of the main reasons why Japanese car producers have moved to Britain and the USA.)

Does a home-base still matter?

We noted previously that there is dispute over the nature of the new market-access regime and what has brought it about. One very contentious disagreement concerns the role of a 'home base' for firms in the context of growing globalization. On one side stand those, such as Robert Reich (1991), who argue that big firms have become largely detached from their countries of origin and with respect to choice of production and trading sites. They are entities unto themselves, dependent only on the continuing supply of highly trained professionals and technicians. To the extent that a country or region wishes to keep or increase its rate of economic growth, therefore, it must invest in the production of the relevant professionals ('symbolic analysts' to Reich). In this way it can maintain its attractiveness to businesses irrespective of where they come from. Indeed, government policies on taxes and training should be neutral as to the 'nationality' of firms, so as to entice and reward firms who invest in your national space. If your 'own' do not, then they should see no benefit. On the other side are those, such as Michael Porter (1990), who have developed a case for why a home base is more important than ever to companies in an era of intensive global competition. To Porter, global competitive advantage lies in the interaction between four dimensions of economic activity: what he terms a 'diamond' model of global competitiveness. The first is what he calls 'factor conditions'. These include infrastructure, finance, and education of the workforce that can only be provided collectively and which underpin a firm's operations beyond national boundaries. Second are 'demand conditions'. A large and growing home market allows some firms to enjoy economies of scale that translates into an advantage in global competition. To the extent that this market is sophisticated and critical in its consuming, so much the better in laying the groundwork for successful forays into global markets. Third, to be globally competitive an industry must have globally competitive suppliers. Better suppliers make for better-quality and lower-price finished products. But good relations depend upon trust, the exchange of information, and the pooling of expertise that are examples of the externalities resulting from sustained contact. Relative proximity helps to produce the complementarity within and between industries that Porter refers to as 'industry clusters'. Finally, rivalry between competitive firms, suppliers and customers honed on home ground is seen as crucial to global competitiveness. It is domestic rivalry that drives investment, innovation, and the pressure to upgrade technology and the quality of labour. Thus, the Japanese *keiretsu* (giant holding companies such as Mitsubishi) in the car and consumer electronics industries maintain their competitive edge in global markets by constantly looking at one another's strengths and weaknesses in the search for new strategies to enhance firm competitive advantage.

A better approach is to link firm strategies to the changing geography of production. The major assumption now is that each type of firm has a particular 'establishment structure' laid out over space, ranging from highly localized clusters of craftsmen/small firms to the globe-spanning networks created by transnational corporations. There is no single model of firm locational behaviour. But if firms in the past were oriented more to regional and national markets with most links to suppliers and consumers confined within national boundaries, most firms now at least aspire to participate in geographically extensive markets. A new spatial division of labour is emerging in which the regional specialization in different sectors or industries of organized capitalism (or Fordism) is being challenged and supplemented by a powerful set of economic processes that rely to a much greater extent on functional separation of production processes (management, research, assembly, component production, etc.) between locations (Massey, 1984; Urry, 1985). This has happened because new transportation and communication technologies have provided a 'permissive' environment in which firms can decentralize manufacturing and primary-production activities yet still maintain central control (e.g. networked computers, telecommunications, air travel). There is now the possibility of intensive interaction without geographical proximity. The main push for restructuring operations has come from the increasingly competitive business environment resulting from a less regulated and more globalized market-place. One important consequence is that labour markets once internal to firms and regional centres of production have been progressively replaced by a geographical segmentation of labour markets external to firms and vulnerable to the switching of low-wage (and other) operations from place to place but with pools of skilled labour and the sunk costs of established facilities restricting the mobility of some functions and activities. This suggests, as Gertler (1986, p. 77) expresses it: 'that the greater economies of "scale" and "agglomeration" are now realizable in a situation in which production is spatially dispersed'.

This by no means signals the end of regional specialization. Rather, it represents the emergence of new locational strategies that better fit the needs of new industries (such as microelectronic assembly) and some old industries (such as the toy and clothing industries). The evolution of transnational strategies of production does not require the demise of older ones. The major change since the 1970s has been the development of globally competitive strategies, particularly by the largest firms, in which the world is a single investment surface for production and marketing. This challenges the older multidomestic strategies in which subsidiaries cater to different national and world-regional markets (Porter, 1986). American firms have been leaders in the shift to a 'borderless world' (Coca-Cola, McDonald's, Disney, Mattel, Ford, IBM, General Motors and Boeing come to mind). Implicit in this approach is a sequence of organizational-geographical moves as firms shift from (1) exporting to (2) foreign sales outlets to (3) foreign production to (4) the world as an 'investment surface' with production spread around over a large number of locations in different countries. Firms have learned that they can improve their profitability if they use their economies of scope and coordination (returns to complexity and managerial capability in producing multiple products) to compete effectively for global markets against local producers who may have advantages in economies of scale and local connections. Brand names, financial clout and managerial savvy can overcome these barriers. But this can happen only if foreign markets are opened up to competition, which is what the market-access regime is all about.

The example of the semiconductor industry

Semiconductors are the basic components in electronics technologies. Electronics technologies are at the heart of the information revolution that has both allowed and encouraged the transformation of the world economy since the 1970s. World-wide, the semiconductor business conducts over US$90 billion worth of transactions and the world market has been growing at around 15 per cent per year since the early 1980s. More than 30 per cent of all semiconductors are exported, so it is a good candidate for showing the workings of the market-access regime. In the 1980s the Japanese makers broke through in world markets, weakening the grip of American producers, and further threatening the already fragile European industry. This challenge was particularly strong in the core market for standard memory chips (the DRAM or dynamic random-access memory chips) as innovation was rapid and Japanese manufacturers cut prices steeply. It is the DRAM market that has the largest economies of scale and thus can in the long term underwrite other types of production. Not surprisingly, semiconductors became a key trade issue. By 1991, DRAMs accounted for over a quarter of the world semiconductor market and production was dominated by Japanese firms. US firms had the lead in more specialized chips and custom-designed integrated circuits. The Europeans were behind on all fronts. One outcome was increased concentration in the hands of relatively few firms. For example, only 11 companies produce over 95 per cent of the DRAMs on the open market. In 1991, 6 of the top 10 semiconductor producers were Japanese, 3 were American, and number 10 was European. The second 10 included 5 more Japanese firms, 2 from the USA, 2 from Europe, and Samsung from South Korea (Cowhey and Aronson, 1993, p. 129).

The semiconductor industry began in the United States. Though organized roughly in accord with the 'American model', Department of Defense policies boosted the home market. In Japan the Ministry of International Trade and Industry (MITI) accorded semiconductors the highest priority and worked to establish a major presence in global markets. In practice, Japanese success was the result of the marriage of the *keiretsu* system of big companies at the centre of affiliated networks to a national industrial policy. In particular, the Japanese firms enjoyed ready access to capital and government policy reduced the risks of overextension and fostered domestic competition to achieve the best results in global markets. Out of this Japanese challenge came a whole new approach to global trade and investment in semiconductors. The first element was two US–Japan Semiconductor Agreements (1986 and 1991) which opened up Japan to foreign (US) firms and monitored charges of Japanese 'dumping' of semiconductors in the USA at below world prices. The second was the move of both Japanese, European, and US producers into a number of international corporate alliances. This is particularly critical. To anticipate future costs of research and development, 'Firms are turning to ICAs [international corporate alliances] to build common, global infrastructures for the next generation of technologies. Alliances also allow firms to

105

reduce the cost and risk of fielding extensive product lines' (Cowhey and Aronson, 1993, p. 162).

The net effect is a perfect example of the market-access regime of trade and investment. The new structure of the semiconductor industry is a hybrid of Japanese, American, and European models (pillar 1). International agreements are the main means by which the industry is regulated (pillar 2). The mix of hardware and software makes it hard to say where the product ends and servicing begins. What is clear is that both must be available on a world-wide basis for a product to be competitive (pillar 3). Since 1986 specialized industry codes have steadily displaced older industry-wide ones (pillar 4). Where chips were once traded relatively freely but there was not much foreign direct investment, there are now major networks across national boundaries based on international alliances. Semi-conductor firms have gone global in their organization as well as in their search for markets (pillar 5). Finally, firms are building global and world-regional advantages in order to give them leverage over home and foreign markets (pillar 6). Product and technology flows are now so globalized that closing markets (national and regional) would doom affected producers to limited market share and to not participating in new rounds of innovation and product development.

Production networks and regional motors

From the business point of view, the response to the competitive pressures of the market-access regime has been to acquire greater flexibility through technological change, reorganizing labour relations, and establishing links with other firms. One solution has predominated: the creation of networks among producers. This has several origins. One lies in attempts by large firms to reduce the size of their work forces and to hive off activities to other firms. This process of **vertical disintegration** can be cost saving if a unionized labour force is replaced by a non-union one, for example, and it also taps into the specialized skills of suppliers and sub-contractors. A second source lies in the attempt to penetrate foreign markets and build a global presence by collaborating with other firms (including erstwhile competitors). Thus, Honda builds cars in the USA, using at least some US suppliers, and Toyota enters a joint venture with General Motors, to use just two familiar examples from the past 15 years. The focus on production networks highlights the central role of geographical shifts in investment and production as a response to changes in the competitive environment experienced by firms in many economic sectors with the advent of the market-access regime (e.g. Clark, 1993; Malmberg and Maskell, 1997).

vertical disintegration
The tendency for specialized firms to be created as sub-contractors and suppliers within industries formerly dominated by large, functionally integrated firms.

Firms and markets

Before discussing production networks in some detail it is important to establish the understanding of firms and markets upon which the later discussion rests (Nelson and Winter, 1982; Belussi, 1996). Firms are seen as collective agents, that is as organizations endowed with capacities that guide their use of technology and labour. Most importantly, they can learn. Through experience firms follow paths that prove to be most efficacious. This may not be the optimal from an idealized point of view. As argued by Dosi (1988), from this point of view firms are organizations in which capital, labour and knowledge are combined and put to work. Firms differ from one another not only in size (usually measured by number of employees or annual turnover in sales) but also in terms of trajectories of development, risk propensities, production functions and spatial histories (connections to one or several places). As a result, firm diversity and idiosyncrasy are the rule. Firms follow evolutionary strategies to adjust to changing conditions by adopting new technologies, developing new products and shifting production and other facilities to take advantage of cost differences between locations. This evolutionary view of firms differs from the usual (so-called neo-classical) view adopted in most economic geography texts in which firms are seen as atomistic agents with full information in a world of pure markets (with no entry barriers) and all have exactly the same resources, technological capability and market power with deviations regarded as market 'imperfections'.

The view of markets is likewise different. In the orthodox view the market is an abstract space in which prices are set by firms and consumers. In the evolutionary vision, firms must 'bet' on the future but because of various constraints (uncertainty, limited information, differing expectations) can never choose a profit-maximizing strategy in advance. Their bets – on product, technology and locational decisions – are then subject to selection resulting from competitiveness in price and quality among firms in the same industry. This is the market for a given product. It provides feedback (in the form of profits or losses) on the relative success of previous bets. This information can then be used by firms to modify their strategies. From this point of view, therefore, firms change their environment as well as being subject to predetermined conditions so the organization of firms and industrial structure is the outcome of firm strategies as well as initial market conditions (demand from consumers, degree of competitiveness within a sector or industry, etc.) (Jacquemin, 1987; Nelson, 1995). As a result, industrial structures will be subject to continual adaptation and transformation.

Production networks

A number of types of networks among firms can be distinguished (Harrison, 1994, pp. 134–41; more generally, see Scott, 1986; Scott and Storper, 1986; Walker, 1988; Christopherson, 1989; Benko and Dunford, 1991; Bonacich *et al.*, 1994; Gereffi and Korzeniewicz, 1993; Storper and Scott, 1992). One type occurs with craft-based industries. These industries are themselves organized around *projects* more than firms, *per se*. In construction, publishing, film and recording, architecture and software engineering highly skilled work forces are employed by firms but share knowledge easily across firm boundaries. Consequently, such industries tend to cluster to take advantage of the externalities implicit in such sharing. External economies include such factors as a labour pool with relevant skills, a broad network of suppliers, excellent educational and training support and, perhaps most importantly, access to 'venture capital' knowledgeable about the nature of the business (more generally on the role of venture capital in production networks, see Florida and Smith, 1993). But they also include intangibles such as a 'culture' accepting of innovation, tolerance of failure, local reinvestment, collaboration, promotion on merit, and openness to new enterprises. Two other types of network are more firm related. The first are small firm industrial districts such as those often associated with so-called 'Third Italy', but also found elsewhere. These are local integrated networks of producers with different firms specializing in different phases of the production process but competing for work with other local firms when new projects come along. This mix of cooperation and competition has elicited much interest and comment, possibly out of all proportion to the relative importance of this type of network in the larger scheme of business restructuring. These networks fared well during the period of reduced firm profitability, so others want to know the secret of their success. Evidence suggests that they rely in equal measure on external economies of scale in production (collaborative production, local government financing, craft traditions, pools of skilled labour, etc.) and on what can be called non-traded interdependencies – a long history of social collaboration, institutionalized cooperation and agreement on social conventions governing everyday interfirm relations. In the 1990s, however, many Italian industrial districts show signs of both increased concentration of firms and a crisis in production brought on by both competitive pressure from cheaper foreign manufacturers and the increased efficiency of large mass-production firms in Northwest Italy.

The third type of network is that of agglomerated big firm-based production systems, such as that of Toyota and its ring of suppliers around Toyota City in Nagoya, Japan, Boeing and its suppliers around Tacoma-Seattle in the USA and FIAT and its suppliers around Turin in Northwest Italy. In some cases the suppliers pre-existed the emergence of the big firms, in others (as in Japan and South Korea) the suppliers were financed by the dominant company (Shiba and Shimotami, 1997). Since the 1970s, however, the main process stimulating this kind of network has been the vertical disintegration of the big firms themselves. Whether territorially connected to them or not, large firms can now achieve improved flexibility by using sub-contractors to carry out aspects of production that used to be performed within the boundaries of the firm (Harrison, 1994, p. 135). A high-tech industrial complex such as Silicon Valley is somewhere in-between the 'classic' industrial district and the agglomerated big-firm production system, sharing features of both (Saxenian, 1994).

Finally, there is the type of network represented by strategic alliances between firms. This is one of the most important innovations of the market-access regime, even though there are certainly historical precursors for joint operations by distinct companies. What is new is the extent to which strategic alliances take place between international competitors. 'Each partner brings to the marriage its own specialty – technology, financial power, access to government regulators or procurement officials – and its own constellation of small firm suppliers' (Harrison, 1994, p. 138). What each gains is the knowledge and connections intrinsic to the other to further their efforts at conquering global markets for their products. One logical consequence of alliances would be merger or acquisition of one partner by the other. But some national laws and customs set limits to this (for example, American and Japanese laws restrict foreign acquisition of home-grown firms) and the goal of flexibility is best met by maintaining or recreating alliances rather than engaging in fully fledged mergers.

The new international division of labour

The first attempt at relating the changing character of business organization to the economic geography of production posited the emergence of a new international division of labour (NIDL). From this point of view big multinational corporations were seen as creating a new global economic geography. One of the architects of this viewpoint, Stephen Hymer (1976), imagined that these corporations:

> create a world in [their] own image by creating a division of labor between countries that corresponds to the division of labor between various levels of the corporate hierarchy. [They] tend to centralize high-level decision-making occupations in a few key cities (surrounded by regional sub-capitals) in the advanced countries . . . confining the rest of the world to lower levels of activity and income.

(quoted in Dicken 1992a, pp. 223–4)

As a result, the new international division of labour was seen as reflecting the hierarchical social division of labour within the big firms themselves. Hymer claimed that a tight **space–process relationship** was coming about in which control and operational activities within firms would be completely separated. In terms of Table 3.8 the close space–process relationship would take the form A to B to C. But the table reveals other possibilities: D, E, and F. Logically, a major metropolis (or **world city**) could dominate all levels in the hierarchy of functions in absolute terms even with the *addition* of foreign operations. This is the case reported for the English electronics industry by Sayer (1985), in which the London-Southeast region dominates the entire industry. The advent of networks was largely responsible for confounding the simple story of the new industrial division of labour. Production networks allowed much more complex geographies of production than those predicted by Hymer's simple hierarchy. Big firms have changed their internal structures and external relations in ways that undermine their own internal hierarchies. So the analogy between internal (organizational) and external (geographical) hierarchies now seems overdrawn.

More importantly, much of the explosion of foreign direct investment of the past thirty years has involved within-core and not core to periphery/semi-periphery flows on a world scale. For example, the Western European share of total US

space–process relationship
The idea that different types of firm activity are carried out at different locations within a hierarchy of places, from world cities to various peripheries.

world city
One of the cities that dominate world finance and serve as headquarters to transnational corporations. Typically, London, New York and Tokyo are identified as the leading tier of world cities, although other cities such as Paris, Frankfurt, Chicago and Zurich also have important global roles.

109

Table 3.8: 'Hymer's stereotype', in which the space–process relationship takes the form A → B → C

	Type of area		
Level of corporate hierarchy	**Major metropolis (for example, New York)**	**Regional capitals (for example, Brussels)**	**Periphery (for example, South Korea, Ireland)**
1 Long-term strategic planning	A		
2 Management of divisions	D	B	
3 Production, routine work	F	E	C

Source: Sayer (1985), Table 1, p. 13.

manufacturing investment abroad accounts for a steadily expanding share of the total: from 26 per cent in 1956 to 43 per cent in 1966 and 53 per cent by 1987. This suggests how important market access has become, relative to the search for cheap labour in assembly processes (Schoenberger, 1990, pp. 380–1). Increased competition for market shares in the USA has forced a search for markets elsewhere that cannot be served by an export strategy. Supplier performance, alliances, and service to customers, as well as avoidance of tariffs and other trade barriers, have dictated moving close to potential markets. Moving to the global periphery to minimize labour costs is a much less important activity in terms of total investment.

Geographies of production networks

An approach which goes beyond the NIDL focuses on the elements of networked production systems and how they come together to produce different kinds of economic geographies specific to sectors and types of interfirm network (Storper and Harrison, 1991). In this construction, a production network has an input–output structure (a set of interacting firms), a governance structure (a hierarchy of power), and a territoriality (a concentrated or dispersed geography of production). These elements define the forms that networks take and the ways that they work. In some industries there are major returns to agglomeration because of the relative importance of sub-contractors and localized external economies of scale, whereas in others dispersed networks indicate that external economies can be captured over space. In each case very different rules of governance are involved: in the former, power is widely shared (depending on the degree of dominance by a core large firm) and in the latter either power-sharing between equals (as in intercorporate alliances) or asymmetric relations (between a headquarters and a branch plant, for example) tend to prevail.

Whatever the precise outcome, lying at the heart of the geography of the world economy under market-access conditions are the trade-offs specific to different sectors between the benefits/costs to firms of conducting economic transactions over space, on the one hand, and the benefits/costs to firms of clustering together, on the other (Scott, 1996; see Table 3.9). The former come down to the costs of production involved in overcoming distance in the transactions (bringing together inputs, serving markets, etc.) implicit in production. The latter involves the cost-saving role of locating close to suppliers, sub-contractors, competitors and specialized pools of labour. In one scenario, perhaps pertinent for resource-based industries, resource-dependent industries, wholesaling, and retailing in which either transport costs and/or direct access to customers still figure prominently in firm locational calculi, producers will seek out low-cost locations relative to basic inputs and/or markets. There is little or no incentive for firms to cluster together. The result will be locational patterns conforming closely to the distribution of resources and population (1). Even in such cases, however, the situation facing firms is not the same as it once was. In particular, international transport costs have declined by about two-thirds in real terms (after taking inflation into account) over the period 1950–90. New types of ship design, containerization, improved methods for routing shipments, and the almost universal availability of faxes have produced massive improvements in the efficiency of the shipping and land transport industries. So, even for industries in which there is major 'weight loss' during production (a major reduction in the amount of output relative to amounts of inputs), the 'friction' of distance is no longer the constraint on making decisions about where to locate plants that it undoubtedly once was.

A second scenario, in which externalities are significant but where transaction costs are high, defines the situation of industrial districts and high technology complexes (such as California's Silicon Valley). As noted previously, there are important differences within this category relating to the relative importance of large firms versus small ones and craft industries. But nevertheless they all share an important focus on externalities while requiring heavy inputs of outside resources and an orientation to external consumer markets. Here intensive relations between firms encourage agglomeration but there are limits to which this can go before costs associated with serving outside markets and obtaining outside resources start to go up (2). A third scenario, essentially that of branch-plant industrialization, sees the possibility of firms consuming externalities at a distance because of low transaction costs. This is the case referred to by Gertler (1986) in the previous discussion. In this way externalities can be internalized within a firm (or interfirm alliance) and then realized through dispersal of functions to locations where they can achieve cost advantages (lower wage bills, etc.) (3). A fourth scenario is literally one where anything can be located anywhere. This is the fantasy of some prophets of globalization who predict the 'end of geography' as telecommuting and a radical decentralization of production characterize all sectors. This is a state of spatial entropy in which there would be no spatial limits on access to externalities. As yet, this appears to be more fantasy than reality (4). More likely is a case in which, though externalities can be obtained at a distance, transaction costs mandate attraction to markets or inputs (5).

Most important to the evolving world economy is the sixth scenario. In this, spatial transaction costs are assumed to be moderate (on average) but externalities are high (6). There is thus a major incentive to cluster. Such clusters represent the major novelty in recent times. They are the concentrations of innovative, knowledge-based and high-value producing industries (both manufacturing and services) that are increasingly driving the world economy. This is why Allen Scott (1996, p. 400) refers to them as the 'regional motors' of the world economy. The major reason for this claim is:

Table 3.9: Spatial transaction costs versus externalities: six scenarios

		Spatial Transaction Costs		
		Low	**Medium**	**High**
	Low	(4)	(5)	(1)
Externalities				
	High	(3)	(6)	(2)

Source: Based on Scott (1996).

that contemporary forms of economic production and organization are rife with externality effects, having their roots in the augmenting levels of flexibility, uncertainty, product destandardization, and competitiveness that are some of the hallmarks of contemporary capitalist enterprise.

But in addition to the increasing returns to agglomeration in many leading sectors (such as high-technology industry, design-intensive consumer goods, and business and financial services), many transactions are still intensely sensitive to the effects of distance:

> while spatial transaction costs have fallen dramatically across a wide front in recent decades, allowing many firms ready access to global markets, there still remain important kinds of transactions that are extremely sensitive to the effects of distance. External economies tend to be well developed in the interaction networks constituted by just such transactions as these and, in order to secure them, producers agglomerate together in geographic space.

The main geographical implication is that with increasing world economic integration, the leading production activities will become more concentrated in metropolitan areas and their hinterlands (see Scott, 1993). These are places with long established competitive advantages, built up by trial and error over the years more than as the result of some single overriding locational advantage relative to other locations. This geographical path dependence seems set to continue, when in the old refrain 'the rich get richer and the poor get the blame'. Giant metropolitan areas such as Tokyo, São Paulo, New York, Mexico City, Shanghai, Los Angeles, Mombai (Bombay), and Seoul with populations in excess of 10 million in 1990, not to mention at least 25 other urban agglomerations in excess of 5 million apiece, constitute the dynamic centres of the world economy as national boundaries lose some of their grip on channelling the processes governing economic growth. Those employed in occupations and sectors with a global market reach who operate from these centres are able to cash in personally on the larger demand only they can satisfy because of *where they are*. London re-insurers, Hollywood actors, New York lawyers, Silicon Valley software engineers have the capacity to extract higher incomes because they embody major specialized activities of where they are from. They have global access in sectors for which there is high global demand.

What is also clear is that the flip-side of this concentration of economic growth based on serving global markets is the increased marginalization of large parts of the world and their populations (Agnew and Corbridge, 1995, Chapter 7). These are not necessarily at great distance from the 'motors' themselves. Indeed, internal to the dominant metropolitan areas are rich and poor districts housing the increasingly polarized income groups that the new world economy seems to be bringing in its train (e.g. Harrison and Bluestone, 1990; Levy and Murnane, 1992; Smeeding *et al.*, 1990). Many of the poor, when they find employment, find it in providing services for the more affluent, which is increasingly one of the driving forces behind local economic growth. Services for local consumption are responsible for much of the economic growth in large cities (Krugman, 1995). Yet, this growth depends on the incomes generated by the goods- and service-producing activities of industries oriented to national and global markets. Indeed, the growth of smaller cities and areas surrounding major cities is increasingly dependent on the growth of the networked economy. For example, much of the growth of employ-ment and incomes in Britain in the 1980s was concentrated in an arc of 'growth areas' extending from Cambridge to Bournemouth that can be seen as beneficiaries of the explosive growth of the financial service industries in London (Champion and Townsend, 1990). Computer hardware manufacturing and service, accounting, billing, paperwork, and other 'back-office' functions have decentralized into Lon-don's hinterland. In the new 'services economy', in which both the most lucrative and the poorest paying jobs involve providing services to others (from banking and finance to fast-food and house-cleaning), globalization causes London to cast a new shadow over its hinterland. Another fascinating example comes from Italy where information on exports from provinces shows an unmistakable geographical pat-tern of some growing regions and others largely missing out on the possibilities offered by the lowering of national barriers to trade (Celant, 1996).

One theoretical implication is unmistakable: the increased globalization of the world economy is leading not to a spreading out of economic growth or a homogenization of global space but to heightened differences between regions and localities. Some so-called world cities, such as London, Tokyo, and New York, are becoming centres (among other things) of business and financial services (Daniels, 1991; Knox and Taylor, 1995). Other regions, like the Third Italy, the American Midwest, and Taiwan (if in different ways) have a focus on manufacturing. Some regions remain primarily agricultural (also in very different ways), whereas others are the sites of low-wage assembly or back office functions. There is no single universal model of business organization that accounts for all of them. In an increasingly competitive world economy they are all the result of adaptive responses to pressures on states and firms to change their ways of doing business.

Summary

This chapter has laid out the historical-geographical perspective that lies behind the rest of the book. We sought to show how the processes that produce the geography of economic activities evolve over time as the world economy evolves. Particular attention has been paid to the specific features of the world economy and its evolution. These are:

1. the single world market
2. the state system
3. the three geographical tiers: core, periphery, semi-periphery
4. temporal patterns and hegemony
5. subordination and resistance
6. alternative adaptations.

The second section of the chapter has been devoted to examining the changing relationship between states and the contemporary world economy under conditions of disorganized capitalism. The main conclusions are that states must now operate in a world in which (1) economic power is no longer best thought of as an attribute of states, and (2) states and societies/national economies are no longer mutually defining entities. However, we have stressed the continuing importance of states as regulators of economic activities. Historical experience suggests that a trend towards a globalized world economy can be reversed if there is an increase in economic nationalism or protectionism if a major economy such as that of the USA or of the European Community undermines the fairly open trading system that presently exists. Even a thoroughly 'globalized' world economy will have political divisions that will take a geographical form.

The third part of the chapter introduced the idea of a 'market-access' economy, how it differs from the previously dominant free-trade regime, and the critical role of certain regional 'motors' to the emerging world economy. The importance of transnational companies and strategic alliances between them was emphasized, along with the increasingly complex international division of labour in many economic sectors. The location of economic activities appears to involve a decreased reliance on the costs of assembling the factors of production (spatial transaction costs) – largely because of a dramatic decline in transportation costs – and an increased reliance on both traded and non-traded interdependencies (externalities) which encourages a clustering of specialized activities. Rather than encouraging a spreading out of manufacturing and service industries across the world, therefore, the market-access model of increased global economic interdependence produces a remarkable regional clustering of many activities. This reflects the competitive advantage achieved by certain locations (often for historic or 'path dependence' reasons) more than patterns of raw material availability or labour cost advantages. For certain sectors (e.g. pulp and paper or apparel, respectively) these factors maintain their relevance. But leading 'high tech' sectors (such as information and bio-technologies) and many other sectors (such as stock trading and insurance, metal-working) are wedded closely to particular locations.

Parts 2 and 3 of this book are taken up with exploring the specific economic-geographical consequences of the evolution of the world economy as described in the first and third parts of this chapter. Part 4 is concerned with some of the manifestations of the globalization/localization nexus at the centre of the today's world economy that pose challenges to state management raised in the second part: the growth of regional trading blocs and decentralist reactions to the changing world economy.

Key Sources and Suggested Reading

Agnew, J. A. 1987. *The United States in the World Economy: A Regional Geography*. Cambridge: Cambridge University Press.

Agnew, J. A. 1992. The United States and American hegemony. In P. J. Taylor (ed.), *The Political Geography of the Twentieth Century*. London: Belhaven Press.

Agnew, J. A. 1993. Trading blocs or a world that knows no boundaries? In C. H. Williams (ed.), *The Political Geography of the New World Order*. London: Belhaven Press.

Agnew, J. A. and Corbridge, S. 1995. *Mastering Space: Hegemony, Territory and International Political Economy*. London: Routledge.

Amsden, A. 1989. *Asia's Next Giant: South Korea and Late Industrialization*. New York: Oxford University Press.

Arthur, W. B. 1989. Competing technologies, increasing returns, and lock-in by historical events. *The Economic Journal*, **99**, 116–31.

Bain, J. S. 1959. *Industrial Organization*. New York: Wiley.

Bairoch, P. 1982. International industrialization levels from 1750 to 1980. *European Journal of Economic History*, **11**, 269–333.

Bayoumi, T. 1989. *Saving-Investment Correlations*. Washington, DC: IMF Working Paper 89/66.

Beenstock, M. 1983. *The World Economy in Transition.* London: Allen and Unwin.

Belussi, F. 1996. Local systems, industrial districts and institutional networks: towards a new evolutionary paradigm of industrial economics? *European Planning Studies,* **4**, 5–26.

Benko, G. and Dunford, M. (eds) 1991. *Industrial Change and Regional Development: The Transformation of New Industrial Spaces.* London: Belhaven Press.

Berry, B. J. L. 1991. *Long Wave Rhythms in Economic Development and Political Behavior.* Baltimore: Johns Hopkins University Press.

Bonacich, E. et al. 1994. *The Globalization of the Garment Industry in the Pacific Rim.* Philadelphia: Temple University Press.

Bourgin, F. 1989. *The Great Challenge: The Myth of Laissez-Faire in the Early Republic.* New York: George Braziller.

Bradshaw, M. 1991. *The Soviet Union: A New Regional Geography?* London: Belhaven Press.

Brenner, R. 1977. The origins of capitalist development: a critique of neo-Smithian Marxism. *New Left Review,* **104**, 25–91.

Celant, A. 1996. Italy's foreign trade: economic and territorial characteristics. In A. Vallega et al., *The Geography of Disequilibrium: Global Issues and Restructuring In Italy.* Rome: Società Geografica Italiana.

Champion, A. G. and Townsend, A. R. 1990. *Contemporary Britain.* London: Edward Arnold.

Christopherson, S. 1989. Flexibility in the US service economy and the emerging spatial division of labour. *Transactions of the Institute of British Geographers,* **14**, 131–43.

Clark, G. L. 1993. Global interdependence and regional development: business linkages and corporate governance in a world of financial risk. *Transactions of the Institute of British Geographers,* **18**, 309–25.

Cohen, S. S. and Zysman, J. 1987. *Manufacturing Matters: The Myth of the Post-Industrial Economy.* New York: Basic Books.

Cooke, P. 1988. Flexible integration, scope economies, and strategic alliances: social and spatial mediations. *Society and Space,* **6**, 281–300.

Corbridge, S. E. and Agnew, J. A. 1991. The US trade and budget deficits in global perspective: an essay in geopolitical economy. *Society and Space,* **9**, 71–90.

Cowhey, P. F. and Aronson, J. D. 1993. *Managing the World Economy: The Consequences of Corporate Alliances.* New York: Council on Foreign Relations Press.

Crook, C. 1991. Sisters in the wood: a survey of the IMF and the World Bank. *The Economist,* 12 October, survey.

Crook, C. 1992. Fear of finance: a survey of the world economy. *The Economist,* 19 September, survey.

Curtin, P. et al. 1978. *African History.* Boston: Little, Brown.

Daniels, P. W. 1991. A world of services? *Geoforum,* **22**, 359–76.

Destler, I. M. 1992. *American Trade Politics.* Second edition. Washington, DC: Institute for International Economics.

Deudney, D. and Ikenberry, G. J. 1991/92. The international sources of Soviet change. *International Security,* **16**, 74–118.

Dicken, P. 1992a. *Global Shift: The Internationalization of Economic Activity.* Second edition. London: Paul Chapman.

Dicken, P. 1992b. International production in a volatile regulatory environment: the influence of national regulatory policies on the spatial strategies of transnational corporations. *Geoforum,* **23**, 303–16.

Dosi, G. 1988. Sources, procedure and microeconomic effects of innovation. *Journal of Economic Literature*, **26**, 1–12.

Dunning, J. H. 1979. Explaining changing patterns of international production: in defence of the eclectic theory. *Oxford Bulletin of Economics and Statistics*, **41**, 269–96.

Dunning, J. H. (ed.) 1997. *Governments, Globalization, and International Business*. New York: Oxford University Press.

The Economist 1992. Zero inflation: how low is low enough? 7 November, 23–26.

The Economist 1993. The global firm: R.I.P. 6 February, 69.

Edwards, C. 1985. *The Fragmented World. Competing Perspectives on Trade, Money and Crisis*. London: Methuen.

Eichengreen, B. 1996. *Globalizing Capital: A History of the International Monetary System*. Princeton: Princeton University Press.

Epstein, G. 1992. Political economy and comparative central banking. *Review of Radical Political Economics*, **24**, 1–30.

Ettlinger, N. 1991. The roots of competitive advantage in California and Japan. *Annals of the Association of American Geographers*, **81**, 391–407.

Florida, R. and Smith, D. F., Jr. 1993. Venture capital formation, investment and regional industrialization. *Annals of the Association of American Geographers*, **83**, 434–51.

Friedman, B. 1989. *Day of Reckoning: The Consequences of American Economic Policy*. New York: Random House.

Gann, L. H. and Duignan, P. 1978. *The Rulers of British Africa, 1870–1914*. Stanford: Stanford University Press.

Gereffi, G, and Korzeniewicz, M. (eds) 1993. *Commodity Chains and Global Capitalism*. Westport, CT: Greenwood Press.

Gertler, M. 1986. Discontinuities in regional development. *Society and Space*, **4**, 71–84.

Gill, S. 1990. *American Hegemony and the Trilateral Commission*. Cambridge: Cambridge University Press.

Glyn, A. *et al.* 1990. The rise and fall of the Golden Age. In S. A. Marglin and J. B. Schor (eds), *The Golden Age of Capitalism: Reinterpreting the Postwar Experience*. Oxford: Clarendon Press.

Grant, R. J. *et al.* 1993. Global trade flows: old structures, new issues, empirical evidence. In C. F. Bergsten and M. Noland (eds), *Pacific Dynamism and the International Economic System*. Washington, DC: Institute for International Economics.

Hall, T. D. 1986. Incorporation in the world-system: toward a critique. *American Sociological Review*, **51**, 390–402.

Hansson, K. 1952. A general theory of the system of multilateral trade. *American Economic Review*, **42**, 59–88.

Harris, N. 1986. *The End of the Third World: Newly Industrialising Countries and the End of an Ideology*. London: I. B. Tauris.

Harris, N. 1996. *The New Untouchables: Immigration and the New World Worker*. London: Penguin.

Harrison, B. 1994. *Lean and Mean: The Changing Landscape of Corporate Power in the Age of Flexibility*. New York: Basic Books.

Harrison, B. and Bluestone, B. 1990. Wage polarization in the U.S. and the 'flexibility' debate. *Cambridge Journal of Economics*, **14**, 351–73.

Hayes, S. L. and Hubbard, P. M. 1990. *Investment Banking: A Tale of Three Cities*. Cambridge, MA: Harvard Business School Press.

Heim, C. E. 1991. Dimensions of decline: industrial regions in Europe, the U.S., and Japan in the 1970s and 1980s. Paper presented at the Social Science History Association, Annual Meeting, New Orleans, November.

Herzog, H. W. and Schlottmann, A. M. (eds) 1991. *Industry Location and Public Policy*. Knoxville: University of Tennessee Press.

Hewett, E. A. and Gaddy, C. G. 1992. *Open for Business: Russia's Return to the Global Economy*. Washington, DC: Brookings Institution.

Hoekman, B. M. and Kostecki, M. M. 1996. *The Political Economy of the World Trading System: From GATT to WTO*. New York: Oxford University Press.

Hutton, W. 1995. *The State We're In*. London: Jonathan Cape.

Hymer, S. 1976. The multinational corporation and the law of uneven development. In H. Radice (ed.), *International Firms and Modern Imperialism*. London: Penguin.

Jackson, R. H. 1990. *Quasi-States*. Cambridge: Cambridge University Press.

Jacquemin, A. 1987. *The New Industrial Organization*. Oxford: Clarendon Press.

Johnson, A. W. and Earle, T. 1987. *The Evolution of Human Societies: From Foraging Group to Agrarian State*. Stanford: Stanford University Press.

Johnston, R. J. 1992. The rise and decline of the corporate-welfare state: a comparative analysis in global context. In P. J. Taylor (ed.), *The Political Geography of the Twentieth Century*. London: Belhaven Press.

Knox, P. L. and Taylor, P. J. (eds) 1995. *World Cities in a World-System*. Cambridge: Cambridge University Press.

Kopytoff, I. 1987. *The African Frontier: The Reproduction of Traditional African Societies*. Bloomington: Indiana University Press.

Krugman, P. 1990. *The Age of Diminished Expectations: U. S. Economic Policy in the 1990s*. Cambridge, MA: MIT Press.

Krugman, P. 1995. The localization of the world economy. *New Perspectives Quarterly*, Winter, 34–38.

Laird, S. and Yeats, A. 1990. Trends in non-tariff barriers of developed countries, 1966–1986. *Weltwirtschaftliches Archiv. Review of World Economies*, **126**, 299–325.

Lesser, A. 1961. Social fields and the evolution of society. *Southwestern Journal of Anthropology*, **17**, 40–8.

Levy, F. and Murnane, R. 1992. U.S. earnings levels and earnings inequality: a review of recent trends and proposed explanations. *Journal of Economic Literature*, **30**, 1333–81.

Leyshon, A. and Thrift, N. 1997. *Money/Space: Geographies of Monetary Transformation*. London: Routledge.

Lipietz, A. 1986. Behind the crisis: the exhaustion of a regime of accumulation. A 'regulation school' perspective on some French empirical works. *Review of Radical Political Economics*, **18**, 13–32.

Maddison, A. 1983. A comparison of levels of GDP per capita in developed and developing countries, 1700–1980. *Journal of Economic History*, **43**, 27–41.

Malmberg, A. and Maskell, P. 1997. Towards an explanation of regional specialization and industry agglomeration. *European Planning Studies*, **5**, 25–41.

Martin, R. L. and Rowthorn, B. (eds) 1986. *The Geography of De-Industrialisation*. London: Macmillan.

Mason, M. 1992. *American Multinationals and Japan: The Political Economy of Japanese Capital Controls, 1899–1980*. Cambridge, MA: Harvard University Press.

Massey, D. 1984. *Spatial Divisions of Labour*. London: Methuen.

Nakane, C. 1970. *Japanese Society*. London: Weidenfeld and Nicolson.

Nau, H. R. 1990. *The Myth of America's Decline: Leading the World Economy into the 1990s*. New York: Oxford University Press.

Nelson, R. 1995. Recent evolutionary theorizing about economic change. *Journal of Economic Literature*, **33**, 1089–98.

Nelson, R. and Winter, S. 1982. *An Evolutionary Theory of Economic Change*. Cambridge, MA: Belknap Press of the Harvard University Press.

O'Brien, P. 1982. European economic development: the contribution of the periphery. *Economic History Review*, **35**, 1–18.

O'Brien, R. 1992. *Global Financial Integration: The End of Geography*. New York: Council on Foreign Relations Press.

Parboni, R. 1988. U. S. economic strategies against Western Europe: from Nixon to Reagan. *Geoforum*, **19**, 45–54.

Pauly, L. W. and Reich, S. 1997. National structures and multinational corporate behavior: enduring differences in the age of globalization. *International Organization*, **51**, 1–30.

Perez, C. 1983. Structural change and assimilation of new technologies in economic-social systems. *Futures*, **10**, 357–75.

Picciotto, S. 1991. The internationalization of the state. *Review of Radical Political Economics*, **22**, 28–44.

Polanyi, K. 1957. The place of economies in societies. In K. Polanyi *et al.* (eds), *Trade and Markets in the Early Empires*. Glencoe, IL: Free Press.

Porter, M. E. 1986. *Competition in Global Industries*. Cambridge, MA: Harvard Business School Press.

Porter, M. E. 1990. *The Competitive Advantage of Nations*. New York: Free Press.

Reich, R. E. 1991. *The Work of Nations: Preparing Ourselves for 21st Century Capitalism*. New York: Knopf.

Rodriguez, E. and Griffith-Jones, S. (eds) 1992. *Cross-Conditionality, Banking Regulation and Third World Debt*. London: Macmillan.

Roosevelt, A. C. 1992. Secrets of the forest: an archaeologist appraises the past – and the future – of Amazonia. *The Sciences*, November/December, 22–8.

Rose, R. 1984. *Understanding Big Government*. London: Sage.

Ruggie, J. G. 1983. International regimes, transactions and change: embedded liberalism in the postwar economic order. In S. D. Krasner (ed.), *International Regimes*. Ithaca, NY: Cornell University Press.

Sayer, A. 1985. Industry and space: a sympathetic critique of radical research. *Society and Space*, **3**, 3–29.

Saxenian, A. 1994. *Regional Advantage: Culture and Competition in Silicon Valley and Route 128*. Cambridge, MA: Harvard University Press.

Schoenberger, E. 1990. US manufacturing investments in Western Europe: markets, corporate strategy, and the competitive environment. *Annals of the Association of American Geographers*, **80**, 379–93.

Scott, A. J. 1986. Industrial organization and location: division of labor, the firm, and spatial process. *Economic Geography*, **62**, 215–31.

Scott, A. J. 1993. *Technopolis: High-Technology Industry and Regional Development in Southern California*. Berkeley and Los Angeles: University of California Press.

Scott, A. J. 1996. Regional motors of the global economy. *Futures*, **28**, 391–411.

Scott, A. J. and Storper, M. J. (eds) 1986. *Production, Work, Territory*. London: Allen and Unwin.

Sen, A. 1981. Public action and the quality of life in developing countries. *Oxford Bulletin of Economics and Statistics*, **43**, 111–23.

Shiba, T. and Shimotami, M. (eds) 1997. *Beyond the Firm: Business Groups in International and Historical Perspective*. New York: Oxford University Press.

Skocpol, T. 1976. France, Russia, China: a structural theory of social revolution. *Comparative Studies in Society and History*, **18**, 181–96.

Smeeding, T. *et al.* (eds) 1990. *Poverty, Inequality, and Income Distribution in Comparative Perspective: The Luxembourg Income Study*. London: Harvester Wheatsheaf.

Smith, R. and Walter, I. 1996. *Global Banking*. New York: Oxford University Press.

Smith, T. 1979. The underdevelopment of development literature: the case of dependency theory. *World Politics*, **31**, 247–88.

Storper, M. J. and Harrison, B. 1991. Flexibility, hierarchy and regional development: the changing structure of industrial production systems and their forms of governance in the 1990s. *Research Policy*, **20**, 407–22.

Storper, M. J. and Scott, A. J. (eds) 1992. *Pathways to Industrialization and Regional Development*. London: Routledge.

Terlouw, C. P. 1992. *The Regional Geography of the World-System: External Arena, Periphery, Semiperiphery, Core*. Utrecht: Netherlands Geographical Studies, No. 144.

Tilly, C. (ed.) 1975. *The Formation of National States in Western Europe*. Princeton: Princeton University Press.

Tylecote, A. 1992. *The Long Wave in the World Economy: The Current Crisis in Historical Perspective*. London: Routledge.

Tyson, L. D. 1992. *Who's Bashing Whom? Trade Conflict in High Technology Industries*. Washington, DC: Institute for International Economics.

UNCTAD 1996. *World Investment Report 1996*. Geneva: United Nations Conference on Trade and Development.

Urry, J. 1985. Social relations, space and time. In D. Gregory and J. Urry (eds), *Social Relations and Spatial Structures*. London: Macmillan.

Vernon, R. 1966. International investment and international trade in the product cycle. *Quarterly Journal of Economics*, **80**, 190–207.

Walker, R. 1988. The geographical organization of production-systems. *Society and Space*, **6**, 377–408.

Wallerstein, I. 1974. *The Modern World-System: Capitalist Agriculture and the Origins of the European World-Economy in the Sixteenth Century*. New York: Academic Press.

Wallerstein, I. 1979. *The Capitalist World-Economy*. Cambridge: Cambridge University Press.

Wallerstein, I. *et al.* 1979. Cyclical rhythms and secular trends of the capitalist world-economy. *Review*, **2**, 483–500.

Webb, M. C. 1991. International economic structures, government interests, and international coordination of macroeconomic adjustment policies. *International Organization*, **45**, 309–42.

Webber, M. J. and Rigby, D. L. 1996. *The Golden Age Illusion: Rethinking Postwar Capitalism*. New York: Guilford Press.

Williams, R. 1983. *The Year 2000: A Radical Look at the Future and What We Can Do to Change It*. New York: Pantheon.

World Bank 1991. *World Development Report*. New York: Oxford University Press.

Zysman, J. 1983. *Government, Markets and Growth: Financial Systems and the Politics of Industrial Change*. Ithaca, NY: Cornell University Press.

PART 2

THE RISE OF THE CORE ECONOMIES

In the next four chapters we trace the emergence of the world's core economies, following their different paths towards increasing scale and complexity with case histories that illuminate many of the patterns, models and theories outlined in Part 1. We seek to show that the world's economic landscapes, however unique or exceptional they may seem, are now part of a single, overarching world economy. In Chapter 4 we describe the way in which this world economy came to be centred on Europe, how it came to be consolidated by the emergence of merchant capitalism, and how the nature and organization of merchant capitalism came to be reflected in particular kinds of urban and regional change. In Chapter 5 we describe the very different trajectories that have marked the ascent of Europe, North America, Japan, and Russia within the world economy, emphasizing the spatial changes consequent upon the emergence and evolution of industrial capitalism. In Chapter 6 the globalization of the core economies is described, and in Chapter 7 the spatial implications of the latest form of economic organization on the capitalist countries of the world's core regions are examined. But, although the emphasis throughout this part of the book is on the interaction of dominant forms of economic organization and major dimensions of spatial change, there is an important sub-theme. This is the role of human agency in shaping and differentiating the mosaic of regional landscapes. What is done, where, and how – under any form of economic organization – reflects human interpretations of how resources should be used. As Ron Johnston[1] has noted, these interpretations 'are shaped through cultural lenses (which may be locally created, or may be imported); they reflect reactions to both the local physical environment and the international economic situation; they are mediated by local institutional structures; they are influenced by historical context; and they change that context, and hence the environment for future operations', p. 446).

[1]Johnston, R. 1984. The world is our oyster. *Transactions, Institute of British Geographers*, **2**, 463–59.

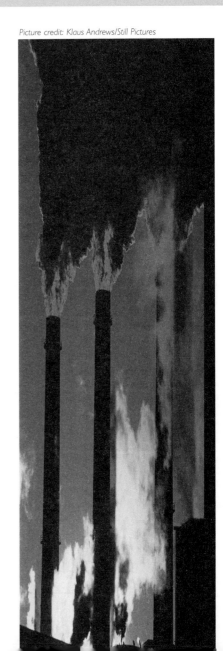

Picture credit: Klaus Andrews/Still Pictures

CHAPTER 4

PRE-INDUSTRIAL FOUNDATIONS

Picture credit: Paul Knox

In this chapter, we trace the emergence of an embryonic world economy centred on Europe and describe the way in which Europeans became, as Robert Reynolds put it, the 'leaders, drivers, persuaders, shapers, crushers and builders' (1961, p. vii) of the rest of the world's economies and societies. It was as a result of these changes that the core areas of Europe forged the template for the economic geography of the modern world. It must be recognized from the outset, however, that pre-industrial economic development was by no means exclusively a European phenomenon. The early trajectories of other parts of the world often eclipsed that of Europe and were sometimes important in influencing events in Europe itself. We begin, therefore, with a brief review which spans the origins and diffusion of the first, crucial 'revolution' in the development of agricultural systems, the rise of ancient empires, the establishment of urban systems, and the spread of feudalism as the dominant mode of production. Our purpose here is not so much to attempt to provide a thumbnail sketch of early economic history as to point to the emergence and spatial implications of certain fundamental socio-economic forces.

BEGINNINGS

Relatively little is in fact known about the first transitions from primitive hunting-gathering minisystems to larger-scale, agriculturally based world-empires and world economies. Despite significant advances in the accuracy of archaeological research, we still have to rely on speculation as much as established facts. It is generally agreed, however, that the transition began in the Proto-Neolithic period (between 9000 and 7000 BC), when a series of innovations among certain hunter-gatherer peoples established the *preconditions for agriculture*. These innovations included (1) the use of fire to process food, (2) the use of grindstones, and (3) the improvement of basic tools for catching, killing and preparing animals, fish, birds and reptiles. Given these preconditions, it was a relatively straightforward transition to a simple system of 'long-fallow' agriculture, which involved sowing or planting familiar species of wild cereals or tubers on scorched land. No special tools are required for such a system, nor are weeding or fertilization necessary, provided that cultivation is shifted to another burned plot after a few crops have been taken from the old one.

Meanwhile, the domestication of cattle and sheep had begun. By Neolithic times (7000 to 5500 BC), farming had developed to the point where stock breeding and seed agriculture were established techniques of food production. The switch from hunting and gathering to food production seems to have occurred very slowly, however: it was not a revolutionary change which suddenly transformed local practices. Archaeological evidence from a Neolithic village in western Asia, for example, shows that the wild legumes which were the major food item in 7500 BC were gradually replaced by cultivated grains over a span of almost 2000 years (Flannery, 1969). Ester Boserup (1981) suggested that there was little incentive to switch to food production until population densities began to increase and/or wild food sources became scarce because hunting and gathering often provided adequate levels of subsistence with relatively low workloads. From this perspective, then, *demographic conditions as well as technological innovations were a critical precondition for economic change.*

Hearth areas

The weight of available archaeological evidence suggests that the transition to food production took place *independently* in several different agricultural 'hearths':

1. The earliest hard evidence comes from the Middle East, in the foothills of the Zagros Mountains of what are now Iran and Iraq, where radio-carbon analysis has dated the remains of domesticated sheep at around 8500 BC (Clark, 1977). In addition, evidence of early Neolithic activity has been found in other parts of the Middle East, particularly around the Dead Sea Valley in Palestine and on the Anatolian Plateau in Turkey.
2. A second early Neolithic hearth area was in South Asia, along the floodplains of the Ganges, Brahmaputra and Irawaddy rivers.
3. Later, from around 5000 BC, a third hearth area seems to have emerged in China, around the Yuan River valley in Western Hunan.
4. Finally, there is evidence of independent agricultural organization in four regions of the Americas: the southern Tamaulipas area and the Tehaucán Valley in Central America, coastal Peru, and the North American Southwest. In these regions, however, agricultural development not only came later but it was painfully slow, with widespread food production coming to dominate the exploitation of wild plants and game only after AD 1000.

Meanwhile, the agricultural 'revolution' had been diffused from the Middle East. By 5000 BC it had begun to spread eastwards, to southern Turkmenia, and westwards, via the Mediterranean and the Danube, into Europe; by 3000 BC it had reached the Sudan and Kenya (via the Nile), much of India (via Afghanistan and Baluchistan) and had penetrated Europe as far as the British Isles and southern Scandinavia. By 1500 BC the last European stronghold of pure hunter–gatherer economies was the zone of tundra and coniferous forest stretching eastwards from the Norwegian coast.

Archaeological evidence is inevitably rather patchy, however, so that the patterns of diffusion from agricultural hearth areas remain a topic of considerable academic debate. Thus, for example, Carl Sauer's influential treatise on *Agricultural Origins and Dispersals* (1952) argued that hearth areas logically had to fulfil six criteria:

1. Plentiful natural food supplies: 'People living in the shadow of famine do not have the means or time to undertake the slow and leisurely experimental steps ... [to develop] a better and different food supply' (p. 21).
2. A diversity of species in order to provide a large reservoir of genes for hybridization. This implies well-diversified terrain and varied climatic conditions.
3. Freedom from the necessity to drain or irrigate land. This implies, however, that river valley sites were unlikely hearth areas.
4. Natural vegetation dominated by woodland, which is much easier to open up than grassland.
5. A population whose economy and technology were oriented towards gathering rather than hunting.
6 A sedentary rather than nomadic population, in order to facilitate the protection of growing crops from 'all manner of wild creatures that fly, walk and crawl to raid fruits, leaves and roots' (p. 23).

On this basis, Sauer deduced that the first hearth area must have been in South Asia, and that Middle Eastern agriculture was a later, more sophisticated outgrowth. Whether or not such deductions are vindicated by future archaeological findings is of less importance to us here, however, than the eventual *outcomes* of the transition to food production:

- Most important of all for the long-term evolution of the world economy were changes in social organization that resulted from the establishment of settled agriculture. The previous communal social order was steadily replaced by a kin-ordered system that laid the basis for a new, stratified social structure. Kin groups emerged as a 'natural' way of assigning rights over resources and organizing the production and storage of food, but they also generated new social institutions to deal with the ownership of property and the formal exchange of goods.
- The increased volume and reliability of food supplies allowed much higher population densities and encouraged the proliferation of settled agricultural villages. Together with the new social institutions of kin-ordered societies, this, in turn, facilitated the development of non-agricultural crafts, such as pottery, weaving, jewellery and weaponry. Such specializations in their turn encouraged the beginnings of barter and trade between communities, sometimes over substantial distances.

The framework of early urbanization

These outcomes of the agricultural revolution were effectively the preconditions for another 'revolutionary' change in the economic and spatial organization of the world: the emergence of cities and city systems. As with the evidence on the agricultural transition, our knowledge of the earliest cities is partly a function of where archaeologists have chosen to dig and partly a function of fortuitous factors like the durability of building materials and artifacts. It now seems firmly established, however, that *urbanization developed independently in different regions, more or less in the wake of the local completion of the agricultural transition.* Thus the first region of independent or 'nuclear' urbanism, from around 3000 BC, was in the Middle East, in the Mesopotamian valleys of the Tigris and Euphrates and the Nile Valley (together making the so-called Fertile Crescent). By 2500 BC cities had appeared in the Indus Valley, and by 1000 BC they were established in northern China. Other areas of nuclear urbanism include Central America (from around AD 1500). Meanwhile, of course, the original Middle Eastern urban hearth had generated successive urban world-empires, including those of Greece, Rome and Byzantium.

Explanations of these first transitions to city-based economies have emphasized several factors. Boserup (1981), for instance, stressed the role of local concentrations of population; while Jane Jacobs (1969) interpreted the emergence of cities mainly as a function of trade; and the classical archaeological interpretation rests on the availability of an agricultural surplus large enough to facilitate the emergence of specialized, non-agricultural workers (Childe, 1950; Woolley, 1963).

Another important factor was the emergence of 'primitive accumulation' through the exaction of tributes, the control of fixed assets, and/or the control of labour power – usually through some form of religious persuasion or despotic coercion. Once established, a parasitic élite provided the stimulus for urban development by investing its appropriated wealth in displays of power and status. This not only created the kernel of the monumental city but also required an increased degree of specialization in non-agricultural activities – construction, crafts, administration, the priesthood, soldiery, and so on – which could only be organized effectively in an urban setting.

This kind of expansion, however, could only be sustained in the most fertile agricultural regions, where the peasant population could produce enough to support not only the parasitic élite but also the growing numbers of non-productive workers. In this context, the development of irrigation seems to have been a critical factor. It not only intensified cultivation and increased productivity; it also required the kind of large-scale cooperation that could only be organized effectively in a hierarchical, despotic society. Yet, even in the most fertile and intensively farmed regions, rank-redistributive economies could only expand beyond a certain point if overall levels of productivity could be increased: through harder work, improvements in technology, or improvements in agricultural practices. All three of these solutions will have required more non-agricultural specialists and so will have reinforced the incipient process of urbanization:

> administrators and, perhaps, an army to oversee the harder work (their actions may have been accompanied by the élite taking to itself the ownership of land) in the first, craftsmen to create the tools in the second, and also, probably, miners and others to provide the raw materials; and 'researchers' to develop the new strains and the new technology (notably irrigation) in the third. Thus the demands for more production are reflected in the urban node as well as in the countryside, and continued growth of the

society, to meet the never-satisfied demands of an expanding élite and its associates, leads to self-propelling urban growth.

(Johnston, 1980, p. 52)

Such developments are ultimately limited by the size of the society's resource base, however. The obvious response – the enlargement of the resource base through territorial expansion – also tended to reinforce and extend the process of urbanization. Thus,

> To enable successful colonial activity, the structure of the society would need to be reorganized and several new functions created. Among the latter, the most important in the new areas would be administration – both civil and military – probably accompanied by religious colonization of the new subject population; to ensure contact between the colonies and the 'heartland', and the movement of surplus production back to the élite centre, a transport infrastructure would have to be established and maintained, and the ability to move substantial quantities of goods created.

(Johnston, 1980, p. 53)

All these changes would involve the creation of city-based jobs. Furthermore, whereas small-scale colonial expansion could be organized from one centre and controlled by a single élite group, expansion beyond easy reach of the main settlement (beyond, perhaps, a day or two's journey) would require 'the establishment of secondary settlements, to act as the nodes for parts of the controlled territory, as intermediate centres in the flow of demands from élite to producers and of goods in return' (Johnston, 1980, p. 54). *As long as growth was maintained, therefore, the empire would have to be continually enlarged, with an increasing number of urban control centres.* Hence the expansion of the Greek and Roman empires, which laid the foundations of an urban system in Western Europe (Fig. 4.1).

It would be wrong, however, to draw a picture of the steady growth, expansion and succession of ancient and classical empires. Urbanized economies were a precarious phenomenon, and many lapsed into ruralism before being revived or recolonized. In a number of cases, this was a result of demographic setbacks (associated with wars and epidemics). Such setbacks left too few people to maintain the social and economic infrastructure necessary for urbanization. An early example of this kind of relapse occurred in the Indus Valley, where the urban economy was displaced by Aryan pastoralists in the middle of the second millennium BC. Elsewhere, it was changes in resource/population ratios which precipitated the breakdown and decay of urban economies. The demands of repair and upkeep of irrigation systems, for example, on top of the need for increasing productivity resulting from population growth, sometimes put overwhelming strains on the available peasant labour. After a while, investments were neglected, armies grew small, and the strength and cohesion of the empire were fatally undermined.

This kind of sequence seems to have been the root cause of the eventual collapse of the Mesopotamian empire and may also have contributed to the abandonment of much of the Mayan empires more than 500 years before the arrival of the Spanish. Similarly, the population of the Roman empire began to decline in the second century AD, allowing the infiltration of 'barbarian' settlers and traders from the German lands of east-central Europe.

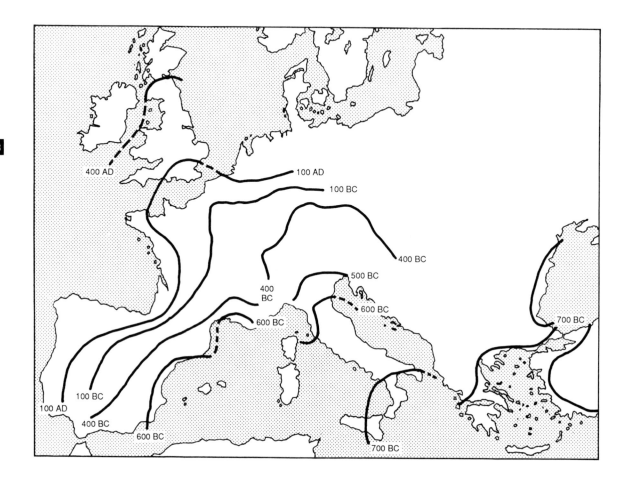

Figure 4.1

The urbanization of the classical world.

Source: Carter (1983), Fig. 2.2, p. 21

feudal systems
Forms of economic organization based on a mode of production wherein the surplus product (i.e. the outputs of productivity in excess of subsistence levels) is appropriated through a hierarchy of socio-political ranks by institutionalized coercion. In classical feudal systems, land is allocated to vassals by lords, in return for military service and/or labour on the lord's estate.

Rural consolidation

Although the emergence of urbanization provided an important framework for future development, it was the reorganization and consolidation of rural areas that provided the immediate platform for the critical transition to merchant capitalism and the emergence of a European world economy. At the heart of this rural consolidation was the evolution of the elaborate feudal systems of medieval Europe, China, India and Japan.

In economic terms, **feudal systems** were characterized by being almost wholly agricultural, with 80 or 90 per cent of the work-force being engaged in mixed arable and pastoral farming, and most of the rest occupied in basic craft work. Moreover, most production was for immediate needs: very little of a community's output ever found its way to wider markets. The basis of the feudal system was the feudal estate, owned by lay or ecclesiastical lords, who delegated parcels of land to others in return for allegiance and economic obligations, the latter being fulfilled mainly in the form of money dues. The lords in their turn would normally owe allegiance and homage to higher lords from whom they held delegated grants of land.

In other words, a feudal system would at least initially be associated with the existence of a chain of dependent tenures and of a corresponding hierarchy of territorial lords composed, at the bottom, of local notables and, at the top, of great magnates owning immense stretches of land.

(Dunford and Perrons, 1983, pp. 91–2)

The labour power which ran each estate consisted of a peasant population, most of whom were serfs (descended from slaves and therefore not free in public law) or tenants whose freedom of movement, freedom of marriage, freedom to leave property to their heirs and freedom to buy goods and sell their labour were closely circumscribed by public law. It was this socio-political institution – the peasantry – that was the key to feudal economic systems in that it formed the basis on which feudal lords were able to accumulate wealth, through a combination of labour services, rents in kind, taxes, seigneurial dues and payments for the use of essential services – milling, baking, olive pressing, and so on – monopolized by the lords.

By 1000 AD the countryside of most of Western Europe had been consolidated into a series of feudal agricultural sub-systems which were largely autonomous. Every estate was more or less self-sufficient in the raw materials for simple industrial products. Some of the members of every rural household would be capable of specialized, non-agricultural, part-time activities such as cloth-making or basketry; and nearly every community supported a range of specialist artisans and craft workers. In addition, most regions were able to sustain at least some small towns, whose existence hinged mainly on their role as ecclesiastical centres, defensive strongholds, and administrative centres for the upper echelons of the feudal hierarchy. *Improbably, it was this economic landscape – inflexible, slow-motion and introverted – which nurtured the essential preconditions for the by-passing of feudalism and the rise of merchant capitalism in Western Europe: the resurgence of trade and the revival of cities.*

SUMMARY: EMERGING IMPERATIVES OF ECONOMIC ORGANIZATION

Before moving on to examine the transition to merchant capitalism and the emergence of the European world-system, it is useful to pause briefly in order to review some of the organizing principles which seem to have been important in delineating the formative stages of pre-industrial economic geography:

- Major changes in patterns of economic activity were gradual and incremental, even in 'hearth' or 'core' regions.
- Such changes were generally preceded by the development of critical innovations, particularly in technology and economic organization.
- Such innovations were a necessary but not sufficient condition to bring about radical change: institutional and socio-political changes were also necessary in order to exploit them.
- Demographic factors were also critical. Insufficient absolute numbers of potential workers sometimes hindered economic development, while changes in

diminishing returns, law of

The tendency for productivity to decline, after a certain point, with the continued application of capital and/or labour to a given resource base. A simple example is provided by agricultural productivity: a large farm will yield progressively higher levels of output with the addition of more farm hands and more machinery, but there will be a point at which productivity decreases as some of the labour and machinery is underemployed, people get in one another's way, and coordination of activities becomes costly.

the balance between a population and its local resource base could be important in precipitating *either* progressive *or* regressive economic change.

- The law of **diminishing returns** provided an early impetus for territorial expansion. In addition to the obvious spatial consequences in terms of establishing dominant/subordinate territorial relationships, colonization was pivotal in the development of hierarchical urban systems and improved transportation. Colonization also stimulated the development of militarism, which itself induced important changes in spatial organization: through the influence of defensive sites for key settlements, for example. Not least, the environmental and social circumscription which translated into the law of diminishing returns was responsible for the emergence of a new geo-political phenomenon – the state.

These organizing principles will be recurring themes as we go on to examine successive epochs of economic development; they must also be reconciled, together with the principles which emerge from our examination of subsequent epochs, within any comprehensive theoretical approach to regional economic development.

THE EMERGENCE OF THE EUROPEAN WORLD-SYSTEM

This section deals with a long period: from the first stirrings of the transition from feudalism to merchant capitalism in the thirteenth century, through the creation of the European world-system in the sixteenth and seventeenth centuries, to the proto-industrialization of the early eighteenth century which laid the foundations for the Industrial Revolution. A comprehensive and detailed review of the period can be found in the works of Cipolla (1981), De Vries (1976), and Wallerstein (1980); our purpose here is merely to point to the emergence, interaction and spatial implications of the salient aspects of economic change. We must begin our examination of the emergence of the European world-system, however, with an obvious but somewhat neglected question: why Europe?

Why Europe?

In the twelfth century, before Europe embarked on the path of capitalist development that was to shape – directly or indirectly – virtually the entire global economy, there were several well-developed 'economic worlds' in the Eastern Hemisphere. One was the Mediterranean region, whose principal elements included Byzantium, the Italian city states and Muslim North Africa. A second was the Chinese Empire. The central Asian land mass from Russia to Mongolia was a third; the Indian Ocean/Red Sea complex was a fourth; and the Baltic area was on the verge of becoming a fifth.

Why did Europe become the locus of innovatory economic change? In particular, why not China? China had approximately the same total population as Europe and for a long time – well into the fifteenth century – was at least as far advanced in science and technology. Chinese ironmasters had developed blast furnaces that allowed the casting of iron as early as 200 BC. Iron ploughs were introduced in the sixth century, the compass in the tenth century, and the water clock in the eleventh century. The Chinese were also significantly more advanced than the Europeans in medicine, papermaking and printing, and the production of explosives. In addition, because China had retained an imperial system, it held a potentially telling advantage: its centralized decision-making, extensive state bureaucracy, well-developed internal communications and unified financial system were well suited to economic development and territorial expansion.

China's failure to take off in the way Europe did must be attributed in part to its failure to pursue economic opportunities overseas. The Chinese had in fact matched early European exploratory successes by spanning the Indian Ocean from Java to Africa in a series of lucrative and informative voyages; but they simply lost interest in further exploration. One explanation for this lack of a colonizing mission is that they saw their own 'world' as the only one that mattered. Another is that they were distracted by the growing menace of Mongol nomad barbarians and/or Japanese pirates. A third explanation is that the centralized power structure of imperial China did not contain enough different interest groups for whom overseas exploration was an attractive proposition.

This last point is seen by some to be part of a broader set of structural constraints associated with the imperial form. The administration and defence of a huge population and land mass are held to have been a drain on attention, energy and wealth which might otherwise have been invested in capital development. The imperial system also meant that cultural and social élites tended to be focused on the arts, humanities and self promotion *vis-à-vis* the imperial bureaucracy. The centralization of decision-making, meanwhile, is seen as having been insensitive to the economic potential of China's estimated 1700 city states and principalities (Jacobs, 1984). There is a link too between China's imperial framework and its failure to develop military technology (after having gained a flying start) in the way that enabled Europeans to turn exploration into domination: quite simply, the Imperial court suppressed the spread of knowledge of gunnery because it feared internal bandits and domestic uprisings.

> So China, if anything seemingly better placed prima facie to move forward to capitalism in terms of already having an extensive state bureaucracy, being further advanced in terms of the monetization of the economy and possibly of technology as well, was nonetheless less well placed after all. It was burdened by an imperial political structure.
>
> (Wallerstein, 1974, p. 63)

Another important difference in the trajectories followed by China and Europe was that European agriculture had become focused on the production of cattle and wheat, whereas Chinese agriculture was dominated by rice production. Because rice production requires relatively little land, China did not feel such a great need for territorial expansion. Conversely, Europe's reliance on wheat and cattle provided a strong impetus for territorial expansion and exploration, while the more extensive use of animal power in Europe meant that 'European man possessed in the fifteenth century a motor, more or less five times as powerful as that possessed by Chinese men' (Chaunu, 1969, p. 336). Finally, some writers have emphasized the lack of autonomy of Oriental towns compared to their European counterparts. As we shall see, the legal and political autonomy of European towns was a crucial 'pull' factor in attracting rural migrants whose labour and initiative were central to the emergence of merchant capitalism. So much, then, for China; it remains for us to explain just how Europe became the hub of the embryonic world economy.

The crisis of feudalism in Europe

The transition from feudalism to merchant capitalism in Europe remains an issue of considerable debate, largely because we do not know enough about the details or timing of the critical economic and social changes that took place between 1300 and 1450. As a result, a variety of theoretical interpretations have emerged, each emphasizing different elements in the transition (for a review, see Peet, 1991). It is generally agreed, however, that the overall context for the transition was a phase of economic, demographic and political 'crisis' brought about by the combination of steady population growth, modest technological improvements and limited amounts of usable land.

As a result of improvements in ploughing techniques, harnesses and basic equipment in the early feudal period, wheat yields rose significantly, leading to a steady rise in population over the twelfth and thirteenth centuries. In response, the feudal economy kept up by reclaiming rough pasture-land and woodland; and when this began to prove difficult (from around 1250) the response was to improve crop rotations and shorten the period of fallow.

There were limits, however, to such adjustments. The number of cattle that could be kept, for example, was fixed by climatic constraints which limited the quantity of available winter forage; and this in turn imposed a limit on the supply of fertilizer for arable farming. In the absence of further advances in agrarian technology, food shortages were an inevitable outcome; and in their wake, just as inevitably, came epidemics such as those of the Black Death (bubonic plague) in the 1340s, 1360s and 1370s. These problems were compounded by climatic fluctuations: the cold winters and late springs of the fourteenth century aggravated the food shortages, while some exceptionally hot summers helped to swell the population of the black rat, the host to the rat flea, one of the two vectors of bubonic plague.

Another aggravating factor was the beginning of the Hundred Years War in 1335–45, which put many economies on a war footing. The result was a marked increase in taxation. This, in turn, initiated a downward economic spiral as levels of consumption fell, causing liquidity problems for noble treasuries and eventually leading to a rise in prices. This led to further rounds of tax increases, which provoked a political climate of endemic discontent. The combined result of these pressures was 'not only to exhaust the goose that laid the golden eggs for the castle, but to provoke, from sheer desperation, a movement of illegal emigration from the manor' (Dobb, 1963, p. 21).

The destination of all these fugitives from feudalism was the town, where different laws and tax systems prevailed. The late medieval European town, writes Cipolla (1981, p. 146):

was the 'frontier', a new and dynamic world where people felt they could break their ties with an unpleasant past, where people hoped they would find opportunities for economic and social success, where sclerotic traditional institutions and discriminations no longer counted, and where there would be ample reward for initiative, daring and industriousness.

That the towns should appear so attractive was not simply the result of the legal status of their inhabitants, however. They had, ironically, begun to prosper at the height of feudal economic development. In order to meet the nobility's more sophisticated and ostentatious requirements, seigneurial incomes had been increasingly realized in the form of cash. *This obliged peasants to sell part of their produce on the market in order to pay rents and taxes, and generally facilitated the trading of commodities.* There developed an embryonic pattern of regional trade in basic industrial and agricultural produce, and even some long-distance, international trade in luxury goods such as spices, furs, silks, fruit and wine. One consequence of this trade was an increase in the size and vitality of towns, as more and more merchants and craft workers emerged to cope with the demands of the system. This urban vitality was a major agent in the eventual crisis of feudalism. It helped to highlight the relative inefficiency of the self-sufficient feudal estate and transformed attitudes towards the pursuit of wealth.

The resurgence of trade and expansion of towns under merchant capitalism

Increased trade and urban growth were both a cause and an effect of the transition from feudalism. But they were also the hallmarks of the new economic order. As the feudal system faltered and disintegrated, it was replaced by an economy that was dominated by market exchange, in which communities came to specialize in the production of the goods and commodities which they could produce most efficiently in comparison with other communities (Fig. 4.2). The key actors in this system were the merchants who supplied the capital required to initiate the flow of trade – hence the label merchant capitalism.

In marked contrast to feudalism and earlier rank-redistributive and primitive subsistence-oriented systems, merchant capitalism was a self-propelling growth system, in which the continued expansion of trade was vital: without it, neither merchants nor those dependent on their success – producers, consumers, financiers, etc. – could maintain their position, let alone advance it.

> Mercantile success required the merchants to buy as cheaply as possible, and to sell as expensively as possible; it also demanded that they trade in as large a volume of goods as possible. ... This created a contradiction, however, for the producers were also consumers (though not of the goods they produced), so that if the prices they received were low, they could not afford to buy large quantities of other goods and thus satisfy the demands of the merchant class as a whole. A consequence of this was a great pressure on producers to increase the volume of goods offered for sale, which meant increasing their productivity, while merchants put pressure on consumers to buy more, even if this meant them borrowing money in order to afford their purchases. Both processes ... involved producers raising loans which they had to repay with interest;

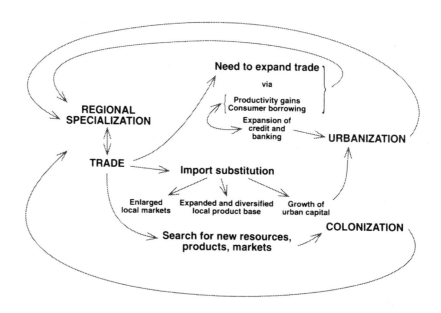

Figure 4.2
The rise of merchant
capitalism and the
changing space
economy

to achieve the latter, they had to produce more (or, if they were employees rather than independent workers, to work harder).

(Johnston, 1980; pp. 33–4)

The regional specializations and trading patterns that provided the foundations for early merchant capitalism were predetermined to a considerable degree by the long-standing patterns which had been developed by the traders of Venice, Pisa, Genoa, Florence, Bruges, Antwerp and the Hanseatic League (which included Bremen, Hamburg, Lübeck, Rostock and Danzig; see Fig. 4.3) from the twelfth century. As the self-propelling growth of merchant capitalism took hold, centres of trade multiplied in northern France and the lower Rhineland, new routes across Switzerland and southern Germany linked the commerce of Flanders more closely to that of the Mediterranean, and sea lanes – across the English Channel, North Sea and Baltic – began to integrate the economies of the British Isles, Scandinavia and the Hansa territories with those of the continental core. Very quickly, a trading system of immense complexity came to span Europe from Portugal to Poland and from Sweden to Sicily. This trading system was based not on the luxury goods of earlier trade routes but on bulky staples such as grains, wine, salt, wool, cloth and metals.

The increased volume of trade fostered a great deal of urban development as merchants began to settle at locations that were of particular significance in relation to major trade routes, and as local economies everywhere came to focus on market exchange. But, once the dynamics of trade had been initiated, the key to urban growth was a process of *import substitution*. Although some things were hard to copy because of the constraints of climate or basic resource endowment, many imported manufactures could be copied by local producers, thus increasing local employment opportunities, intensifying the use of local resources and increasing the

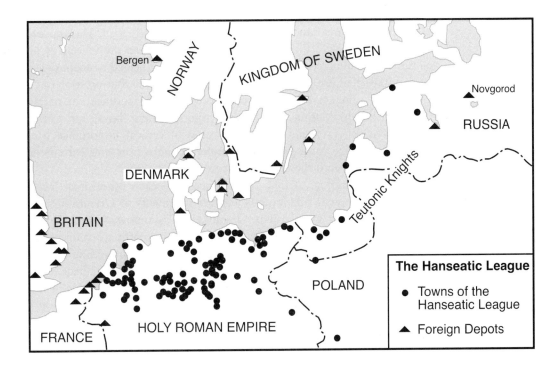

Figure 4.3
Towns and cities of
the Hanseatic League.
Source: Hugill (1994), Fig. 2.5,
p. 50

amount of local investment capital available. As Jane Jacobs (1984) argued, cities which replaced imports in this way could then afford *new* types of goods being produced in other cities. The newly imported innovations, in their turn, might then be replaced with local production, opening up the market for still more innovations from elsewhere. So the cities of Europe, as Jacobs observes:

> were forever generating new exports for one another – bells, dyes, buckles, parchment, lace, carding combs, needles, painted cabinet work, ceramics, brushes, cutlery, paper, sieves, riddles, sweetmeats, elixirs, files, pitchforks, sextants – and then replacing them with local production, to become customers for still more innovations. They were developing on one another's shoulder.
>
> (1984, p. 50)

As a result, patterns of trade and urban growth were very volatile; and long-term local success within the new economic order became increasingly dependent on:

- sustained improvisation and innovation
- repeated episodes of import substitution
- the discovery and control of additional resources and new kinds of resources.

Consolidation and expansion

In the fifteenth and sixteenth centuries, a series of innovations in business and technology contributed to the consolidation of merchant capitalism. These included innovations in the organization of business and finance: banking, loan systems, credit transfers, company partnerships, shares in stock, speculation in commodity futures, commercial insurance, courier/news services, and so on. The importance of these innovations lay not only in the way that they oiled the wheels of industry, agriculture and commerce but also in the way that they helped to encourage savings and to facilitate their use for investment. Furthermore, the routinization of complex commercial and financial activity brought with it the codification of civil and criminal legislation relating to property rights (patent laws, for example): a development which is seen by some as being of critical importance because it provided an incentive for a sufficient number of innovators and entrepreneurs to channel their efforts into the embryonic capitalist economy.

Meanwhile, technological innovations succeeded each other at an accelerated rate. Some of these were adaptations and improvements of Oriental discoveries – the windmill, spinning wheels, paper manufacture, gunpowder, and the compass, for example. But in Europe there was a real passion for the mechanization of the productive process as a means of increasing productivity. In addition to improvements based on others' ideas there emerged a welter of independent engineering breakthroughs, including the more efficient use of energy in water mills and blast furnaces, the design of reliable clocks and firearms, and the introduction of new methods of processing metals and manufacturing glass.

The advantages conferred by these breakthroughs were jealously guarded by the centres of innovation – northern Italy up to the fifteenth century; England and Holland in the sixteenth and seventeenth centuries – while competitors in other regions went to considerable lengths to acquire new technology at the first opportunity. Thus, for example, the Venetian government strictly prohibited the emigration of caulkers; and the Grand Duke of Florence gave a reward for the return, dead or alive, of emigrants from key positions in the brocade industry. The French actually kidnapped skilled iron workers from Sweden; while many governments were happy to provide shelter and handsome rewards from migrant craftsmen who had knowledge of new techniques. These early examples of a 'brain drain' were complemented by the practice of temporary migration in the opposite direction in order to acquire new expertise, sometimes legitimately, sometimes covertly. But the most important vector for the diffusion of technological innovations came with the invention of the printing press using movable type. Within 20 years of its introduction in Mainz, printing shops had spread throughout Europe, opening up vast new possibilities in the fields of knowledge and education.

It was the combination of innovations in *shipbuilding, navigation and naval ordnance*, however, that literally had the most far-reaching consequences for the evolution of the European space-economy (Hugill, 1994). By the fourteenth century, European shipwrights were building ships skeleton first, at a vast saving of labour in comparison with previous methods. In the course of the fifteenth century, the full-rigged ship was developed, enabling faster voyages in larger and more manoeuvrable vessels that were less dependent on favourable winds. Meanwhile, the quadrant (1450) and the astrolabe (1480) were developed and a systematic knowledge of Atlantic winds had been acquired. By the mid-sixteenth century, England, Holland and Sweden had perfected the technique of casting iron guns, making it possible to replace bronze cannon with larger numbers of more effective guns at lower expense. Together, these advances made it possible for the merchants of Europe to establish the basis of a world-wide economy in the space of less than 100 years.

Mercantilism and territorial expansion

As we have already seen in relation to China, however, economic strength and technological ability do not necessarily lead to overseas expansion. What, then, translated Europe's economic power and technological superiority to a broader arena? Fig. 4.4 summarizes the most important factors. The large number of impoverished aristocrats produced by western European inheritance laws and by expensive Crusades and local wars was one important factor. Discouraged from commercial careers by sheer snobbery and encouraged by a culture which romanticized the fighting man, these poverty-stricken gentlemen provided a plentiful supply of adventurers who were willing to die for glory and even more willing to exercise greed and cruelty in the name of God and Country. This points to two other important factors: the evangelical zeal of the Church and the political competitiveness of the monarchies.

Above all, however, overseas expansion was impelled by the *logic of merchant capitalism and the law of diminishing returns*. Self-propelling growth could only be sustained as long as productivity could be improved, and after a point this required food and energy resources which could only be obtained by the conquest – peaceful or otherwise – of new territories. Similarly, merchant capitalism required new supplies of gold and silver to make up for the leakage through trade with Byzantium, China and Arabia.

Collectively, these motivations found expression in the dogma of Mercantilism. Most European countries adhered to this dogma from the sixteenth century to the early eighteenth century. The basis of Mercantilism was that national wealth was to be measured in terms of gold or silver and that the fundamental source of economic growth was a persistently favourable balance of trade. This was the economic 'logic' that justified not only overseas colonization but also the coercion of plantation labour and the prohibition of manufacturing in the colonies. It was also the logic which, on the domestic front, promoted thrift and saving as a means of accumulating capital for overseas investment. It required a high degree of economic regulation, sponsorship and protection by the government.

There is no need for us to reiterate here the pattern and sequence of European expansion and conquest (though it is worth noting that the overall thrust, overseas from Atlantic Europe rather than inland to the east, was essentially because the technological superiority of the Europeans was not as marked on land as it was on the seas: Asians could counterbalance technological inferiority with weight of numbers until after the mid-seventeenth century, when European technology succeeded in developing more mobile and rapid-firing guns). Europeans soon destroyed most of the Muslim shipping trade in the Indian Ocean and captured a large share of the intra-Asian trade. By bringing Japanese copper to China and India, Spice Island cloves to India and China, Indian cotton textiles to Asia and Persian carpets to India, European merchants made good profits and with them paid for some of their imports from Asia.

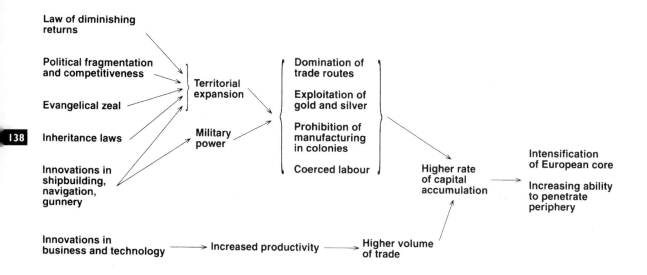

138

Figure 4.4

The emergence of a European-based world-system

It was the gold and silver from the Americas, however, that provided the first major economic transformation, allowing Europe 'to live above its means, to invest beyond its savings' (Braudel, 1972, p. 268). In effect, the bullion was converted into effective demand for consumer and capital goods of all kinds – textiles, wine, food, furniture, weapons, ships – thus stimulating production throughout the economic system and creating the basis for a 'Golden Age' of prosperity for most of the sixteenth century. Meanwhile, overseas expansion made available a variety of new and unusual products – cocoa, beans, maize, potatoes, tomatoes, sugar cane, tobacco and vanilla from the Americas, tea from the Orient – which opened up large new markets to enterprising merchants.

As European traders came to monopolize intra-oriental trade routes, they were literally able to control the flow and pattern of trade between potential rivals. Because of this monopoly, European traders could identify foreign articles with a tested profitable market and ship them home to Europe, where skilled workmen could learn to imitate them. Once Europeans had begun manufacturing these products, it was their goods which were shipped to the rest of the world:

> For example, Europeans long prized the shawls which were made in the north of India in the Kashmir region; much later Scotchmen [sic] were making imitations of those shawls by the dozens per day; called 'Paisley' shawls, they swept the Kashmir shawls off the general market. Europeans admired the very hard vitrified china of the Chinese, and for a long while bought it to sell to other peoples, taking it from China and distributing it. But then the Europeans began to make it in France and elsewhere, and shortly true Chinese china had become a rare article on the world market while Europe was making and selling enormous amounts of its own 'china'. For a good while Europeans bought cottons of a very fine quality from India for markets in Africa, Europe, and America, but before too long they had imitated them in England and were shipping cheaper machine-made cottons back to India where they ruined the Indian cotton-weaving industry in its own home.

(Reynolds, 1961, pp. 45–6)

For Europe, the benefits of overseas expansion thus extended well beyond the basic acquisition of new lands and resources. In addition to the bullion and the opportunities for import substitution, overseas expansion also stimulated further improvements in technology and business techniques, thus adding a further dimension to the self-propelling growth of merchant capitalism. New developments were achieved in nautical mapmaking, naval artillery, shipbuilding and the use of sail; maritime insurance emerged as one of the growing number of tertiary industries; and the whole experience of overseas expansion provided a great practical school of entrepreneurship and investment. Most important of all, perhaps, was the way that the profits from overseas colonies and trading overflowed into domestic agriculture, mining and manufacturing. This contributed to an accumulation of capital that was undoubtedly one of the main preconditions for the emergence of industrial capitalism in the eighteenth century.

The world outside Europe: transoceanic rim settlements

Outside Europe, the most important features of the economic landscape to emerge as a result of merchant capitalism were the gateway towns and entrepôts that were established along the coastal rims of the Americas, Africa and South Asia. These *transoceanic rim settlements* (Fig. 4.5) were of three main kinds (Haggett, 1983):

1. *Trading stations*, such as Canton (China), Madras (India) and Goa, which grew up as the points of contact between Europe and the – as yet – relatively autonomous economic worlds of the Orient. Few Europeans lived in these towns and cities, and only in India was it possible to exercise any secure measure of political control over the large hinterland areas that served the ports.
2. Entrepôts and colonial headquarters for tropical plantations, such as Rio de Janeiro (Brazil), Georgetown (British Guiana), Port of Spain (Trinidad), Penang (Malaysia), Lagos (Nigeria), Lourenço Marques (Mozambique) and Zanzibar. Substantial numbers of European settlers were required for administrative and military purposes, whereas the indigenous population provided field labour and manual labour in the towns. The colonial plantation system made intensive demands on labour, however, and when the indigenous supply was insufficient it was made up by enforced movements of slave labour from other regions, thus creating distinctive ethnic cleavages among the populations of many colonies.
3. Gateway ports for the 13 farm-family colonies on the northeastern seaboard of America (similar settlements were later established in South Africa, Australia and New Zealand). Although there were several distinctive groups – the Tidewater Colonies of Virginia (e.g. Jamestown, Baltimore), the New Towns of New England (e.g. Boston, Newport), the

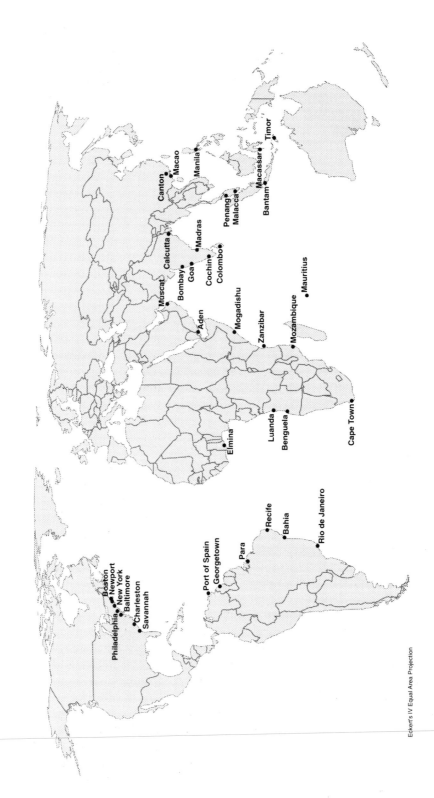

Eckert's IV Equal Area Projection

Figure 4.5
Transoceanic rim settlements of the mercantile era

Middle Colony towns (e.g. New York, Philadelphia) and the Colonial Towns of the Carolinas (e.g. Charleston, Savannah) – they were essentially a direct extension of the West European urban system, peopled by West Europeans and oriented much more to their homelands than their hinterlands.

141

The shifting locus of economic power

The dominant feature of the changing economic geography of Europe in the sixteenth and seventeenth centuries was a dramatic shift in the focus of economic activity from the Mediterranean to the North Sea. At the end of the fifteenth century, the Mediterranean was the most highly developed region in the world, with central and northern Italy as the hub of economic activity. During the sixteenth century, the relative prosperity of the Mediterranean was further enhanced as Spain and Portugal benefited immensely from the influx of treasure from the Americas. By the end of the seventeenth century, however, the Mediterranean had become a backward region in relation to the levels of prosperity generated by the Dutch economy; while England, hitherto a very marginal economy in European terms, stood poised to threaten the position of the Dutch as world leaders.

Between these two extremes of stagnation/regression and dynamic expansion was the experience of France, Scandinavia, Germany and much of the rest of continental Europe, where there was a general penetration of economic development and maturing of local economies: a consolidation of merchant capitalism which helped to maintain the coherence of the European economy during a period of volatile change in its spatial organization. In detail, therefore, the changing centre of gravity of the European economy involved a complex of overlapping, interlocking and interacting regional struggles and transformations. The basic processes involved, however, were more general, and they can be illustrated with reference to the decline of Spain and Italy and the rise of Holland and England.

Spain

Spain provides a good example of the importance of import substitution. Quite simply, Spain declined because it had never been fully 'developed' to begin with; it had only been rich. The increased demand generated by its acquisition of bullion from the Americas did not stimulate domestic production as much as it might have done because of bottlenecks in the productive system – the restrictive practices of guilds and the lack of skilled labour, for example – and the complacent attitude of the Spanish élite. In 1675 Alfonso Nuñez de Castro wrote:

> Let London manufacture those fabrics of hers to her heart's content; Holland her chambrays; Florence her cloth; the Indies their beaver and vicuña; Milan her brocades; Italy and Flanders their linens, so long as our capital can enjoy them; the only thing it proves is that all nations train journeymen for Madrid and that Madrid is the queen of Parliaments, for all the world serves her and she serves nobody.

(quoted in Cipolla, 1981, p. 25)

The treasure of the Americas provided Spain with purchasing power but ultimately it stimulated the development of England, France, Holland and the rest of Europe. Meanwhile, Spain's artificial prosperity had induced the government to pursue a persistently warmongering policy, which represented a serious drain on the treasury. In the course of the seventeenth century then, as the influx of bullion from Spain's colonies declined (partly through depleted mines), the momentum of the economy evaporated, leaving insufficient entrepreneurs and artisans to counterbalance an overabundance of bureaucrats, lawyers and priests and a mounting national debt.

Italy

Italy's decline was more complex, but its beginning can be dated more accurately: the end of the fifteenth century, when for almost 50 years Northern Italy became the battlefield for an international conflict involving Spain, France and Germany. As a result there were not only famines and epidemics but also severe disruptions to trade at a time when trade was beginning to expand elsewhere. Buoyed up by the international boom in demand during the late sixteenth century, the economy made something of a recovery; but it was a recovery based on traditional methods of organization which meant, among other things, that competition and innovation were suppressed by the renewed strength of craft guilds. Between 1610 and 1630 a series of *external events* led to the collapse of some of the Italians' major markets – the decline of the Spanish economy, disruptive wars in the German states, and political instability within the Turkish empire. At the same time, many of Italy's competitors had been able to substitute domestic products for Italian imports. At this point, the self-propelling growth of merchant capitalism broke down. Unable or unwilling to respond through innovation and increased productivity, Italian entrepreneurs began to disinvest in manufacturing and shipping. By the end of the seventeenth century Italy was importing large quantities of manufactures from England, France and Holland while exporting agricultural goods for which the terms of trade were poor: oil, wheat, wine and wool. Foreign trade had thus been transformed from an 'engine of growth' to an 'engine of decline'.

The Netherlands

The Netherlands' 'economic miracle' of the seventeenth century was launched from a fairly solid platform of trading and manufacturing. Although overshadowed in the early phases of merchant capitalism by the prosperity of nearby Bruges and Antwerp, Holland (and Amsterdam in particular), had steadily developed an entrepôt function for Northern Europe (importing flax, hemp, grain and timber, and exporting salt, fish and wine), around which it had begun to establish a manufacturing base. From the stability of this base, the Dutch successfully rebelled against Spanish imperialism and emerged, in 1609, with political independence and religious freedom.

Thereafter, a combination of factors helped the Dutch to become leaders of the world economy for more than 150 years. One was the 'modernity' of Dutch institutions: relatively few restrictive guilds, a small nobility of landowners, and a relatively weak church after the departure of the Spanish. Another was the vigorous pursuit of Mercantilist policies, including not only a strong colonial drive and a massive commitment to merchant shipping but also an uncompromising stance towards competitors. For example, the Dutch blockaded Antwerp's access to the sea from 1585 to 1795, taking over its entrepôt trade and its textile industry. In relation to their drive to dominate trade, the Dutch were able to turn their geographical situation to great advantage, both developing ocean ports and exploiting the inland waterways that penetrated the heart of continental Europe. They were also able to benefit from a highly developed and very innovative shipbuilding industry whose output completely overshadowed that of the rest of Europe. Finally, the Dutch were the major beneficiaries of the flight of skilled craftsmen, merchants, sailors, financiers and professionals from the fanaticism and intolerance of the Spanish in Flanders and Wallonia.

England

England, at the end of the fifteenth century, was distinctly backward, with a small population (around 5 million, compared to more than 15 million in France, 11 million in Italy and 7 million in Spain) and a weakly developed economy. The only significant comparative advantage the English held was the manufacture of woollen cloth; though Mediterranean supplies and German merchants had long dominated the European trade in woollen textiles. The first real break for the English economy came in the first half of the sixteenth century, when Italian production and trade collapsed because of war and its ensuing disasters, leaving the English literally to capitalize on a sustained increase in woollen exports – a trend which was further enhanced by the progressive deterioration of English currency resulting from Henry VIII's extravagant military expenditures. The boom was halted in the mid-sixteenth century, however, by the recovery of the Italian textile industry and by the war between the Dutch and the Spanish, which disrupted English exports.

By this time, though, English entrepreneurial and expansionist ambitions had become established and were articulated through a strong mercantilist philosophy. Like the Dutch, the English were able to take advantage of their geographical situation, at least in relation to transoceanic trade. They had also developed a strong navy and gave high priority to establishing a large merchant fleet and to acquiring colonial footholds. Like the Dutch, they also benefited from the skills of immigrants driven from France and the Low Countries by religious persecution. Innovation, improvisation and import substitution also played their part in ensuring a rapid escape from the mid-century economic crisis and, indeed, in building an economy to challenge that of the Dutch. The development of iron artillery in the 1540s, for example, enabled the English to arm their merchant ships, privateers and warships more extensively *and* at lower cost. Meanwhile, the exploitation of coal as a substitute for the relatively sparse and rapidly diminishing timber reserves not only helped the English to avoid an energy crisis but also helped to develop new processing techniques. 'Concentrating on iron and coal', writes Cipolla (1981, p.

290), 'England set herself on the road that led directly to the Industrial Revolution.'

Summary

At this stage it is useful to review the major organizing principles which had been important in delineating the evolving space-economy up to the eve of the industrial revolution. First, note that the observations we made in relation to early economic systems (p. 129) appear as recurring elements in subsequent economic epochs. Thus we can confirm the gradual and incremental nature of major economic change (though we should note that increasing economic sophistication and territorial integration make for a more rapid diffusion of short-run changes).

We can also confirm the continuing importance of innovations in technology and business organization (though we should note the innovative process to this point was carried out in small steps, by way of the gradual cumulation of improvements rather than by distinct bursts of invention which, as we shall see, have characterized economic change since the industrial era).

The importance of institutional and socio-political factors was also a recurring theme (as, for example, in the constraints of a centralized imperial system on the evolution of the Chinese economy, in the stimulus provided by European laws on property rights, and the role of European governments in implementing Mercantilist policies). Similarly, we must acknowledge the continuing interaction between demographic change and economic development and, finally, the continuing impetus for territorial expansion provided by the law of diminishing returns. In addition, though, we can identify several new dimensions of spatial-economic organization:

- The emergence of a true 'world economy', involving long-distance economic interaction based on a sophisticated division of labour.
- The progressive elaboration of the world economy, with competitive ('price-setting') markets penetrating into more and more space and more and more commodities, was *uneven*. Some sectors, nations, and regions expanded more quickly than others, and some spheres of opportunity and lines of communication were penetrated more quickly than others, so that its early spread was in a selective, spatially discontinuous fashion.
- The pattern of specialization and the nature of economic interaction within the world economy resulted in the emergence of *core* areas, characterized by such mass-market industries as had emerged (e.g. textiles, shipbuilding), international and local commerce in the hands of an indigenous bourgeoisie, and relatively advanced forms of agriculture; *peripheral* areas, characterized by the monoculture of cash crops by coerced labour on large estates or plantations; and *semi-peripheral* areas, characterized by a process of de-industrialization but retaining a significant share of specialized industrial production and financial control.
- The spatial organization of the European space-economy was based around a cluster of core areas in northwestern Europe – southeastern England and Holland, together with the Baltic states, the Rhine and Elbe regions of Germany, Flanders and northern France. Peripheral regions included northern Scandinavia, Britain's Celtic fringe, east-central

Europe and all of the trans-oceanic rim settlements and colonies. The semi-periphery consisted of the Christian Mediterranean region which had been the advanced core area at the beginning of the merchant capitalist era.

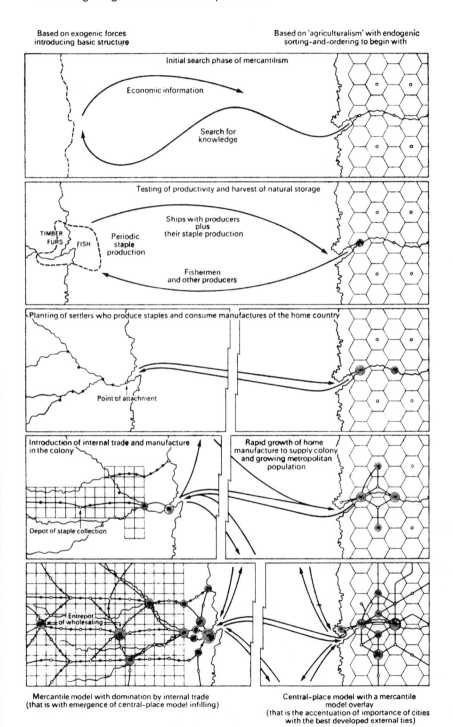

Based on exogenic forces
introducing basic structure

Based on 'agriculturalism' with endogenic
sorting-and-ordering to begin with

Initial search phase of mercantilism

Economic information

Search for
knowledge

Testing of productivity and harvest of natural storage

TIMBER
FURS
FISH

Periodic
staple
production

Ships with producers
plus
their staple production

Fishermen
and other producers

Planting of settlers who produce staples and consume manufactures of the home country

Point of attachment

Introduction of internal trade and manufacture
in the colony

Depot of staple collection

Rapid growth of home
manufacture to supply colony
and growing metropolitan
population

Entrepôt
of wholesaling

Mercantile model with domination by internal trade
(that is with emergence of central-place model infilling)

Central-place model with a mercantile
model overlay
(that is the accentuation of importance of cities
with the best developed external ties)

Figure 4.6
Colonialism and urban
settlement patterns
(after Vance).

Source: Vance (1970), p. 151

● The articulation of the European world economy also produced a distinctive pattern of settlement and urbanization. Merchant capitalism was reflected in the urban landscape by a strengthening of the hierarchical system of settlements and the development of a *central place system*. The overseas territorial expansion associated with merchant capitalism was also reflected in a distinctive urban landscape, as illustrated in Fig. 4.6. Johnston once again provides a succinct description:

> In the initial stages of mercantile exploration no permanent settlement is established in order to obtain the required products (fish, timber and furs). Then the colony is settled by agriculturalists; the export of their products moves through local articulation points to the colonial port, and thence to the port in the homeland, which grows in size and status relative to its inland competitors. As settlement of the colony expands further inland, so both of the ports increase in size, railways replace rivers as the main traffic arteries within the colony, and internal gateways develop to articulate the trade of areas some distance from the port, while in the homeland places near to the original port benefit from the imports and a new outport is built to handle the larger volume of trade and the bigger vessels. (1980, p. 74)

● The emergence of the European world economy brought about a system of internal dynamics that involved three important mechanisms of spatial change:

1. The switching of investment from one area to another by merchants in response to the shifting comparative advantages enjoyed by local producers. These shifts in comparative advantage, in turn, were associated with technological innovations and improvements, institutional changes, currency fluctuations, and so on. A good illustration of such change is provided by the rapid and almost continuous redrawing of the map of textile production in early modern Europe (Dunford and Perrons, 1983).

2. Import substitution. Communities able to achieve repeated episodes of import substitution, as Jacobs (1984) pointed out, benefit from five aspects of economic development:

   (a) enlarged markets for new imports and innovations
   (b) an expanded and more varied employment base
   (c) new applications of technology to increase rural productivity
   (d) a spillover of employment to rural areas as older, expanding enterprises are crowded out of cities
   (e) growth of city capital.

3. Militarism and geopolitical change.

Key Sources and Suggested Reading

Boserup, E. 1981. *Population and Technology*. Oxford: Blackwell.

Braudel, F. 1972. *The Mediterranean and the Mediterranean World in the Age of Philip II*. Translated by S. Reynolds. New York: Harper & Row.

Carter, 1983. *An Introduction to Urban Historical Geography*. London: Edward Arnold.

Chaunu, P. 1969. *L'expression européenne du XIIIe au XVe siècle*. Collection Nouvelle Clio 26. Paris: Presses Universitaires de France.

Childe, V. G. 1950. The urban revolution. *Town Planning Review*, 21, 3–17.

Cipolla, C. 1981. *Before the Industrial Revolution. European Society and Economy, 1000–1700*. 2nd edn. London: Methuen.

Clark, C. 1977. *World Prehistory in New Perspective*. Cambridge: Cambridge University Press.

De Vries, J. 1976. *Economy of Europe in an Age of Crisis, 1600–1750*. Cambridge: Cambridge University Press.

Dobb, M. 1963. *Studies in the Development of Capitalism*. London: Routledge & Kegan Paul.

Dunford, M. and Perrons, D. 1983. *The Arena of Capital*. London: Macmillan.

Flannery, K. V. 1969. Origin and ecological effects of early domestication. In P. J. Ucko and G. W. Dimbleby (eds), *The Domestication and Exploitation of Plants and Animals*. London: Duckworth.

Haggett, P. 1983. *Geography: A Modern Synthesis*. 3rd edn. London: Harper & Row.

Hugill, P. 1994. *World Trade Since 1431*. Baltimore: Johns Hopkins University Press.

Jacobs, J. 1969. *The Economy of Cities*. New York: Random House.

Jacobs, J. 1984. Cities and the wealth of nations. *Atlantic Monthly*, March, 41–66.

Johnston, R. J. 1980. *City and Society*. Harmondsworth: Penguin.

Peet, R. 1991. *Global Capitalism: Theories of Societal Development*. New York: Routledge.

Reynolds, R. 1961. *Europe Emerges: Transition Toward an Industrial World-Wide Society*. Madison: University of Wisconsin Press.

Sauer, C. 1952. *Agricultural Origins and Dispersals*. New York: American Geographical Society.

Tilly, C. 1992. *Coercion, Capital, and European States*. Cambridge, MA: Blackwell.

Vance, J. Jr. 1970. *The Merchant's World: The Geography of Wholesaling*. Englewood Cliffs, NJ: Prentice Hall.

Wallerstein, I. 1974. *The Modern World-System: Capitalist Agriculture and the Origins of the European World-Economy in the Sixteenth Century*. New York: Academic Press.

Wallerstein, I. 1980. *The Modern World-System II: Mercantilism and the Consolidation of the World-Economy 1600–1750*. London: Academic Press.

Woolley, L. 1963. The urbanisation of society. In J. Hawkes and L. Woolley (eds), *History of Mankind*, vol. 1 *Prehistory and the Beginnings of Civilisation*. Paris: UNESCO, Chapter 3.

Picture credit: Paul Knox

EVOLUTION OF THE INDUSTRIAL CORE REGIONS

From the second half of the eighteenth century, industrialization brought a new pattern and tempo to the economic organization of the world economy, and new dimensions to the world's economic landscapes. Today, economic geography within the tri-polar core of the world economy is dominated by the legacy of industrial capitalism. The economic geography of the world's peripheral regions has, meanwhile, been shaped by their role in sustaining the industrial expansion of the core economies. In short, there are few of the world's economic landscapes that are not largely a product, directly or indirectly, of the industrial era. In this chapter, we outline the evolving economic geography of the industrial core regions, analysing the major processes involved in the relative ascent and decline of nations and regions within the industrial core areas.

THE INDUSTRIAL REVOLUTION AND SPATIAL CHANGE

The transition during the late eighteenth and early nineteenth centuries from merchant capitalism to industrial capitalism as the dominant mode of production is conventionally ascribed to the industrial revolution. The industrial revolution, in turn, is conventionally depicted as a revolution in the techniques and organization of manufacturing, based on a series of innovations in the technology of production (e.g. Kay's flying shuttle (1733), Hargreaves's spinning jenny (1765), and Cartwright's machine loom (1787)), and in transport technology and engineering (particularly the development of canal and railway systems). But technological advance was really part of a wider economic, social and political transition, whose origins and preconditions are to be found in the era of merchant capitalism.

The most important context for technological advance was the existence within merchant capitalism of *industry* organized on capitalist lines by entrepreneurs employing wage-labour and producing commodities for sale in regional and national markets. In addition, the *capital* that had been accumulated through trading provided the means for entrepreneurs to finance investment in the capital-intensive but highly productive technology of the industrial revolution.

From these roots, machine production and the organizational setting of the factory – **machinofacture** – emerged as the central characteristics of industrialization. While machinery provided the basis for higher levels of productivity, it was the factory setting that enabled this productivity to be exploited to the fullest possible extent. This was achieved through specialization – the assembly-line division of labour – and internal economies of scale. At the same time, the concentration of workers in big industrial units generated urban environments which themselves represented a new dynamic force for economic, social and political change.

machinofacture
A form of organization of industrial production that is capital-intensive, with labour tending machines rather than operating them directly. It was the basis for the regime of accumulation that preceded Fordism.

149

Regimes of accumulation and modes of regulation

Thus emerged a distinctive regime of accumulation – a systematic process of production, income distribution and consumption based on the organization of capitalist firms. Regimes of accumulation tend to evolve in response to the opportunities and constraints of production, transportation and communications technologies. At the same time, this evolution is associated with a succession of *technology systems* that are imprinted, differentially, onto the world's economic landscapes. These technology systems can be thought of as 'A sort of paradigm for the most efficient organisation of production, i.e. *the main form and direction along which productivity growth takes place within and across firms, industries and countries*' (Perez, 1983, p. 361, emphasis added).

This, in turn, brings us to another useful and important concept: the idea of **mode of regulation**. These are specific local and historical structures and institutions that emerge in order to facilitate the operation of successive regimes of accumulation and technology systems within the wider national and international context. Modes of regulation have four principal functions (Dunford, 1990):

mode of regulation
A collection of structural forms (political, economic, social, cultural) and institutional arrangements which define the 'rules of the game' for individual and collective behaviour within a specific regime of accumulation or phase of economic development. The mode of regulation gives expression to, and serves to reproduce, fundamental social relations.

1. Regulating the monetary system and financial mechanisms
2. Regulating wages and collective bargaining
3. Facilitating (or, in some circumstances, constraining) competition, and establishing the relations between the private sector and the public economy
4. Establishing the roles of governments at various spatial scales.

Regimes of accumulation, technology systems and modes of regulation of industrial capitalism are all driven by the need for the *circulation and accumulation of capital*:

> More must be produced, and then sold, to ensure the needed returns on investment. Again, this leads to the contradictory situation whereby forcing down wages so as to increase returns on investment (profits), while at the same time attempting to maintain if not increase prices, produces a reduction in consumption and thus capitalist failure. Wages must rise, to keep pace with prices, in the system as a whole, and this could only be ensured by growth – more production, more productivity, more consumers, more people. The resulting greater profits lead to a greater pool of investment capital, seeking the best returns; to attract this investment, the productivity potential of workers is increased by improving the tools with which they work. Better and more

complex machinery means more production, more and bigger factories, and more goods seeking purchasers. Growth demands more growth.

(Johnston, 1980, p. 37)

Like merchant capitalism before it, industrial capitalism has had to confront the twin obstacles of market saturation and the law of diminishing returns. In response, industrialists have pursued a variety of strategies that have contributed to the dynamics of changing regimes of accumulation (at the level of the firm) and changing modes of regulation (at the level of the nation state). In addition to the constant search for technological advances, these have included:

- the pursuit of new ways of exploiting internal and external economies of scale;
- the exploitation of new, cheaper sources of labour and/or raw materials and energy;
- the penetration of new (i.e. overseas) markets for existing products;
- the development of new products, either through new inventions (e.g. video recorders, microcomputers) or by the 'commodification' of activities previously performed within the household (e.g. food processing and preparation);
- the acceptance of increasingly formalized relations with labour unions and governments, in order to establish a more stable context (economic, social and political) in which to operate.

As a result of all this dynamism, the changes imposed on economic landscapes by the first waves of the industrial revolution have been overwritten by a succession of episodes of industrial development, restructuring, and reorganization. In addition, it is important to recognize the differences that were created between the major industrial regions as a result not only of variations in resource endowment and previous patterns of economic development but also because of variations in the relative timing and interaction of these episodes of industrial change.

MACHINOFACTURE AND THE SPREAD OF INDUSTRIALIZATION IN EUROPE

There was in fact not one industrial revolution but several distinctive transitional phases, each having a different degree of impact, in different ways, on different regions and nations. *As new technologies shifted the margins of profitability in different kinds of enterprise, so the fortunes of specialized places shifted.* These regional differentials, in turn, helped to influence the changing character of capitalism itself.

First wave industrialization: Britain

It is possible to identify three major waves of industrialization in Europe, each consisting of several phases and each highly localized in their impact. The first wave saw the imprint of the first *technology system* of the industrial revolution, based on new iron and cotton textile technologies, using water power, trunk canals, and turnpike roads. Even within the span of this first wave, however, the imprint was highly differentiated. 'Above all,' Pollard emphasizes, 'the industrial revolution was a *regional* phenomenon' (1981, p. 14, emphasis added).

The springboard for the first wave of industrialization, which began in Britain around 1760, consisted of several local hearths of 'proto-industrialization'. These were areas with long-standing concentrations of industry based on a wage-labour force using the most advanced of the available industrial processes. This early industrial activity was highly localized because of industry's need to be near mineral resources and water power and because of the importance of local canal systems. This localization was also a product of the principle of **comparative advantage** whereby industry had been displaced into areas which were least profitable for agriculture.

This pattern of proto-industrialization, with its external economies, infrastructural advantages and well-developed markets, helped to determine the nuclei of industrial development in Great Britain during the first phase of the first wave of industrialization, between 1760 and 1890. These included north Cornwall, eastern Shropshire, south Staffordshire, North Wales, upland Derbyshire, south Lancashire, the West Riding of Yorkshire, Tyneside, Wearside and parts of the central Lowlands of Scotland. Although these sub-regions shared the common impetus of certain key resources and innovations, each retained its own distinctive business traditions and industrial style. To use the terminology of Regulation Theory, each had its own emergent *mode of regulation* and *regime of accumulation*. Much of the required capital was raised locally, labour requirements were drawn (in the first instance) from the immediate hinterland, and industrialists formed themselves into regional organizations and operated regional cartels.

From the start, then, industrialization was articulated at the regional level; and this has been a feature of subsequent phases and waves. The second phase, between 1790 and 1820, reinforced the position of those embryo industrial regions with a coalfield base and saw the emergence of Ulster and South Wales as industrialized regions. Meanwhile, the prosperity of four of the early starters – Cornwall, North Wales, eastern Shropshire and upland Derbyshire – declined markedly as their relative advantages were eclipsed by a combination of three different factors:

1. The exhaustion of minerals, or the discovery of cheaper alternative supplies of them.
2. Inaccessibility to markets because of poor communications and/or relatively remote locations.
3. Lack of size to develop. As Pollard put it, 'A small compass may be an ideal milieu for the first halting steps in an innovation, but beyond a certain point an industrial region has to be of a certain size to remain viable' (1981, p. 20).

The third phase of the 'British' wave, between 1820 and 1850, was dominated by the expansion of the railway system. This did not foster any new industrial regions, but it did widen the market area of the existing industrial regions, drawing more of Britain into the sphere of industrial capitalism.

comparative advantage
The principle used to explain patterns of trade and specialization: if each region or country specializes in those economic activities which they perform *relatively* better than others, each is likely to gain (transport costs and terms of trade notwithstanding).

The second wave: a new technology system and new regimes of accumulation

It was at this point that industrialization began to spread to continental Europe. It should be emphasized, however, that this did not take the form of a straightforward spatial diffusion of industrialization or 'modernization'. By this time, a second *technology system* had begun to emerge, based on coal, steel, heavy engineering, steam power, and railways. Exploiting these new technologies meant:

- drawing on new resources;
- the development of new labour practices (the spread of wage-labour norms);
- the development of new corporate structures (large limited-liability firms that were national rather than local in scope);
- new relationships between governments and industry (increased government regulation of, and investment in, key industries).

Thus there evolved new *regimes of accumulation* and new *modes of regulation*.

Just as the British wave of industrialization was initially based on localized concentrations of proto-industrialization, so the second wave was launched from the proto-industrial regions of Continental Europe. Initially, from around 1850, industrialization was concentrated in the Sambre-Meuse region of Belgium and in the valley of the Scheldt in Belgium and France. Subsequent phases saw the spread of industrialization to the Aachen area, the right bank of the River Rhine around Solingen and Remscheid, the Ruhr, Alsace, Normandy, the upper Loire valley and Swiss industrial district between Basel and Glarus.

Meanwhile, however, the *initial advantage* enjoyed by British industries over their would-be competitors on the Continent meant that later-industrializing regions had to confront a situation in which British industries, having secured comfortable advantages in technology, had come to dominate world markets. Britain also had a series of 'natural' geographical advantages: a compact territory with a large population, favourable conditions for intensive agricultural production, and a rich variety of minerals, including coal. This competitive disadvantage for Continental European industrial regions was compounded by the consequences of the revolutionary and Napoleonic Wars of the early nineteenth century (as it was in the United States by the Civil War of 1861–65). Conscription, conflict and military occupation disrupted production and suppressed industrial expansion, allowing British industries to forge still further ahead on the basis of the new technology system (and, of course, a constantly evolving and adapting mode of regulation).

On the other hand, Continental entrepreneurs and governments did not have to industrialize by trial-and-error in the way that the British had: they could benefit from British experience and they could import British managers, workers, capital and technology. These regions of 'inner' Europe were differentiated one from another not only by their different mix of industries, but also by what economic historian Sidney Pollard calls the *differential of contemporaneousness*, whereby new technologies, ideas and market conditions reached different regions simultaneously but affected them in very different ways *because they were differently equipped to respond to them*. Thus, for example:

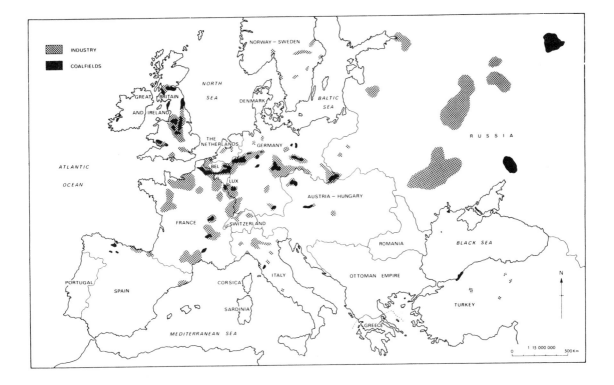

Figure 5.1
Europe in 1875

Source: Pollard (1981), Map 2,
p. xv

Legislation permitting the easy formation of joint-stock companies spread quickly across Europe in the 1850s, and their contribution to overspeculation and wide-spread bankruptcies in the less sophisticated European economies has often been commented on. In banking, the backward economies, using the experience of the pioneers, could bypass some of the difficulties of the latter by enjoying the benefits of more efficient banks, ahead, as it were, of their own stage of economic growth.

(Pollard, 1981, pp. 188–9)

In general terms, however, the cumulative impact of innovations in first- and second-wave industrializers made for convergence: the French Nord began to look and function increasingly like the central belt of the Scottish Lowlands, and the Ruhr began to look and function increasingly like the Sambre-Meuse region. At the same time, there was increasing divergence between those areas which had adopted an industrial base and those which had yet to follow suit. By 1875, the latter still covered a great deal of the map (Fig. 5.1), but many of them were incorporated in the third wave of industrialization between 1870 and 1914.

The third wave: intermediate Europe

The third wave of industrialization included 'Intermediate Europe' – parts of Britain, France, Belgium and Germany that had not been directly affected by the first two waves, together with most of the Netherlands, southern Scandinavia, northern Italy, eastern Austria, and Catalonia. By this time, all European landscapes were beginning to be reorganized in response to the imperatives of a third *technology system*: one based on steamships, world shipping, the internal combustion engine, heavy chemicals, and heavy engineering.

In the regions of 'intermediate' Europe, the imprint of this industrialization was distinctive in several important respects. There was little by way of antecedent development on which to base industrialization (apart from small enclaves in Barcelona, Milan–Turin and Vienna), and the relative amount of capital these regions were required to find in order to support industrialization was ten times greater than during the first wave. This, together with the increasing sophistication of industrial technology and its related services, led central governments to take on ever-increasing responsibilities. The economic role of the state among the later industrializers therefore tends to be more pronounced than in the countries of 'inner' Europe.

Peripheral Europe

The residual territories of Western Europe – most of the Iberian peninsula, northern Scandinavia, Ireland, southern Italy, the Balkans and east-central Europe, which Pollard collectively terms the 'outer periphery' – remained, like the interstices of 'inner' and 'intermediate' Europe, mainly outside the fold of industrial capitalism, to be penetrated to different degrees over the next 50 years.

One of the reasons for the continued peripherality of these regions was that their entrepreneurs and governments often felt compelled to adopt the new technologies and forms of organization that had served the pioneer regions well, despite the reality of very different economic and geographic settings in the periphery. Railway systems provide a simple illustration. Railway networks in pioneer regions had been able to integrate industrial development and operate profitably by carrying regular passenger traffic as well as heavy bulk freight like coal, ore and grain. The extension of railway systems to regions with neither an emerging industrial base nor a sufficiently high density of population (as in Ireland, southern Italy, Spain, and most of east-central Europe) invited heavy losses. The fact that most governments were willing to underwrite such losses reflects the potency of the railways as political virility symbols. What was not foreseen at the time, however, was that, rather than integrating national territories and fostering industrial development, the penetration of the railways to peripheral regions tended to result in their specialization in a subordinate, agricultural role: a special case of Pollard's 'differential of contemporaneousness'.

Another reason for continued peripherality is to be found in the very different nature of urban development within the later-industrializing regions. In Britain, 'inner' Europe and 'intermediate' Europe there had been a symbiotic relationship between urban and industrial development (with cities providing capital, labour, markets, access to transport systems, and a variety of agglomeration economies). In much of peripheral Europe, the 'demonstration effect' of these events led to a very different relationship, largely because of the attitudes of the élite:

> Railways were laid to royal palaces, gas or water mains supplied a narrow layer of
> privileged classes ... innovations intended for mass markets were misused for a

narrow luxury market and either diverted resources, or led to burdensome capital imports. .... Above all, the city became the gate of entry to new technology manufactures from abroad, spreading outward from Naples, Madrid, Budapest or St. Petersburg, to kill off native industry as unfashionable.

(Pollard, 1981, p. 212)

In short, conspicuous consumption precluded import substitution, creating cities that inhibited rather than fostered industrial growth.

155

Dislocation and depression

In the first half of the twentieth century, the economic development of the whole of Europe was punctuated twice by major wars. The disruptions of the First World War were immense. The overall loss of life, including the victims of influenza epidemics and border conflicts which followed the war, amounted to between 50 and 60 million. About half as many again were permanently disabled. For some countries, this meant a loss of between 10 and 15 per cent of the male workforce. In addition, material losses caused a severe dislocation to economic growth: it has been calculated that the level of European output achieved in 1929 would have been reached by 1921 if it had not been for the war.

Economic dislocation in Europe was further intensified by several indirect consequences of the war. In terms of tracing the evolving economic geography of the core regions, two of these were particularly important:

- The *relative* decline of Europe as a producer compared with the rest of the world. Europe accounted for 43 per cent of the world's production and 59 per cent of its trade in 1913, compared with only 34 per cent of production and 50 per cent of trade in 1923. The main beneficiaries were the USA and Japan for manufactures, and Latin America and the British dominions for primary production.
- The redrawing of the political map of Europe. This created 38 independent economic units instead of 26; 27 currencies instead of 14; and 20 000 extra kilometres of national boundaries. The corollary of these changes was a severe dislocation of economic life, particularly in east-central Europe: frontiers separated workers from their factories, factories from their markets, towns from their traditional food supplies, and textile looms from their spinning sheds and finishing mills; while the transport system found itself only loosely matched to the new political geography.

Just as European economies had adjusted to these dislocations, the *stagflation* crisis of 1929–35 – the Great Depression – created a further phase of economic damage and reorganization throughout Europe. It should be emphasized, however, that the effects of the Depression varied a good deal from one sector of the economy to another *and from one region to another*. The image of the 1930s depends very much on whether attention is focused on Jarrow or Slough, on Bochum or Nice, on Glasgow or Geneva.

These contrasts are reflected in the unequal distribution of the unemployed – the 'casualties of peace' – in the United Kingdom. Even at the depth of the Depression, in the winter of 1932–33, a steep regional gradient persisted. Local variations, meanwhile, were sharper still: in parts of South Wales, for example, up to 80 per cent of the civilian labour force was out of work. In northeast England the figure approached 70 per cent in towns like Jarrow; and in Scotland it approached 40 per cent in the Glasgow area. Meanwhile, unemployment never rose above 5 per cent in places such as Guildford, Slough and Romford in southeast England.

Subsequently, recovery came to the depressed regions only slowly. Economic revival was concentrated in the southeast and the Midlands, where industries based on new consumer products – domestic appliances, mass-produced clothes and furniture, etc. – were assured of a large and relatively buoyant market. Nevertheless, the availability of cheap manufactured products also affected patterns of consumption in the depressed regions, producing the contradictory situation noted by George Orwell in *The Road to Wigan Pier*:

> Twenty million people are underfed but literally everyone . . . has access to a radio. What we have lost in food we have gained in electricity. Whole sections of the working class are being compensated, in part, by cheap luxuries which mitigate the surface of life. . . . It is quite likely that fish-and-chips, art-silk stockings, tinned salmon, cut-price chocolate, the movies, the radio, strong tea and the football pools have between them averted revolution.

(Orwell, 1962, pp. 80–1; first published in 1937)

Meanwhile, the coherence of the European economic world began to disintegrate as individual countries attempted to protect their industries with import quotas and restrictions, currency manipulation, and exclusionary trade agreements. The result was a substantial fall in trade, both in absolute terms and as a proportion of output, with the USA and Japan, once again, as the major beneficiaries.

The Second World War and recovery

The Second World War resulted in a further round of destruction and dislocation. The total loss of life in Europe was 42 million, two-thirds of whom were civilian casualties. The German occupation of Continental Europe involved ruthless exploitation. By the end of the war, France was depressed to below 50 per cent of her pre-war level of living and had lost 8 per cent of her industrial assets. The United Kingdom lost 18 per cent of her industrial assets (including overseas holdings) and the USSR lost 25 per cent. Germany herself, however, lost 13 per cent of her assets and ended the war with a level of income *per capita* that was less than 25 per cent of the pre-war figure.

After the war, the political cleavage between Eastern and Western Europe (that resulted from the imposition of what Winston Churchill called the 'Iron Curtain' along the western frontier of Soviet-dominated territory) resulted in a further erosion of the coherence of the European economy and, indeed, of its economic geography. Ironically, it was this cleavage which led to a surprisingly rapid economic recovery in Western Europe: the USA, believing that poverty and economic chaos would foster communism, embarked on a massive programme of aid under the Marshall Plan. This pump-priming action, together with the pent-up backlog of demand in almost every sphere of production, provided the basis for a remarkable recovery. By the early 1950s most of Europe had exceeded pre-war levels of prosperity. By the early 1960s, European central banks were in a position to step in, when necessary, to support the US dollar. As Table 5.1 illustrates, growth rates throughout Western Europe surged forward to impressive levels.

Table 5.1: Growth rates in Europe

	Annual average compound growth rate of real output per capita	
	1913–50	1950–70
Austria	0.2	4.9
Belgium	0.7	3.3
Denmark	1.1	3.3
France	1.0	4.2
Germany (West)	0.8	5.3
Greece	0.2	5.9
Ireland	0.7	2.8
Italy	0.8	5.0
Netherlands	0.9	3.6
Norway	1.8	3.2
Portugal	0.9	4.8
Spain	−0.3	5.4
Sweden	2.5	3.3
Switzerland	1.6	3.0
UK	0.8	2.2
Av. Western Europe	1.0	4.0

Source: Adapted from Pollard (1981), Table 9.2, p. 315.

Centre and periphery in Western Europe

The cumulative effects of the differential impact of successive waves of industrialization and reorganization have often been interpreted in terms of centres of capital accumulation and economic power and peripheries of limited (or suppressed) potential for economic development. Yet different criteria yield different definitions of 'centre' and 'periphery'. The West European centre, for example, has variously been interpreted as the triangular region defined by Lille–Bremen–Strasbourg (the 'Heavy Industrial Triangle'), as an axial belt stretching between Boulogne and Amsterdam in the north and Besançon and Munich in the south, as a T-shaped region whose stem extends down the Rhine to Stuttgart; and so on.

Such definitions can be confusing in their variability, but their major weakness is that they overlook the *interdependence* which exists between

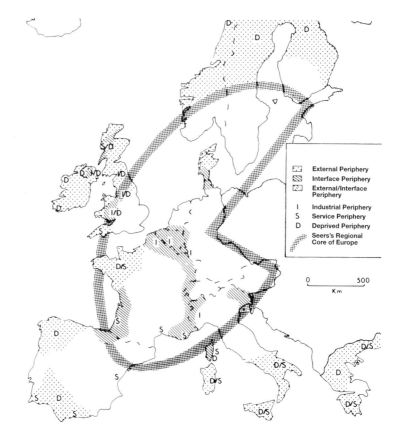

Figure 5.2
Centre and periphery
in Western Europe
(after Seers, 1979; and
Aarebrot, 1982).

Source: Knox (1984) Fig. 5.8,
p. 147

centre and periphery. It is therefore more satisfactory to think in terms of a centre consisting of a number of linkages or *flows* (capital, migrants, taxes, tourists, consumer fashions) *which bind centre and periphery together, reinforcing their unequal but symbiotic relationship.* From this perspective, the European centre has been portrayed as an egg-shape centred on Kassel in West Germany, with its long axis stretching 2700 kilometres from Barcelona in the southwest to Helsinki in the northeast (Fig. 5.2).

This kind of economic dominance tends to go hand in hand with political control and cultural standardization. This, in turn, points to the existence of different *kinds* of peripheral regions as a result of the interplay of economic, political and cultural factors. Rokkan (1980), for example, demonstrated how the changing patchwork of European nation states created a number of peripheral regions as *interfaces* between major politico-economic core territories. These include parts of the European centre, such as Flanders/Wallonia, Alsace-Lorraine, the Bernese Jura and the Austro-Hungarian interface (Fig. 5.2). Rokkan also recognized *external* peripheries, characterized by their relative distance from national centres. These include Brittany, Galicia, Scotland and Wales. At the same time, differentiation along the major lines of cleavage between centre and periphery – economic, political and cultural – are also recognized as important. In terms of economic profiles, for example, peripheral regions can be grouped into *industrial* peripheries (e.g. northern England, northern Italy), *service* peripheries dependent on tourism (e.g. southwestern England, the Italian Riviera, and the French southern Atlantic and Mediterranean coasts) and *deprived* peripheries (southern Italy, the Republic of Ireland, Scotland, Wales, Sardinia and Corsica). Finally, it must be acknowledged that a second 'centre' of the West European space–economy has emerged: that of the United States, which is now a major supplier of capital, technology and consumer trends to the whole of Europe – centre and periphery.

FORDISM AND NORTH AMERICAN INDUSTRIALIZATION

The emergence of North America as the dominant component of the core of the world economy was essentially due to the fact that it had:

- vast natural resources of land and minerals;
- a large and – thanks to immigration – rapidly growing market and labour force;
- sufficent size to breed giant corporations with large research budgets, which helped to institutionalize the innovation process in a way which European industry had never done.

Within North America, the evolving pattern of spatial organization can be interpreted in terms of the interaction of (1) the geography of resources, (2) the introduction of major technological innovations (particularly in transport), and (3) movements of population. Thus:

> Major changes in technology have resulted in critically important changes in the evaluation or definition of particular resources on which the growth of certain urban regions had previously been based. Great migrations have sought to exploit resources – ranging from climate or coal to water or zinc – that were either newly appreciated or newly accessible within the national market. Usually, of course, the new appreciation or accessibility had come about, in turn, through some major technological innovation.

(Borchert, 1967, p. 324)

As Borchert and others have noted, the history of the development and evolution of the American economy found its clearest geographic expression in changing patterns of urbanization. At the time when Europe was experiencing the first waves of industrialization, the spatial organization of the North American economy was focused on the gateway ports of the Atlantic seaboard, each of which controlled a limited hinterland where the economy was dominated by the production of agricultural staples for export to Europe and the consumption of manufactured goods imported from Europe. From the end of the eighteenth century, however, the North American economy began to break loose from this dependent relationship, and within 100 years it had become the dominant component of the world economy, articulated around a closely integrated but highly differentiated urban system.

A major factor in this metamorphosis was the *political independence* of the United States, which was formally achieved in 1783. This stimulated economic development in several ways:

1. Independence from Britain and national political integration under a federal system provided an important stimulus for economic links to be forged between the component parts of the old colonial system.
2. Independence meant that a much greater proportion of investment was financed by American capital, with the result that less of the profits were 'leaked' back to Europe.
3. Independence stimulated a proliferation of government employment, as every county seat and state capital developed the infrastructure of democracy.
4. The territorial expansion of the new nation provided a large, rich resource base.

Both urban and economic development in this period, however, were constrained by the relatively primitive transportation system of the 'sail and wagon' epoch. It was not until the 1840s, when the second *technology system* of industrial capitalism (coal/steel/steam/railways) began to be exploited, that American industrialization took off.

The growth of the manufacturing belt

The acceleration of industrialization in the 1840s was in part the result of the diffusion of industrial technology – particularly the wider industrial application of steam and the accompanying changes in the iron industry – and methods of industrial and commercial organization from the hearth of the industrial revolution in Europe. In addition, the demand for foodstuffs and other agricultural staples, both in North America and abroad, stimulated the growth of industrial capitalism as farmers sought to increase productivity through mechanization and the use of improved agricultural implements. Increasing agricultural productivity, in turn, helped to sustain the growing numbers of immigrants from Europe, thus allowing them to be channelled into industrial employment in North America's mushrooming cities.

The development of the railway system was central to the evolution of this new economic order. Initially, the railways were complementary to the waterways as competitive long-haul carriers of general freight; but by the end of the 'iron horse' epoch the railway network had not only realigned the economic system but also extended it to a continental scale. In 1869, the railway network reached the Pacific when, at Promontory, Utah, the Union Pacific railway, building west from Omaha, met the Central Pacific railway, building east from Sacramento. By 1875 intense competition between railway companies had began to open up the western prairies as far as Minneapolis-St Paul and Kansas City. The significance of this was to be profound:

Not only did this permit American enterprise to exploit fully the commercial advantages and scale economies of large, diversified natural resources and of the revolutionary technologies evolved in those decades, but it generated rapid, large-scale functional and spatial concentration of finance and management unimpeded by world events, creating a '*transcontinental*' business mentality. Wide spatial separation of major resources, cities and markets, and adjacency to the easily penetrated Canadian economy all induced mental thresholds for thinking '*intercontinental*' once imported resources and markets overseas became a necessary ingredient to sustain business activity at home, especially during and after the Second World War.

(Hamilton, 1978, p. 26, emphasis added)

In short, the railways can be seen as the catalyst which allowed regional economies to develop into a continental economy which stood poised to become the leading component of the world economic system.

Meanwhile, the westward extension of the railways inevitably affected the fortunes of the inland gateway cities. Buffalo and Louisville, for instance, slowed their rate of growth and came increasingly to rely on more diversified regional functions. Further west, St Paul and Kansas City grew rapidly to become major wholesaling depots. The development of improved transportation networks also led to adjustments in spatial organization within the northeast, where fierce competition between the railways and water-borne transport, coupled with equally fierce rivalry between neighbouring cities, led to a marked increase in *intra*-regional trade. This helped to lay the foundations of what was to become the Manufacturing Belt.

In essence, the consolidation of the Manufacturing Belt as the continental economic heartland was the result of *initial advantage*. With its large markets, well-developed transport networks and access to nearby coal reserves, it was ideally placed to take advantage of the general upsurge in demand for consumer goods, the increased efficiency of the telegraph system and postal services, the advances in industrial technology and the increasing logic of scale economies and external economies which characterized the late nineteenth century. The overall effect was twofold:

1. Individual cities began to specialize, as producers were able to gear themselves to *national* rather than regional markets:

> Between 1870 and 1890, advances in milling technology and concentration of ownership supported the emergence of Minneapolis as a milling centre. Furniture for the mass market centralized in fewer, larger plants using wood-working machinery. . . . The rise of national brewers between 1880 and 1910 is an example of national market firms encroaching on local-regional firms. The brewers in Milwaukee and St Louis achieved economies of scale in manufacture, used production innovations such as mechanical refrigeration, and capitalized on distribution innovations made possible by the refrigerated rail car and an integrated rail network.
>
> (Meyer, 1983, p. 160)

Similarly, musical instrument manufacture and men's clothing emerged as specialties in Boston; meat packing, furniture manufacture and printing and publishing in Chicago; coach-building and furniture manufacture in Cincinnati; textile manufacture in Philadelphia; and so on. In smaller cities, specialization was often much more pronounced, as in the production of iron and steel and coachbuilding in Columbus, furniture in Grand Rapids, agricultural implements in Springfield, and boots and shoes in Worcester. Overall, there emerged a three-part segmentation of the Manufacturing Belt (Fig. 5.3), with a heavy bias towards consumer-good production in the ports of Baltimore, Boston and New York, a producer-goods axis between Philadelphia and Cleveland, and a western cluster of rather less specialized consumer-oriented manufacturing cities.

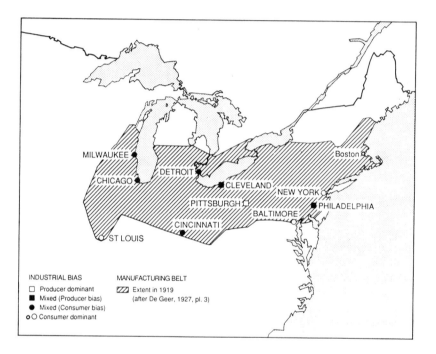

Figure 5.3
The American Manufacturing Belt in 1920 (after Conzen).

Source: Knox *et al.* (1988) Fig. 5.1, p. 117

2. This specialization provided the basis for increasing commodity flows between individual cities, thus binding the Manufacturing Belt together. These linkages, in turn, generated important multiplier effects through wholesaling, finance, ware-housing and transportation, adding to the cumulative process of regional industrial growth and increasing the region's comparative advantage. These advantages meant that the Manufacturing Belt was able to attract a large proportion of any new industrial activities with large or national markets, thus stifling the chances of comparable levels of industrialization in late-developing regions.

This does not mean, of course, that other regions did not become industrialized. Rather, it was the scale and the intensity of industrialization which differed: later-developed regions were able to support an array of locally oriented manufacturers, together with some nationally oriented activities based on particular local advantages or raw materials; but they were rarely able to attract manufacturers of producer-goods for the national market.

Fordism, Taylorism and regional economic change

If the template of North American economic geography had been established by 1920, the full details of industrial capitalism were etched in between 1920 and 1940, when the arrival of truck and automobile transportation triggered a further series of shifts and adjustments. Road and air travel, along with improvements in electronic communications, increased the *capacity* and *efficiency* of the economy and facilitated the functional integration of both businesses and regions at an unprecedented pace. The 1920s were the 'New Economic Era', and the liberal reactions to industrialism, which had characterized the Progressive Era before the First World War, were quickly edged aside by consumerism and boosterism.

The larger companies based in the major metropolitan centres were best placed to exploit the increased capacity and efficiency of the economic system. As they did so, they also exploited new principles of economic organization based on a more intensive division of labour, flow-line assembly and 'scientific' management (also known as 'Taylorism'). The resulting increases in efficiency and productivity meant that many goods could be *mass-produced* at low prices for mass markets. The consequent combination of mass production and mass consumption is generally referred to as a *Fordist* regime of accumulation, after Henry Ford, the automobile manufacturer who led the way in implementing these changes.

This new regime of accumulation required ever-larger companies. A flurry of company mergers soon transformed the business structure of the economy, resulting in a relatively small number of very powerful corporations which now stood poised to dominate the economy. By 1920, over 30 per cent of all jobs and nearly 50 per cent of the country's production was accounted for by just 1 per cent of all firms.

> The Captains of Industry were clearly in charge. Across the country, territorial communities watched effective control over local production slip out of their grasp. Political power came to focus on the national level of territorial integration which, for the time being, effectively bounded the operation of most businesses.
>
> (Friedmann and Weaver, 1979, p. 22)

But the new regime of accumulation associated with this 'New Economic Era' fostered some serious problems. Mechanized agriculture became so 'over-productive' that commodity prices plummeted; while the industrial market became unstable as a result of the labyrinth of holding companies that had been created. In October 1929 the stock market collapsed, triggering the Great Depression in which millions of workers lost their jobs. Because of the regional division of labour that had emerged over the previous 50 years, some areas suffered particularly acute social and economic problems. The political response took the form of a 'New Deal' in which the central government took on much more responsibility both for overall economic growth and for regional economic well-being. Before long, this evolved into a system of public macroeconomic management that came to be known as **Keynesianism** (after the doctrine of British economist John Maynard Keynes). In these changes, we can see that a significant shift occurred in the *mode of regulation* associated with the Fordist regime of accumulation.

With the outbreak of war in Europe in 1939, the entire North American economy entered a phase of accelerated growth; and in the aftermath of the war the

Keynesianism
Specifically, this refers to a doctrine of macro-economic management that is closely associated with the British economist John Maynard Keynes, who advocated the use of fiscal policy (e.g. budget deficits) and the exploitation of economic multiplier effects in order to achieve and maintain full employment. The term is also used to denote the mode of regulation associated with the Fordist regime of accumulation.

United States and Canada emerged not only with stronger and more efficient industries but also with new technologies and with control over new international markets. The United States, in its new, outward-looking role as leader of the capitalist world, was able to dictate the pattern of world affairs through the terms of Marshall Aid, its control of the Organization for European Economic Co-operation, and the Bretton Woods agreement (which established a new framework for international economic relations). By 1960, GNP *per capita* stood at US$2513 in the USA, compared with US$1909 in Canada, US$1678 in Sweden, US$1259 in the UK, US$1200 in West Germany, US$1193 in France, and US$421 in Japan. By this time, however, the economic geography of North America had begun to respond to the imperatives of the new technology systems and regimes of accumulation of advanced capitalism: themes that we explore in detail in Chapter 7.

THE SOVIET ATTEMPT TO JOIN THE CORE

The experience of the former Soviet Union and its Eastern European satellites provide a very different case study of industrial development (though in terms of spatial organization the differences between them and capitalist industrial countries were fewer than might be expected). It should be emphasized at the outset that the economies of the former Soviet Union and its satellites were *not* based on a true socialist or communist mode of production in which the working class had democratic control over the processes of production, distribution and development. Rather, they seem to have evolved as something of a hybrid, in which state power was used by a bureaucratic class to exploit workers and to compete for power and economic advantage in the world economy. Social objectives aside, their experience can be interpreted as the pragmatic response of late-comers whose leadership felt they could only industrialize by disengaging from their semi-peripheral role in the world economy in order to pursue modernization and economic development through highly centralized and rigidly enforced government direction.

Eventually, the constraints imposed by excessive state control, the inherent disadvantages that resulted from (1) the absence of entrepreneurship and competition, and (2) the dissent that resulted from the lack of democracy, combined to bring the experiment to a sudden halt (in 1990). By then, parts of the Soviet world-empire had been able to achieve a relatively advanced stage of industrial development; but none of the survivors of the experiment, including Russia, have achieved better than semi-peripheral status in the world economy.

Revolutionary economic reorganization

Since the time of Peter the Great (around 1700), Tsarist Russia had been attempting to modernize. By 1861, when Alexander II decreed the abolition of serfdom, Russia had built up an internal core with a large bureaucracy, a substantial intelligentsia and a sizeable group of skilled workers. The abolition of feudal serfdom was designed to accelerate the industrialization of the economy by compelling the peasantry to raise crops on a commercial basis, the idea being that the profits from exporting grain would be used to import foreign technology and machinery. In many ways, the strategy seems to have been successful: grain exports increased fivefold between 1860 and 1900, while manufacturing activity expanded rapidly. Further reforms, in 1906, helped to establish large, consolidated farms in place of some of the many small-scale peasant holdings. But the consequent flood of dispossessed peasants to the cities created acute problems as housing conditions deteriorated and labour markets became flooded.

These problems, to which the Tsar remained indifferent despite the petitions of desperate municipal governments, nourished deep discontent which eventually, aggravated by military defeats and the sufferings of the First World War, spawned the revolutionary changes of 1917. It was not the peasantry or the oppressed provincial industrial proletariat, however, that emerged from the chaos to take control. It was the Bolsheviks, a dissatisfied element drawn from the internal core of the nation, whose orientation from the beginning favoured a strategy of economic development in which the intelligentsia and skilled industrial workers would play the key roles.

In the first instance, however, the ravages of war and the upheavals of revolution precluded the possibility of planned economic reorganization of any kind. Thus the centralization of control over production and the nationalization of industry resulted as much from the need for national and political survival as from ideological beliefs. Similarly, it was rampant inflation that led to the virtual abolition of money, not revolutionary purism. By 1920, industrial production was still only 20 per cent of the pre-war level, crop yields were only 44 per cent of the pre-war level, and national income *per capita* stood at less than 40 per cent of the pre-war level.

In 1921, a New Economic Policy was introduced in an attempt to catch up. Central control of key industries, foreign trade and banking was codified under *Gosplan*, the central economic planning commission. But in other spheres – and in agriculture in particular – a substantial degree of freedom was restored, with heavy reliance on market mechanisms operated by 'bourgeois specialists' from the old intelligentsia. Improvement in national economic performance was immediate and sustained, with the result that recovery to pre-war levels of production was reached in 1926 for agriculture and the following year in the case of industry.

Soon afterwards, however, there occurred a major shift in power within the Soviet Union. This power shift swept aside both the New Economic Policy and its 'bourgeois specialists'. They were replaced by a much more centralized allocation of resources: a 'command' economy operated by a new breed of engineers/managers/ *apparatchiks* drawn from the new intelligentsia that had developed among the membership of the Communist Party. With this shift there also came a more explicit strategy for industrial development. Like Japan (see below), the Soviet Union gave national economic and political independence the highest priority; but, in contrast to the Japanese strategy of aggressive international trade, the Soviet Union chose to withdraw from the capitalist world-system as far as possible, relying instead on the capacity of its vast territories to produce the raw materials needed for rapid industrialization. As in Meiji Japan, the capital for creating manufacturing capacity and the required infrastructure and educational improvements was extracted from the agricultural sector. The foundation of Stalin's industrialization drive was the collectivization of agriculture. This involved the compulsory relocation of peasants into state or collective farms, where their labour was expected to produce bigger yields. The state would then purchase the harvest at relatively low prices so that, in effect, the collectivized peasant was to pay for industrialization by 'gifts' of labour.

In the event, the Soviet peasantry was somewhat reluctant to make these gifts, not least because the wages they could earn on collective farms could not be spent on consumer goods or services: Soviet industrialization was overwhelmingly geared towards manufacturing *producer* goods like machinery and heavy equipment. It proved very difficult to regiment the peasants. Requisitioning parties and tax inspectors were met with violence, passive resistance, and the slaughter of animals. At this juncture Stalin employed police terror to compel the peasantry to comply with the requirements of the Five-Year Plans that provided the framework for his industrialization drive. Severe exploitation required severe repression. Dissidents, along with enemies of the state uncovered by purges of the army, the bureaucracy and the Communist Party, provided convict (*zek*) labour for infrastructural projects. Altogether, some 10 million people were sentenced to serve in the zek workforce, to be imprisoned, or to be shot. The barbarization of Soviet society was the price paid for the modernization of the Soviet economy.

Economic and territorial expansion under socialism

The Soviet economy *did* modernize, however. Between 1928 and 1940 the rate of industrial growth increased steadily, reaching levels of over 10 per cent per annum in the late 1930s: rates which had never before been achieved, and which have been equalled since only by Japan and China. The annual production of steel had increased from 4.3 million tons to 18.3 million tons; coal production had increased nearly five times; and the annual production of metal-cutting machine tools had increased from 2000 to 58 400. In short, 'An industrial revolution in the Western sense had been passed through in one decade' (Pollard, 1981, p. 299). When the Germans attacked the Soviet Union in 1941 they took on an economy which in absolute terms (though not *per capita*) had output figures comparable with their own.

The Second World War cost the Soviet Union 25 million dead, the devastation of 1700 towns and cities and 84 000 villages, and the loss of more than 60 per cent of all industrial installations. In the aftermath, the Soviet Union gave first priority to national security. The *cordon sanitaire* of independent East European nation states that had been set up by the Western nations after the First World War was appropriated as a buffer zone by the Soviet Union. Because this buffer zone happened to be relatively well developed and populous it also provided the basis of a Soviet world-empire as an alternative to the capitalist world economy, thus providing economic as well as military security.

But the Soviet Union felt vulnerable to the growing influence and participation of the United States in world economic and political affairs, and Stalin felt compelled, in 1947, to intervene in Eastern Europe. In addition to the installation of the 'iron curtain' which severed most economic linkages with the West, this intervention

Table 5.2: Industrial growth in the Soviet bloc, 1950–78

	1950/52–1958/60		1958/60–1967/69		1971–75		1976–78	
	A	B	A	B	A	B	A	B
Bulgaria	15.5	7.5	11.5	6.4	9.1	6.6	6.9	6.1
Czechoslovakia	8.0	4.6	5.5	5.1	6.6	5.4	5.4	4.6
GDR	10.1	3.6	5.2	5.7	6.5	5.4	5.6	4.8
Hungary	7.1	8.5	8.7	7.8	6.4	6.2	5.5	5.8
Poland	9.4	3.1	8.3	7.0	10.4	7.7	7.3	6.7
Romania	11.9	8.4	13.2	10.7	12.9	6.2	11.0	7.5
Soviet Union	10.7	6.6	8.7	4.9	7.4	5.9	5.1	3.4

Notes:
A = Average annual per cent change in industrial output.
B = Average annual per cent change in labour productivity.

Source: UN Economic Commission for Europe (1972) and (1979).

resulted in the complete nationalization of the means of production, the collectivization of agriculture, and the imposition of rigid social and economic controls. The Communist Council for Mutual Economic Assistance (CMEA, better known as COMECON) was also established to reorganize the Eastern European economies in the Stalinist mould – even to the point of striving for autarky for individual members, each pursuing independent, centralized plans. This proved unsuccessful, however, and in 1958 COMECON was reorganized by Stalin's successor, Kruschev. The goal of autarky was abandoned, mutual trade among the Soviet bloc was fostered, and some trade with Western Europe was permitted.

Meanwhile, the whole Soviet bloc gave high priority to industrialization. Between 1950 and 1955, output in the Soviet Union grew at nearly 10 per cent per annum, though it subsequently fell away to more modest levels (Table 5.2). The experience of the East European countries varied considerably, but, in general, rates of industrial growth were high. Equally important was the structural transformation of industry, for although producer-goods remained dominant, the economic base of all Soviet bloc countries expanded to the point where per capita consumption of food, clothing and consumer goods came to be much closer to that of the capitalist core countries than to that of the peripheral and intermediate countries of the capitalist world economy.

The economic geography of state socialism

As in Western Europe, North America and Japan, industrialization brought about radical changes in the economic landscapes of the Soviet Union and Eastern Europe (for detailed economic geographies, see Bradshaw, 1991; Shaw, 1988; and Turnock, 1988). But did the state socialism, or statism, that guided Soviet bloc industrialization result in qualitatively different economic landscapes? At face value, there were

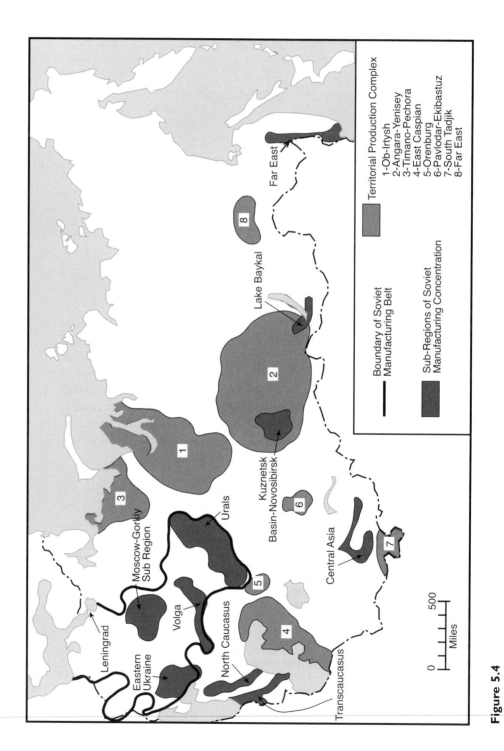

Figure 5.4
Manufacturing regions and territorial production complexes in the former Soviet Union

sound reasons for anticipating substantial differences. Central planning and control of economic development meant that ideological objectives could be translated into administrative fiat, while the absence of a competitive market eliminated risk and uncertainty, precluded the influence of powerful monopolies, and facilitated the rapid dissemination of technological innovations.

In practice, however, spatial organization within the Soviet bloc did not exhibit any really distinctive dimension. As in the industrial core regions of the West and Japan, the industrial landscape came to be dominated by the localization of manufacturing activity (Fig. 5.4), by regional specialization, by centre–periphery contrasts in levels of economic development, and by agglomeration and functional differentiation within the urban system. The reasons for this were several:

1. At the most fundamental level was the unevenness of natural resources and the consequent unevenness of population and economic development inherited from the pre-socialist era.
2. The principles of rationality and the primacy of *national* economic growth took precedence over ideological principles of spatial equality. As a result, Soviet planners applied the logic of agglomeration economies, developing *territorial production complexes*: planned groupings of industries designed to exploit local energy resources and environmental conditions. As developed by the Tenth Five-Year Plan (1976–80), these territorial production complexes were broadly defined (Fig. 5.4), designed to foster broad-scale agglomeration economies among specialized sub-regional territorial production complexes.
3. The extensive bureaucracy required by 'command' economies meant that a pronounced 'control hierarchy' developed:

   > Central places in the spatial control hierarchy will have disproportionate numbers of high-level business and party posts. This, along with the tendency for the world of culture, the arts and education to concentrate spatially, will create local élites, as in Moscow, enjoying living standards substantially better than those of the mass of people.
   >
   > (Smith, 1979, pp. 341–2)

   Conversely, places at the lower end of the control hierarchy offered limited occupational opportunities and limited access to upper-level jobs.
4. Centralized economic planning was unable to redress unwanted spatial disparities because large parts of the system came to be characterized by inertia, insensitivity, conservatism and compartmentalization. As a result, resource allocation was strongly conditioned by *incrementalism*, whereby those places already well-endowed by past allocations got proportionally large shares of each successive round of budgeting.
5. Places and regions which were able to establish an initial advantage (by proximity to market or raw materials, for instance) were generally able to maintain a significant comparative advantage over other regions. David Turnock, reviewing the outcomes of 'socialist' location principles, observed that 'Despite oft-repeated assertions forecasting the impending elimination of backward regions, through appropriate allocations of investment under the system of central planning, growth rates continue to show wide spatial disparities' (1984, p. 316).

We can briefly illustrate both the extent of the resultant unevenness in economic development and the degree to which the economic landscapes of industrial socialism, like those of industrial capitalism, were dominated by centre–periphery contrasts.

Within the former Soviet Union, the major contrast was always between, on the one hand, the richly endowed, relatively densely peopled, highly urbanized core of the manufacturing belt that stretches across Russia from St Petersburg (formerly Leningrad) in the north and eastern Ukraine in the south, through the Moscow and Volga regions to the Urals (Fig. 5.4) and, on the other hand, the rest of the country. Within the latter there are vast reaches where physical isolation and harsh environmental conditions have prevented all but a veneer of modern economic development and where tribal folk still pursue local subsistence economies. Much of the rest of what was Soviet Central Asia and the southern portion of Kazakhstan also lagged well behind the rest of the country in terms of economic development, largely because of the fundamental problem of physical isolation. In addition, however, there were parts of the European portion of the former Soviet Union that remained some way behind the levels of development achieved in the centre. These included Belorussia, Lithuania, eastern Latvia and the Western Ukraine: regions with large rural populations that were systematically excluded from Stalin's industrialization drive.

Such inequalities eventually contributed to the vulnerability of the Soviet system as an alternative mode of economic development. The critical economic failure, however, was state socialism's inflexibility and its consequent inability to take advantage of the new technology system that was developing among core countries: 'Soviet statism failed in its attempt . . . to a large extent because of the incapacity of statism to assimilate and use the principles of informationalism embodied in new information technologies' (Castells, 1996, p. 13).

The dramatic failure of state socialism has led to a period of radical change in the geography of the former Soviet Union. The former states of Yugoslavia and Czechoslovakia have been broken up into smaller entities; East Germany has been absorbed into Germany; and Hungary, Poland and the Baltic states are being drawn rapidly into the European Union's sphere of influence. Russia, meanwhile, is having to reconstitute its economic geography through a chaotic transition towards a market economy. In the process, all local and regional economies have been disrupted, leaving many Russian people to survive through a semi-formal 'kiosk economy'. Many accounts suggest that criminal business has flourished amid the chaos of Russia's transition, with one estimate putting the share of the country's GDP generated by organized crime at nearly 15 per cent (Sukhotin, 1994).

Meanwhile, the overall GDP has fallen dramatically: by 29 per cent in 1991–92, and a further 12 per cent in 1992–93. Material production fell by 63 per cent between 1989 and 1993, as industrial production went into free fall. Foreign capital has flowed into Russia, but it has been targeted mainly at the fuel and energy sector, natural resources, and raw materials (which now account for about half of Russia's total exports) rather than manufacturing industry. As a result, Russia has become increasingly dependent on Siberia and the Far East, which together accounted for two-thirds of Russian exports in 1995.

JAPANESE INDUSTRIALIZATION

The rise of the Japanese economy to join Western Europe and North America at the core of the modern world economy represents a major achievement, and it poses some important questions in relation both to the theory and the reality of economic organization and spatial change. In particular, how was it that a relatively resource-poor country like Japan was successful in industrialization while resource-rich regions elsewhere in Asia and in Latin American were not? In other words, in what way was Japan an exception to the rest of the periphery?

Broadly speaking, the answer lies in the fact that the Japanese economy, although organized along feudal lines until well into the nineteenth century, was autonomous; it had never been penetrated by the capitalism of the core regions. Moreover, the transition from feudalism to capitalism took place as a deliberate attempt to preserve national political and economic autonomy. But, even though Japan was 'lucky' in not having been politically and economically subordinated, the path to progress via industrialization was still obstructed by the core regions' pre-emption of the technology, the infrastructure and the capital for industrial development. This raises a second important question: how were these obstacles overcome? The answer, again in general terms, lies in the combination of a proto-industrial base and a strategy of military aggression, flooding overseas markets with cheap products, and copying and adapting Western technology: a strategy that was achieved at the expense of authoritarian government, widespread exploitation and acute regional disparities.

From feudalism to industrial capitalism

Japan's transition from feudalism to industrial capitalism can be pinpointed to a specific year – 1868 – when the feudal political economy of the Tokugawa regime was toppled by the restoration of imperial power. For over 200 years, just as an industrial system was developing in the western hemisphere, the Tokugawa regime had attempted to sustain traditional Japanese society. To this end, the patriarchal government of the Tokugawa family excluded missionaries, banned Christianity, prohibited the construction of ships above 50 tons, closed Japanese ports to foreign vessels (Nagasaki was the single exception), and deliberately suppressed commercial enterprise. At the top of the feudal hierarchy were the nobility (the *shogunate*), the barons (*daimyos*), and warriors (*samurai*). Farmers and artisans represented the productive base exploited by these ruling classes; and only outcasts and prostitutes ranked lower than merchants.

In terms of spatial organization, the economy was built on a closed hierarchy of castle towns, each representing the administrative base of a local shogun. The position of a town within this hierarchy was dependent on the status of the shogun which, in turn, was related to the productivity of their agricultural hinterland. As a result, the largest cities – which were to become the foundations for subsequent economic growth – emerged among the alluvial plains and the reclaimed lakes and bay-heads of southern Honshu. At the top of the hierarchy was Edo (now Tokyo), which the Tokugawa regime had selected as its capital in preference to the traditional imperial capital of Kyoto, and which, bloated by soldiery, administrators and the entourages of the nobility in attendance at the Tokugawa court, reached a population of around a million by the early nineteenth century. Kyoto and Osaka were next largest, with populations of between 300 000 and 500 000; and they were followed by Nagoya and Kanazawa, both of which stood at around 100 000.

With cities of this size, it was very difficult to suppress commerce and prevent the breakdown of the traditional political economy. As in feudal Europe, the peasantry fled the countryside in increasing numbers in response to a combination of taxation, technological improvements in agriculture, and the lure of the relative freedom and prosperity of the cities. At the same time, the cities evolved into important centres of domestic manufacture: nodes of proto-industrial development that were to become the platform for Japan's subsequent development (Wigen, 1992). Meanwhile, prolonged peace had reduced both the influence and the affluence of the samurai, drawing increasing numbers of them towards commercial and manufacturing activities. Thus 'former peasants mingled with former warriors in secular occupations coordinated as much by market forces as by feudalistic regulations. A class-based commercial society thus developed despite the efforts of the Tokugawa leaders to maintain the pre-industrial, status-oriented society of old Japan' (Light, 1983, p. 158).

By the early nineteenth century, Japan had moved into a period of crises: famines and peasant uprisings, presided over by an introverted and self-serving leadership. In 1853 US Admiral Perry arrived in Edo Bay to 'persuade' the shogunate to open Japanese ports to trade with the USA and other foreign powers. The neo-colonialist threat galvanized feelings of nationalism and xenophobia and precipitated a period of civil war among the shogunate. The outcome was the restoration of the Meiji imperial dynasty in 1868 by a clique of samurai and *daimyo* who were convinced that Japan needed to industrialize in order to maintain national independence.

Under the slogan 'National Wealth and Military Strength', the new élite of ex-warriors set out to modernize Japan as quickly as possible. A distinctive feature of the entire process was the very high degree of state involvement. Successive governments intervened to promote industrial development by fostering capitalist monopolies (*zaibatsu*). In many instances, whole industries were created from public funds and, once established, were sold off to private enterprise at less than cost. Because early manufacturing was motivated strongly by considerations of national security, it was iron and steel, shipbuilding and armaments that were prominent in the early phase of Japanese modernization. The latest industrial technology and equipment were bought in from overseas, and advisers (chiefly British) were bought in to supervise the initial stages of development. Meanwhile, the state indulged in high levels of expenditure on highways, port facilities, the banking system and public education in its attempt to 'buy' modernization. Similarly, the railway system was financed by the state under British direction before being sold to private enterprise.

The Japanese financed this modernization by harsh taxes on the agricultural sector. As a result, there began a sharp polarization between the urban and the rural economies, characterized by the impoverishment of large numbers of peasant farmers. Yet the more productive components of the agrarian sector were able to contribute significantly to Japanese economic growth. Improved technology, better seedstock and the use of fertilizers provided an increase of 2 per cent per annum in rice production during the last quarter of the nineteenth century and the first part of the twentieth century, thereby helping to feed the growing non-farm sector without great dependence on food imports. It is important to note that these increases in agricultural productivity were not absorbed by population growth, as has been the case for most late-comers. The Japanese demographic transition arrived later – after increases in agricultural productivity had helped to finance an emergent industrial sector but in time to provide an expanding labour force and market for industrial products.

Several other factors helped to foster rapid industrialization in Japan in the late nineteenth and early twentieth centuries. One was the cultural order that allowed the Japanese to follow government leadership and accept new ways of life: a recurring theme in modern Japanese economic history. Another was the success of educational reforms: by 1905, 95 per cent of all children of school age were receiving an elementary education. Third, Japanese sericulture (silk production) provided the basis for a lucrative export trade with which to help finance expenditure on overseas technology, materials and expertise. It has been estimated that between 1870 and 1930 the raw silk trade alone was able to finance as much as 40 per cent of Japan's entire imports of raw materials and machinery. Finally, and most important, were the benefits deriving from military aggression. Naval victories over China (1894–95) and Russia (1904–05), and the annexation of Korea (1910) not only provided expanded markets for Japanese goods in Asia but also provided indemnities from the losers (which paid for the costs of conquest) and stimulated the armaments industry, shipbuilding, and industrial technology and financial organization in general.

By the early 1900s, a broad spectrum of industries had been successfully established. Most were geared towards the domestic market in a kind of pre-emptive import-substitution strategy. The textile industry, however, had already begun to establish an export base. Unable to compete with Western nations in the production of high-quality textiles, the Japanese concentrated on the production of inexpensive goods, competing initially with Western producers for markets in Asia. Their success derived largely from labour-intensive processes in which high productivity and low wages were maintained through a combination of (1) exhortations to personal sacrifice in the cause of national independence, and (2) strict government suppression of labour unrest.

Japan advances

During the First World War, Japan was able to become a major supplier of textiles, armaments and industrial equipment on world markets, to almost double its merchant marine tonnage, and to establish a balance of payments surplus. Between 1919 and 1929 this position was consolidated, again under government sponsorship. Steel manufacturing, engineering and textiles were further developed, and aircraft and automobile industries were established. Meanwhile, Japanese innovations began to emerge, weakening the dependence on Western technology and providing an important competitive advantage. This pattern of progress was halted, however, by the stagnation of international trade which followed the stock market collapse of 1929 and the subsequent Depression. Once again, state intervention provided a critical boost. A massive devaluation of the yen in 1931 allowed Japanese producers to undersell on the world market, while a Bureau of Industrial Rationalization was set up to increase efficiency, lower costs, and weed out smaller, less profitable concerns.

Although these interventions helped to sustain Japanese industrialization and improve Japan's overall economic independence, they led directly towards crisis. Western governments – particularly the United States – began seriously to resist the purchase of Japanese goods. At home, the austerity resulting from devaluation and rationalization precipitated social and political unrest. The government response was to indulge in further military expenditure and to adopt a more aggressive territorial policy. In 1931 the Japanese army advanced into Manchuria to create a puppet state. In 1936 a military faction gained full political power and, declaring a Greater East Asian Co-Prosperity Sphere, set about full-scale war with China the following year. In 1939 Japan attacked British colonies in the Far East and by 1940 the Japanese 'had become heavily committed to an industrial empire based on war. In that year, 17 per cent of the entire national output was for war purposes' (Kornhauser, 1982, p. 119). By this time, as the rest of the world quickly realized, Japan had attained the status of an advanced industrial nation. The military leaders overplayed their hand, however, by attacking the United States. With defeat in 1945, Japanese industry lay in ruins. In 1946, output was only 30 per cent of the pre-war level; and the United States, having begun to weaken the power of the *zaibatsu* and to impose widespread social and political reforms, was set to impose punitive reparations.

Postwar reconstruction and growth

Within five years, the Japanese economy had recovered to its pre-war levels of output. Throughout the 1950s and 1960s the annual rate of growth of the economy held at around 10 per cent, compared with growth rates of around 2 per cent per annum in North America and Western Europe (Harris, 1982). After beginning the postwar period at the foot of the international manufacturing league table, Japan had risen to the top by 1963 (Fig. 5.5). By 1980, Japan had outstripped, even in *absolute* terms, all of the major industrial core countries in the production of ships, automobiles and television sets, and only the Soviet Union was producing more steel (Fig. 5.6). The Japanese, in short, have not only achieved a unique transition direct from feudalism to industrial capitalism, they have presided over a postwar 'economic miracle' of impressive dimensions.

Explanations of this 'miracle' have identified a variety of contributory factors. One of the most important, in the first instance, was the reversal of United States policy. Cold War strategy, in response to China's pursuit (in 1949) of a communist path to development, dictated that the punitive stance should be replaced by massive economic aid in order to create a bastion against the spread of communism in East Asia. The Korean conflict (1950–53) helped to reinforce this logic and at the same time stimulated the Japanese economy through US expenditure on Japanese

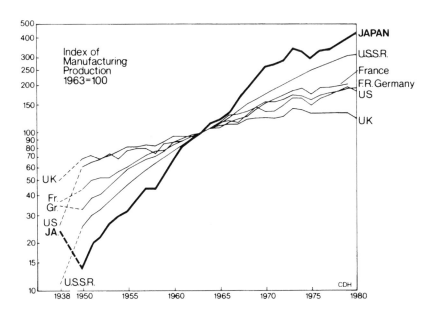

Figure 5.5
Index of manufacturing
production for
selected countries,
1950–80

Source: Harris (1982), Fig. 10,
p. 63

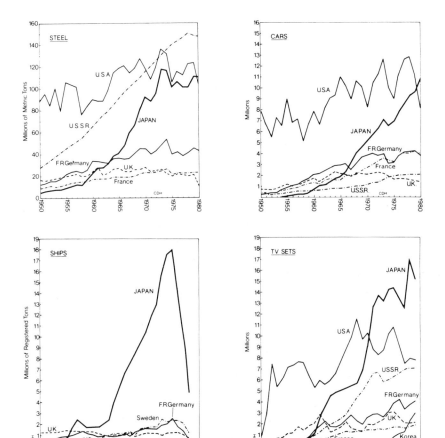

Figure 5.6
Manufacturing
production for
selected countries,
1950–1980

Source: Harris (1982),
Figs. 11–14, p. 64

supplies and military bases. Once under way, the reconstruction of the Japanese economy was able to draw on some of its previously established advantages: a well-educated, flexible, loyal and relatively cheap labour force, a large national market with good internal communications, a good geographical situation for trade within Asia, a high degree of cooperation between industry and government, and a mode of industrial organization – derived from the *zaibatsu* – big enough to compete with the multinational corporations of Western Europe and North America.

In addition, several *new* factors helped to metamorphose reconstruction into spectacular growth. These include:

1. Exceptionally high levels of personal savings (19.5 per cent of personal disposable income in 1980, compared with 4.7 per cent in the United States), which helped to fund high levels of capital investment.
2. The acquisition of new technology: between 1950 and 1969 Japan was able to acquire, for around US$1.5 billion in royalties and licences, a body of thoroughly tested US technology that had cost the United States US$20 billion *per year* in research and development (R & D). More recently, Japanese investment in domestic R & D has overtaken (in relative terms) that of both Western Europe and North America, providing important advantages in production technology and product design. Overall, Japanese investment in technology in 1980 amounted to 6 per cent of her industrial turnover, compared with 1 per cent in the United States.
3. New means of government support. On the one hand, the construction of a rigid and sophisticated system of import barriers – both tariff and non-tariff – has protected domestic markets from overseas competition. On the other hand, the growth of domestic industry was fostered by a multitude of tax concessions and by the provision of investment finance through the Japan Development Bank. Most important of all, however, was the orchestration of industrial growth by the Ministry of International Trade and Industry (MITI). In particular, MITI identified key recovery sectors (e.g. steel, shipbuilding) and potential growth sectors (e.g. automobiles, electronics, computers) and facilitated their development by providing finance, ensuring protection from foreign competition, subsidizing technological development and arranging corporate mergers.

Regional dimensions of Japanese industrialization

The pace and weight of postwar industrialization have dramatically rearranged the economic landscape that existed at the close of the Tokugawa period. In many ways, the changes wrought on the Japanese landscape parallel those which occurred in response to the industrialization of Western Europe and North America. Existing urban centres (the castle towns) grew differentially according to their adaptability as regional industrial, commercial or administrative centres; while new kinds of specialist settlements – ports, mining towns, heavy manufacturing towns and transport centres – emerged and grew rapidly to become major nodes of urbanization. Similarly, the expansion and diversification of the industrial economy imposed a progressive spatial division of labour. The logic of agglomeration and economies of scale made for regional specialization. The concentration of the silk industry in central Honshu, the gravitation of the cotton industry towards the port cities which serve the Asian market, and the location of heavy industry around deposits of raw materials in Hokkaido and northern Honshu, are important examples (for a full account of the emergent economic geography of Japan, see Kornhauser, 1982; Murata, 1980).

Within this overall transformation, one distinctive feature to emerge was the large company town. This, of course, was a reflection of the unique role of the *zaibatsu* in Japanese industrialization. The early leaders among the *zaibatsu* – Mitsui, Mitsubishi and Sumitomo – inevitably came to dominate their host cities (which included Omuta, Niihama, Nobeoka and Nagasaki); while later-established *zaibatsu*, as well as some of the corporate giants spawned by postwar growth, sponsored new company towns in newly industrializing regions: the city of Hitachi, northeast of Mito, for example.

What was most distinctive about the geography of industrialization in Japan, however, was the sheer intensity of development that was crammed into the relatively limited amount of suitable land. The 'megalopolitan' corridor between Tokyo and Kobe represents the embodiment of this development, and it has inevitably brought serious problems: crowding, congestion, environmental pollution and ground subsidence. Meanwhile, the concentration of economic activity in the Tokyo–Kobe corridor has resulted in a relative lack of development elsewhere.

Japan is thus characterized, like Europe and North America, by a *centre–periphery* pattern. In the Japanese case, the periphery consists of northern Honshu, Hokkaido, Kyushu and Shikoku. Like peripheral regions within older core nations, they have experienced the backwash effects of metropolitan development: selective out-migration, restricted investment (both public and private), and limited employment opportunities. In addition, much of the periphery has a climate that most Japanese find severe, thus compounding feelings of deprivation and remoteness.

THE EMERGENCE OF 'ORGANIZED' CAPITALISM

We have seen that the development of the industrial economies of the tri-nodal core brought with it a number of important changes in the nature of economic, social, political and cultural relations (or, put another way, changes in technology systems, regimes of accumulation and modes of regulation), each of which became woven into the urban and regional landscapes of the industrial core regions. Collectively, these changes characterize what has been called 'organized' capitalism. Its principal features include (Lash and Urry, 1987, pp. 3–4):

1. The concentration and centralization of industrial, banking and commercial capital as markets became increasingly regulated; the increased interconnectedness of finance and industry; the proliferation of cartels.
2. The emergence of extractive and manufacturing industry as the dominant economic sector.
3. The concentration of industrial capitalist relations within relatively few industrial sectors and within a small number of nation states.

4. The expansion of empires and the control by the core economies of markets and production in overseas settings.
5. The increasing separation of ownership from control and the elaboration of complex managerial hierarchies.
6. The growth of a new managerial/scientific/technological intelligentsia and of a bureaucratically employed middle class.
7. The emergence of 'Modernism' – a cultural-ideological configuration involving the glorification of science and technical rationality, a machine- and future-oriented aesthetic, and a nationalistic frame of reference.
8. The growth of collective organizations in the labour market: trade unions, employers' associations, nationally organized professions, etc.
9. Regional economic specialization.
10. The dominance of particular regions by large metropolitan areas.
11. An increasing inter-articulation between nation states and large monopolies and between collective organizations and nation states as states increasingly intervene in social conflicts and become involved in welfare state legislation.

Clearly, not all of these changes occurred simultaneously or in the same way everywhere. There are three main factors that determine the timing of these changes and the extent to which a particular national economy develops the characteristics of 'organized' capitalism:

- First is the *point in history at which it begins to industrialize*: the earlier this is, the less 'organized' capitalism will be, because later industrializers need to begin at higher levels of concentration and centralization of capital in order to compete with established industrial economies.
- Second is the *extent to which pre-industrial institutions survive* into the capitalist period: 'Britain and Germany became more highly organized capitalist societies than France and the United States: this is because the former two nations did not experience a "bourgeois revolution" and as a result, guilds, corporate local government, and merchant, professional, aristocratic, university and church bodies remained relatively intact' (Lash and Urry, 1987, p. 5).
- Third is the *size of the country*: for the industry of small countries to compete internationally, resources had to be channelled into relatively few sectors and firms. This, in turn, meant that coordination between state and industry was facilitated.

Lash and Urry argue that German capitalism became 'organized' during the last quarter of the nineteenth century, but that American capitalism was organized fairly early on at the top (e.g. through the concentration of industry, increasing inter-articulation of banks, industry and the state, and the formation of cartels) but very late and only partially at the bottom (e.g. the development of national trade union organizations, working-class political parties, and the welfare state). British capitalism, in contrast, was organized rather early at the bottom but late at the top; while French capitalism only came to be organized, at both top and bottom, after the Second World War.

The characteristics of organized capitalism thus came to be woven into the urban and regional landscapes of the core regions rather unevenly. Meanwhile, they came to represent not just a distinctive set of economic, social, political and cultural relations but also to represent the context – the preconditions – for further transformations of capitalism and the new economic landscapes to emerge with the onset of 'advanced' capitalism (Chapter 6).

The changing role of the state

It is no accident that the rise of competitive capitalism and the evolution of industrial economies took place side by side with the emergence of the modern nation state. Within Europe, it was the system of nation-states, once established in place of the earlier dynastic kingdoms and empires, that fostered the economic, social and political organization required by the industrial revolution. At the same time, strong competitiveness between nation states provided a strong incentive to technological innovation. It is important to bear in mind, however, that few nation states were 'natural' entities developed from distinctive cultural or philosophical bases. Rather, they were *constructed* in order to clothe, and enclose, the developing political economy of industrial capitalism. It follows that the process of building nation states involved the resolution of successive crises that arose from the interaction of territory, economy, culture and government.

One series of crises arose from the long struggle to make state boundaries fit populations with feelings of (or at least the potential for) common identity. This struggle involved (1) states attempting to build 'nations' from a diversity of peoples, and (2) peoples with a common identity, 'nations', attempting to create an autonomous state (Taylor, 1991). The former has often involved the penetration by powerful regions or groups of neighbouring territories with different cultures, languages and economic institutions. As a result, many nation states came into being with inbuilt centre–periphery contrasts, with socio-political tensions compounding economic differentials. We examine the reaction to these developments in Chapter 12.

Another series of crises arose from the increasing degree of organization required by capitalism. The evolution of the industrial core regions posed a succession of problems which resulted in *more* state intervention in a *greater variety* of fields. The initial advantage gained by British manufacturers with the advent of the industrial revolution soon prompted businessmen elsewhere to realize that the old doctrine of *laissez-faire* and free trade only served the interests of the dominant economy. As a result, governments everywhere were looked to as protectors – through tariffs and quotas – against low-priced British imports.

Meanwhile, the coming of the railway involved the state in another sphere – investment in infrastructure – because of the railways' economic and strategic importance. Problems of public health, working conditions, housing and civil disorder induced further kinds of state involvement, as did the need to provide a stable price system for the successful operation of private industry, the need to manage the cyclical fluctuations of the industrial economy, and the need to improve the quality of the work-force and its managers through formal education. Of course, the *capacity* of governments to intervene in all these matters was dependent on economic growth. Nevertheless, it was by no means the wealthiest economies which led the way in terms of state activity, since public expenditure is mediated through the complex arena of politics.

In detail, then, the development of state functions has been complex. It is possible, however, to identify two major trends in the nature of the changes which have taken place:

1. The *centralization* of the functions of states, whereby local and regional activities have been rationalized into centralized national bureaucracies as the organization of government has attempted to keep up with the changing scale of economic organization. With increasingly powerful central bureaucracies, the power of politicians at both local and national levels has been constrained and this, in turn, has led to crises in the legitimacy of political institutions. One response has been the intensification of demands for the devolution of power by the representatives of communities in peripheral areas. Another has been the growth of forms of direct action in the shape of grass-roots pressure groups. These are both examined in more detail in Chapter 12.

2. The dramatic expansion of the *public economy* as governments have become increasingly drawn into the creation of welfare states. In most core countries, the public economy channels a vast amount of resources into everything from defence, health, education and income security to transport, infrastructural development and industrial investment and conditions the whole of the private economy through everything from price guarantees and labour laws to tax structures and import tariffs. Italy, Sweden and the United Kingdom had by the 1970s reached the stage where more than 50 per cent of their GNP was committed to public expenditure, and most of the other industrial core nations are now approaching this figure, having almost doubled their share of their respective national economies since the 1950s.

Indeed, the sheer magnitude of the public economy has come to blur the boundary between the public and the private sectors. In addition to the state ownership of key industries that is common in much of Europe, many governments have been impelled to intervene – for a variety of reasons – to prevent the collapse of private business corporations (examples have included the US government's efforts to rescue Lockheed and Chrysler in the 1970s, and the UK government's efforts in the 1980s to sustain British Leyland). Most important of all, governments everywhere have become the largest single consumer of the goods produced by private sector enterprise. In short, the public economy is pervasive in its effects on economic well-being.

The geography of the public economy

Nearly all of this public sector activity has a geographical expression. One of the most obvious examples is the deliberate bias of regional policy and planning, which we examine in Chapter 12. In this section, we emphasize the spatial bias – often unintentional – that results from other aspects of the public economy.

The geography of public finance is a complex subject, and it is hazardous to attempt detailed comparisons between nations. It is possible, however, to identify major categories of activity and to illustrate their spatial implications with specific examples. In this context, it is convenient to recognize four major categories of government *expenditure*:

1. The salaries of central government employees, including clerks, bureaucrats and other workers in the armed forces, education, public health, the nationalized industries, the police and the courts. Expenditure on these salaries is of course localized in capital cities. In Washington, DC, for example, almost one-third of all earnings come from federal employment. In addition, government salaries can have an important impact in small- or medium-sized cities that have been selected for specific functions – defence installations, for example, or decentralized branches of the bureaucracy (as in the UK government's relocation of social security offices from London to Newcastle upon Tyne).

2. Transfer payments to particular population groups (e.g. the elderly, the unemployed, families with dependent children) and particular industries (e.g. agricultural subsidies and guaranteed prices). These expenditures involve complex flows of monies and are geographically localized only in as much as the 'target' populations and industries are localized.

3. Purchasing and sub-contracting from businesses in the private sector. This includes a wide range of items – buildings, roads, dams, power stations, military equipment, office equipment and publishing, for example – which can make for highly localized impacts. Defence expenditure has been researched in some detail, and it provides good examples of the kind of bias that can result from government purchasing. It seems that the employment generated by defence expenditure generally tends to be sufficiently localized (because it is concentrated in the hands of just a few giant corporations – e.g. General Dynamics, McDonnel Douglas, United Technologies, Boeing, General Electric, Lockheed and Hughes Aircraft in the USA) as to create significant multiplier effects. But the resultant spatial bias seems to bear no consistent relationship to centre–periphery patterns or to dimensions of economic geography. In the UK, defence expenditures after the Second World War came to sustain around 1.25 million jobs – about 5 per cent of all employment – and pay for around 25 per cent of all the R & D activity undertaken in the country. Both the direct spin-off from the employment and the indirect 'seed bed' effects of R & D tended to reinforce the centre–periphery structure of the UK economy, with the Southeast, the Southwest and the East Midlands benefiting most while the likes of Yorkshire and Northern Ireland were under-represented both in terms of job creation and R & D activity. In the USA, on the other hand, defence procurements tended to be biased away from the industrial core. Taking R & D activity as well as prime defence contracts into account, and allowing for the effects of sub-contracting, it was California that benefited most from Department of Defense spending, together with Washington (state), Massachusetts, Maryland, Arizona and Connecticut.

4. Local government expenditure. In many core countries, this soon came to approach the levels of expenditure by central governments (though a large portion of local expenditure is always dependent on revenues provided by central governments in the form of grants and revenue-sharing funds). What is most striking about local government expenditure is that, after fulfilling their statutory obligations, local authorities vary a great deal both in the amount they spend and the categories of their expenditure. This reflects a complex interaction between local resources, local needs, and the local political climate.

In order to gauge the net impact of these expenditures, we have to set them against the geography of *taxation*. Such an exercise is very difficult to achieve at any level of detail, but we can illustrate the kind of biases that can emerge by reference to federal taxes in the USA. Around 40 per cent of all federal revenues are derived from personal income taxes, with another 25 per cent coming from taxes on pension trusts and a further 15 per cent from taxes on corporate profits. *To a large extent, therefore, the geography of federal revenues reflects the geography of income and economic health.*

Yet the *structure* of the tax system can have less straightforward geographical implications. Indeed, the *President's National Urban Policy Report* for 1980 concluded that 'the tax system is perhaps the most pervasive federal influence on the patterns of economic development. Taxes influence the relative cost to businesses of new capital versus existing machinery, of low-wage or lower-skilled workers versus others, and of land in growing areas which is rising in value as opposed to land whose value may be falling' (US Department of Housing and Urban Development, 1982, p. 318). The tax breaks offered under the Investment Tax Credits scheme (which are geared to encourage business investment in new equipment and machinery), for example, tend to benefit growth industries in growth areas; as do the more generous rates of depreciation allowed for new industrial and commercial plants. On the other hand, it should be acknowledged that studies of industrial location have established fairly clearly that tax differences between states are not a significant factor in inducing industrial *relocation*.

Although it is not possible to quantify the local effect of these structural characteristics of the tax system, it is possible to specify the magnitude of the *net* flows of monies between the federal government and individual states. Inter-state differentials in such flows have tended to diminish during the postwar period as the federal tax system has become more uniformly progressive and as state variations in per capita incomes have narrowed. Differences between states remain substantial, however, with a marked disadvantage of states in the Manufacturing Belt and the northeast. This has been interpreted by some as a product of the manipulation of the political 'pork barrel'. Certainly the traditional dominance of the Democrats in the Deep South ensured the seniority of its federal senators and thus the chairmanship of committees of Congress controlling the allocation and distribution of federal funds. Moreover, members of Congress do seek positions on the committees of Congress that control the allocation and distribution of federal funds; they do seek membership on the committees and sub-committees most relevant to the needs of their constituents; and many decisions are blatantly taken for electoral reasons. But political and electoral variables are weak predictors of the overall geography of federal outlays.

What is important about patterns of federal expenditure in the present context is that the *marginal impact* of the flow of federal funds seems to have been much greater in the South and West. By improving the infrastructure of communications, transportation, sewage and facilities and energy, the federal government helped to establish the *preconditions* for the development of new industries in the Sunbelt. By direct and indirect investment in electronics research, semiconductors, computers, aeronautics and scientific instruments in the South and West, the federal government enabled the Sunbelt to capture some of the most dynamic activities of the advanced capitalist economy.

Increasing global interdependence

Although the industrial core regions have been the focus of this chapter, the historical process of industrialization must be seen in a global perspective. Quite simply, the ascent of the industrial core regions could not have taken place without the foodstuffs, raw materials and markets provided by the rest of the world. In order to ensure the availability of the produce, materials and markets on which they were increasingly dependent, the industrial core nations vigorously pursued a second phase of colonialism and imperialism, creating a series of 'trading empires'.

As soon as the industrial revolution had gathered momentum, European nation states embarked on the inland penetration of midcontinental grassland zones in order to exploit them for grain or stock production – although the detailed pattern and timing of this exploitation were heavily conditioned by innovations such as the railways, barbed wire and refrigeration. Hence the settlement of the prairies and the pampas in the Americas, the veld in Africa, and the Murray-Darling and Canterbury Plains in Australasia. The emigration which fuelled this colonization was itself a major factor in the economic development of the core regions, siphoning off the 'surplus' population that was generated by demographic transition and swollen by the rationalization of rural economies. Meanwhile, as the demand for tropical plantation products increased, most of the tropical world came under the political control – direct or indirect – of one or another of the industrial core nations.

The outcome of this expansionism was that the colonies and client states of the industrial core began to specialize in the production of those foodstuffs and raw materials (1) for which there was an established demand in the industrial core, and (2) for which they held a comparative advantage. *This specialization, in turn, established a complex pattern of interdependent development that was articulated, above all, in patterns of international trade.* From the start, however, this expanded and more closely integrated international system was unevenly balanced. On the one hand, the influence of the core countries on the cultural and institutional organization of the peripheral countries has moulded their economic organization to fit core-oriented needs and core-inspired philosophies of 'development' (Chapter 9). On the other, a variety of barriers and imperfections have blunted the effectiveness of international trade as an 'engine of growth'. While the economies of *some* of the semi-peripheral and peripheral nations were able to achieve considerable momentum as a result of the stimulus provided by rapidly increasing levels of demand transmitted from the industrial core regions, many were not. In very general terms, the type and profitability of activity in semi-peripheral and peripheral nations were determined by effective distance from the industrial core regions which represented their major market. Within the resultant zones of specialization, the beneficial effects of trade were conditioned by a variety of factors, including variations in climate, topography, pre-existing systems of agriculture, and population densities. In practice:

> The trade impetus to growth was . . . immensely important for Argentina and Uruguay in Latin America, South Africa and Zimbabwe (formerly Southern Rhodesia) in Africa, Australia and New Zealand, and, to a lesser extent, in Sri Lanka (formerly Ceylon). Elsewhere, there was a significant impact, *but this was inadequate to get sustained development going*, for example on the west coast of Africa. For countries such as India, Pakistan, Bangladesh, Iraq and Iran, *the export trades were too small relative to the total population* to provide much impetus for development, except in very restrict areas.
>
> (Chisholm, 1982, p. 88, emphasis added)

In short, the once-and-for-all benefits of specialization and international trade enabled some regions and some nations to ascend within the world-system while enhancing the position of those at the top. For the rest, the subsequent prospects of economic growth through trade have been further diminished by the in-built differential between themselves and the better-off in terms of access to capital, since this translates into a differential in the use of technology; and this, in turn, keeps productivity at relatively low levels. As a result, the amount of labour power required to produce a given quantity of exports from the industrial core will generally be much less than that needed to produce an equivalent value of exports from peripheral countries. From this point of view, the international trade system is characterized by *unequal exchange*.

Attempts to short-circuit this built-in handicap by borrowing capital with which to purchase new (but not always appropriate) technology have almost always resulted in a *debt trap* as compounded interest on loans has outpaced increases in the rate of productivity.

In addition, many peripheral countries have been affected by another built-in handicap: the differential elasticities of demand for their products *vis à vis* those of their trading partners in the industrial core. In general, elasticities of demand for the primary products that have become the staples of the periphery are low, so that even fairly large price reductions in overseas markets elicit only a modest rise in demand. Similarly, demand for these products will only increase very slightly in response to increases in the purchasing power of consumers in the core nations. Conversely, elasticities of demand for manufactured good are generally very high. The net result is that the *terms of trade* have tended to work to the cumulative advantage of the industrial core.

In other words, although the world economy has come to be characterized by interdependent relationships as a result of the spatial division of labour, it is the periphery that has carried the burden of *dependency*, and this is one of the themes that we explore in detail in Part 3.

PRINCIPLES OF ECONOMIC GEOGRAPHY: LESSONS FROM THE INDUSTRIAL ERA

It will be clear from the case studies in this chapter that the increasing complexity of economic organization and spatial change make it difficult to generalize about organizing principles or characteristic features. What we can do, however, is to emphasize once more the recurring importance of certain elements and the emergence of others in interacting to produce the economic landscapes of the industrial core regions. Among those which carried over from previous eras are:

- the distribution of natural resources – with iron ore and coal now exercising a central role;
- demographic change – particularly (1) the timing of the demographic transition in relation to industrialization and (2) the role of large-scale migrations in relation to changing labour markets;
- technological change – including improvements and innovations in transport and communications;
- colonialism and territorial expansion as responses to the law of diminishing returns;
- changes in institutional and socio-political settings;
- changes in the spatial distribution of investment in response to the shifting comparative advantages enjoyed by producers in different areas;
- import substitution as a mechanism of ascent within the world economy;
- militarism and geopolitical change as a mechanism of ascent within the world economy.

In addition, the industrial era saw the emergence of several new dimensions of spatial-economic organization and the increased prominence of others:

- the extension of the world economy to a global scale with a corresponding extension of the spatial division of labour and the consequent intensification of the interdependencies between core, semi-periphery and periphery;
- the replacement of 'liberal' merchant capitalism with a competitive and, later, an increasingly organized form of industrial capitalism characterized by distinct, specialized regional economies organized around growing urban centres;
- the eclipse of the European core of the world economy by the ascent of the United States and Japan;
- the temporary partitioning of the world economy with the emergence of a Soviet world-empire;
- the emergence of distinctive centre–periphery contrasts within the industrial core territories of the world economy;
- the agglomeration of industrial activity as a result of the logic of economies of scale and the multiplier process;
- the modification of urban systems by the addition of new kinds of towns and cities – mining towns, heavy manufacturing centres, power centres and transport nodes – and the rapid growth of larger pre-industrial cities as they benefited disproportionately (because of their established markets,

entrepreneurship, trading links and commercial infrastructure) from the various growth impulses that characterize industrialization;

- the imprint of cyclical fluctuations in the pace and nature of economic activity;
- the 'differential of contemporaneousness' in regional economic development: a phenomenon linked to the process of technological diffusion and changing technology systems;
- the adaptation of private firms to the changing opportunities and constraints of different technology systems, resulting in evolving 'regimes of accumulation', from simple manufacturing, through machinofacture, to Fordism;
- the adaptation of wider society to these changing regimes of accumulation, resulting in another evolutionary process: that of changing modes of regulation, in which the increasing intervention of governments in economic development was the single most important development; and
- the emergence of an 'organized' form of global capitalism, founded on the power and authority of independent nation states, characterized by a sophisticated interdependence of firms, industries, regions, and governments, and forming the basis for core–periphery relationships at various geographic scales.

Key Sources and Suggested Reading

Borchert, J. 1967. American metropolitan evolution. *Geographical Review*, **57**, 301–32.

Bradshaw, M. 1991. *The Soviet Union: A New Regional Geography?* London: Belhaven.

Castells, M. 1996. *The Information Age: Economy, Society and Culture.* Volume 1, *The Rise of the Network Society.* Oxford: Blackwell.

Chisholm, M. 1982. *Modern World Development.* Totowa, NJ: Barnes and Noble.

Dunford, M. 1990. Theories of regulation. *Society and Space*, **8**, 297–321.

Friedmann, J. and Weaver, C. 1979. *Territory and Function.* Berkeley: University of California Press.

Hamilton, F. E. I. 1978. Multinational enterprise and the European Economic Community. In F. E. I. Hamilton (ed.), *Industrial Change: International Experience and Public Policy.* London: Longman, 24–41.

Harris, C. 1982. The urban and industrial transformation of Japan. *Geographical Review*, **72**, 50–89.

Hugill, P. 1993. *World Trade Since 1431: Geography, Technology, and Capitalism.* Baltimore: Johns Hopkins University Press.

Johnston, R. J. 1980. *City and Society.* Harmondsworth: Penguin.

Knox, P. L. 1984. *The Geography of Western Europe: A Socio-Economic Survey.* Beckenham: Croom Helm.

Knox, P. et al., 1988. *The US: A Contemporary Human Geography.* London: Longman.

Kornhauser, D. 1982. *Japan: Geographical Background to Urban and Industrial Development.* 2nd edn. London: Longman.

Lash, S. and Urry, J. 1987. *The End of Organized Capitalism.* Cambridge: Polity Press.

Light, I. 1983. *Cities in World Perspective.* New York: Macmillan.

Massey, D. and Meegan, R. 1989. Spatial divisions of labour in Britain. In D. Gregory and R. Walford (eds), *Horizons in Human Geography.* Totowa, NJ: Barnes and Noble, 244–57.

Meyer, D. R. 1983. Emergence of the manufacturing belt: an interpretation. *Journal of Historical Geography*, **9**, 145–74.

Murata, K. 1980. *An Industrial Geography of Japan*. London: Bell and Hyman.

Orwell, G. 1962. *The Road to Wigan Pier*. Harmondsworth: Penguin; first published by Victor Gollancz, 1937.

Perez, C. 1983. Structural change and assimilation of new technologies in the economic social systems, *Futures*, October, 357–75.

Pollard, S. 1981. *Peaceful Conquest: The Industrialization of Europe, 1760–1970*. Oxford: Oxford University Press.

Rokkan, S. 1980. Territories, centres and peripheries. In J. Gottmann (ed.), *Centre and Periphery*. London: Sage, 163–204.

Shaw, D. 1988. *Russia in the Modern World*. Oxford: Blackwell.

Smith, D. 1979. *Where the Grass is Greener*. Harmondsworth: Penguin.

Sukhotin, I. 1994. Stabilization of the economy and social contrasts. *Problems of Economic Transition*, November 1994, 44–61.

Taylor, P. 1991. *Political Geography: World-Economy, Nation-State, and Locality*. 2nd edn. London: Longman.

Turnock, D. 1984. Postwar studies on the human geography of Eastern Europe. *Progress in Human Geography*, **8**, 315–45.

Turnock, D. 1988. *The Human Geography of Eastern Europe*. Beckenham: Croom Helm.

UN Economic Commission for Europe 1972. *Economic Survey of Europe in 1971*. United Nations: New York.

UN Economic Commission for Europe 1979. *The European Economy in 1978*. United Nations: New York.

US Department of Housing and Urban Development 1982. *The President's National Urban Policy Report 1982*. Washington, DC: US Government Printing Office.

Wigen, K. 1992. The geographic imagination in early modern Japanese history. *Journal of Asian Studies*, **51**, 3–29.

CHAPTER 6

Picture credit: Paul Knox

THE GLOBALIZATION OF PRODUCTION SYSTEMS

After the Second World War, the economies of the industrial core regions began to enter a substantially different phase in terms of *what* they produced, *how* they produced it, and *where* they produced it. This phase is sometimes referred to as 'advanced capitalism'. It typically involves a combination of ingredients, 'most especially: the accelerated internationalization of economic processes; a frenetic international financial system; the use of new information technologies; new kinds of production; different modes of state intervention; and the increasing involvement of culture as a factor in and of production' (Thrift, 1995, p. 19). It evolved in response to the increasing inflexibility of the old system of Fordist industrial capitalism. Faced with the saturation of domestic consumer markets, increasing overseas competition, increasing costs of unionized labour and of governmental welfare provision, the industrial corporations of the core economies began to pursue new and more flexible strategies. They reorganized themselves, redeployed their operations, and revised their relationships with labour unions and governments. The result has been the deindustrialization of the core economies, the industrialization of certain semi-peripheral countries, and the expansion of business and financial services on a global scale.

THE TRANSITION TO ADVANCED CAPITALISM

The shift to advanced capitalism has been a result of the cumulative interaction of several processes. As in all of the previous major economic transitions we have described, the importance of these processes has been revealed only after a period of *crisis* for the old order. In this case, the crisis was thrown into focus by a phase of stagflation and intensified by a sudden increase in the price of oil. Meanwhile, just as in previous major economic transitions, the processes themselves have drawn on a number of *preconditions* developed during the preceding era (i.e. during 'organized', Fordist industrial capitalism). Here we can recognize three main factors:

(1) new, enabling technologies in transport and telecommunications, (2) changing patterns of demand and consumption, and (3) corporate integration.

Prelude: the crisis of Fordism

The crisis for the Fordist regime of industrial capitalism emerged abruptly in the early 1970s, throwing into reverse the postwar industrial boom. This reversal is clearly illustrated by the performance of the US economy. In overall terms, the US economy performed exceptionally well from the late 1940s right through to the early 1970s. Real disposable income per capita rose from just over US$2200 (in constant 1972 dollars) in 1947 to over US$3800 in 1972. The 1960s were particularly prosperous, with economic growth averaging over 4 per cent per year, thus expanding the GNP by 50 per cent over the decade. Meanwhile, the average family obtained a real increase of over 30 per cent in its disposable income: these were the years of J. K. Galbraith's 'affluent society' (Galbraith, 1977).

After the early 1970s, however, US economic growth averaged only 2.2 per cent, while productivity in the private business sector, having increased at around 3.3 per cent per year in the 1960s, fell away to 1.3 per cent per year in the 1970s. 'By 1979, the typical family with a $20 000 income had only 7 per cent more real purchasing power than it had a full decade earlier. The years had brought a mere $25 more per week in purchasing power for the average family' (Bluestone and Harrison, 1982, p. 4). Unemployment, having remained steady at round 4.5 per cent until the early 1970s, almost doubled over the next 5 years, levelling off at around 10 per cent by the mid-1980s. The rate of inflation doubled from around 2.5 per cent per year in the 1960s to over 5 per cent per year in the mid-1980s. Meanwhile, in 1971, the US economy had moved, for the first time this century, into a negative trade balance with the rest of the world: a performance repeated in 19 of the next 21 years.

The 'system shock' precipitated by the rise in oil prices in 1973 as a result of the OPEC cartel has been widely cited as a major cause of this downturn (in 1973–74, petroleum prices quadrupled as a result of the cartel's actions). The evidence is inconclusive, however. Similarly, it has been difficult to establish the 'guilt' of other popular scapegoats, such as the role of labour unions in obtaining wage increases in excess of productivity. Rather, the crisis of the 1970s must be seen as the product of a conjunction of trends whose origins can be traced to the 1960s or before. Hamilton (1984) identified several such trends:

1. The *stagflation* phase of economic long waves. With the downswing of the Kuznets cycle (see p. 13) there was a slowing down of economic growth and a steady fall in profits, particularly in the industrial core countries of the OECD. This was associated with falling levels of demand for capital goods, particularly transport, building, mining and factory equipment (e.g. ships, vehicles, machinery, machine tools) and, hence, steel. Overall, rates of growth in the OECD countries fell from an annual average of 5 to 6 per cent between 1963 and 1973 to around 2.5 per cent between 1973 and 1978 and less than 1 per cent between 1979 and 1982.

At the same time, rising levels of inflation, associated with the upswing of the Kondratiev cycle (see p. 12), generally served to reduce profits and hampered capital accumulation. This resulted in greater dependency on financing investment via the banking sector. This, in turn, meant high interest rates that retarded technological investment and so hindered competitiveness. Inflation also raised labour costs, thus increasing the urgency of technological investment (particularly automation) at a time when capital was expensive. The result was the widespread depression of both capital-intensive industries (e.g. steel, shipbuilding, vehicles, appliances) *and* labour-intensive industries (e.g. textiles, clothing, footwear).

2. Increased international monetary instability, which took two major forms:

   (a) overvaluation of exchange rates as a result of the transition (in the early 1970s) from fixed exchange rates to floating exchange rates. Where currencies were overvalued (e.g. the currencies of oil and/or gas producers such as the Netherlands, Norway and the United Kingdom, and, more recently, the US dollar), the loss of international competitiveness resulted in

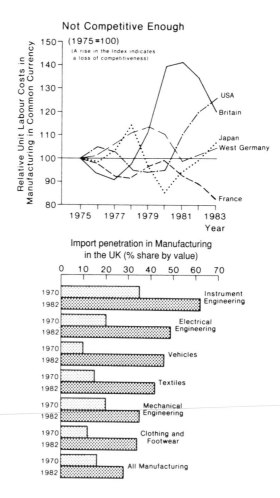

Figure 6.1

Forces in the deindustrialization of the United Kingdom: dramatic loss of competitiveness (1978–83) and consequent import penetration, converting the country from a net exporter to a net importer of manufactures.

Source: Hamilton (1984), Fig. 2, p. 352

import penetration and a consequent decline in industrial capacity
(Fig. 6.1). In the United States, import penetration in clothing and textiles
increased from 34 per cent in 1980 to 55 per cent in 1986; import
penetration in shoes increased from 50 per cent to 81 per cent, in
computers from 7 to 25 per cent; and in automobiles from 35 to 40 per
cent.

(b) problems of indebtedness among NICs and some underdeveloped countries
following massive borrowing from the 'petrodollar' surpluses created in the
OPEC countries (see p. 49). In addition to the international financial
instability associated with uncertainty caused by debt rescheduling and
fears over national bankruptcies, this created a strong incentive for NICs
and underdeveloped countries to increase their exports – of cheap
manufactured goods as well as traditional staples – to the core regions in
order to obtain the necessary foreign exchange. This, in turn, increased the
competitive pressure on the labour-intensive sectors of the core economies.

3. The strengthening, throughout the 1960s, of social values associated with
social welfare provision (e.g. retirement pensions, health care, anti-poverty
programmes) and environmental protection. Although this created new
markets for some products and services, it also raised some industrial costs and
contributed to a higher tax burden on both consumers and producers.

4. The introduction of innovations and technological changes in response to
escalating energy and labour costs created feedback effects that depressed
demand in 'traditional' industrial activities. Energy-saving designs in transport
and heating, for example, reduced the demand for steel; while innovations in
microelectronics reduced the demand for electro-mechanical products.

5. A resurgence of political volatility that reduced the extent of stable business
settings and so inhibited several dimensions of world trade, including East–
West trade (until 1989) and trade involving much of Central America, the
Middle East and Southeast Asia.

Towards flexibility: enabling technologies and economies of scope

The resolution of the crisis of Fordism and the emergence of advanced capitalism
have been made possible, in part, by the availability of 'permissive' technology of
two kinds:

1. *Circulation*: improvement in transport and communications technologies
(wide-bodied cargo jets, containerization, fax, electronic mail networks,
computerized business systems, communications satellites, optical fibre
networks, etc.) that have reduced the time and costs of circulation, bringing a
wider geographic market within the scope of an increasing range of business
activities. This 'global reach' has also been advanced through the economic
development of peripheral areas and the standardization of products across
cultures (the latter itself being a function of the development of
communications media).

import penetration
This describes the result of a significant share of domestic markets for a particular product or service being lost by domestic firms in the face of competition from foreign sources.

191

deskilling
A reduction in the range and level of skills within a local labour market that is the result of two trends: the increased mechanization and computerization of production processes (including management and management-support functions), and the geographic consolidation and localization of higher-skilled activities in world cities, major control centres, and centres of innovation (which leaves other labour markets with a preponderance of routine jobs in local offices and branch plants).

2. *Production*: improvements in production technologies (electronically controlled assembly lines, automated machine tools, computerized sewing systems, robotics, etc.) have allowed for a finer degree of specialization in many production processes, and facilitated a routinization of many operations. This, in turn, has led to the **deskilling** of many production systems while at the same time increasing the *separability* (and therefore the spatial fragmentation) of their constituent parts. This has made it easier for managers to take advantage of new sources of cheaper and less militant labour. Advances in the manufacture and use of synthetic materials have also extended the locational capability of many industries, since raw materials have traditionally been the most restrictive of all factors of production.

The impact of these enabling technologies is difficult to overstate. During the past two decades the global network of computers, telephones, and televisions has increased its information-carrying capacity a million times over. Telecommunications satellites allow for hundreds of thousands of simultaneous telephone conversations and fax messages. Fig. 6.2 shows the tremendous increase that has occurred in the flow of long-distance calls within the United States and international messages originating from the United States. International air cargo increased from 9.3 million tons of freight and 21.5 ton-kilometres in 1976 to 14.7 million tons of freight and more than 43 billion ton-kilometres just 10 years later. On some routes, air cargo now accounts for between 15 and 25 per cent of all trade, by value. The emergence of a tri-polar world economy has encouraged the development of round-the-world shipping services that could link the three main cargo-generating zones of Europe, North America and East Asia. Thus there emerged, in the mid-1980s, the flexible, intermodal, and global services offered by several container lines: Evergreen (Taiwan), United States Lines (US), Senator (Germany) and Maersk (Denmark). In addition, some shipping lines such as American President Line and P&O Line have acquired inland trucking operations,

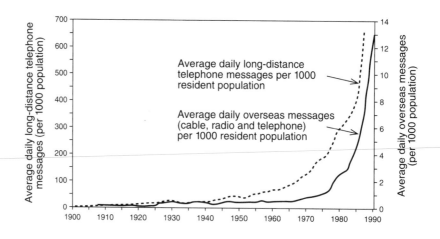

Figure 6.2

Average long-distance and overseas messages, United States, 1900–90 and railway companies have become involved in ocean-borne transport (e.g. the merger of CSX and Sea-Land in the United States).

The 1970s' technological divide

All of the key information technologies that underpin the emerging informational economy of the 1990s are part of a massive diffusion of commercial and civilian applications since the 1970s. The giant leap forward in the diffusion of microelectronics in all machines came in 1971 with the invention by an Intel engineer, Ted Hoff, of the microprocessor, that is the computer on a chip. The advent of the microprocessor turned the electronics world, and indeed the world itself, upside down when it began to diffuse in the mid-1970s. The microcomputer was invented in 1975 and the first successful commercial product, Apple II, was introduced in April 1977, around the same date that Microsoft started to produce operating systems for microcomputers, The Xerox Alto, the matrix of many software technologies for 1990s' personal computers, was developed at PARC labs in Palo Alto in 1973. The first industrial electronic switch appeared in 1969, and digital switching was developed in the mid-1970s and commercially diffused in 1977. Optic fibre was first industrially produced by Corning Glass in the early 1970s. Also by the mid-1970s, Sony started to produce VCR machines commercially, on the basis of 1960s' discoveries in America and England that never reached mass production. And last, but not least, it was in 1969 that the US Defense Department's Advanced Research Projects Agency (ARPA) set up a new, revolutionary electronic communication network, that would grow during the 1970s to become the current Internet. It was greatly helped by the invention by Cerf and Kahn in 1974 of TCP/IP, the interconnection network protocol that ushered in 'gateway' technology, allowing different types of network to be connected. The modem was invented by two Chicago students, Ward Christensen and Randy Suess, in 1978, when they were trying to find a system to transfer microcomputer programs to each other through the telephone to avoid travelling in the Chicago winter between their distant locations.

(Castells, 1996, pp. 42, 44, 47, and 353)

We should also note that the introduction of new circulation and production technologies has in many cases created a powerful second-order effect: economies of scope, the capacity to provide entirely new products and/or services through the same production or service network. Thus, for example, the computerized records developed by airlines have lent themselves to an increased scope of business that includes hotel reservations and rental cars. Similarly, the credit records of major retailing firms have provided a base for them to exploit economies of scope. In the United States, Sears, exploiting its access to over 40 million credit customers, added insurance services (through Allstate), investment services (through Dean Witter) and (for a time) real estate services to its retailing and mail-order activities.

Another important aspect of enabling technologies concerns their capacity for flexibility during periods of intense competition or changing market conditions. This example is from the insurance industry in the United States:

> In the mid-1960s, when the insurance industry was stable and heavily regulated, insurance companies automated their back-room activities to obtain dramatic gains in productivity in handling premium billings and collections. As wildly fluctuating interest rates hit the industry in the mid-1970s, companies had to change their product rapidly to attract premiums and to offset the effects of customers borrowing against their policies at low interest rates. *Only those companies that had flexibly designed computer and control systems could deploy their products rapidly enough to obtain a competitive edge* .... Smaller companies could not afford the huge initial costs of needed technologies and sold out or merged with larger companies who could benefit from their distribution networks. A flexibly automated back room became a key element in survival and competitive success.
>
> (Quinn, 1987, p. 137, emphasis added)

Finally, we must recognize the *differential of contemporaneousness* (see also p. 152) in the broad spatial impact of these enabling technologies. Following Sachar and Öberg (1990), we can see that new technologies have different implications for different regions within the world economy:

- In the core countries, high technology creates new jobs, particularly in business and financial services. It creates new products, facilitates new production and distribution processes and new forms of corporate organization, but reduces the need for employment in manufacturing.
- In semi-peripheral countries, high technology brings an increase in manufacturing employment, increases in productivity, and an overall improvement in their competitiveness in the international economy.
- In peripheral countries, new technologies are often too expensive to acquire or deploy. As a result there is a relative decline in both productivity and international competitiveness. To the extent that new technologies are deployed, their main effect is to displace jobs in labour-intensive sectors, thus adding to sprawling **informal** urban economies and putting pressure on the public sector to absorb labour in government-sponsored jobs.

Changing patterns of demand and consumption

informal sector
Economic activities that are undertaken without any formal systems of regulation or remuneration. In addition to domestic labour, these activities include strictly illegal activities such as drug-peddling and prostitution as well as a wide variety of legal activities such as casual labour in construction crews, on docks or on farms; domestic piece work; street trading; scavenging; and providing personal services such as shoe-shining or letter-writing.

Shifting patterns of consumer demand have also been an important precondition for the evolution of advanced capitalism. Within core countries, the Fordist mode of industrial capitalism, based as it was on mass production coupled with mass consumption, began to be the victim of its own success as mass markets for many of its staple products – from cars to refrigerators to leisure wear – came close to being saturated. As the affluent societies of the core countries came to fulfil more and more of their wants, market saturation could only be avoided by skilful marketing campaigns and continuous modifications in products and packaging. Even so, mass-produced goods were less and less effective in satisfying one of their main roles in affluent societies: that of **positional goods**, possessions that serve as meaures of socio-economic status. As more and more people were able to acquire mass-produced positional goods – a nice home, a new car, a television, audio equipment, etc. – so more people sought the distinction of custom-made, stylish, high-design and fashionable products. Social distinctions, previously marked by the ownership of a basic set of consumer goods on a sliding scale of size/quality, now had to be established via the symbolism of ensembles of *aestheticized commodities*. The problem for producers following the Fordist mode of accumulation was not just that their mass-produced products were rapidly losing their appeal to the most affluent consumers but also that their strategies and processes of production were too inflexible to cater to the many different (and rapidly changing) market niches for aestheticized commodities. The result was that many firms began to adopt more flexible forms of production, using new production and circulation technologies in order to exploit a variety of niches within the overall market. One of the classic success stories in this respect is the Benetton company (see Box on p. 196).

Although the postwar era of steadily increasing affluence in the core economies came to an abrupt end with the stagflation crisis and OPEC oil price increases of the mid-1970s, the shift away from mass markets to niche markets did not slow. Rather, it intensified as a new materialism took root. The chief actors in this change were the 'baby-boomers' whose formative years had been spent in the postwar economic boom. Their reaction to mass consumption was a counter-cultural movement with a collectivist approach to the exploration of freedom and self-realization. The failure of this counter-cultural movement (in particular, the failure of the sit-ins, protest marches, general strikes, student–worker alliances and civil disorder of 1968) meant that self-realization slid into self-centred and narcissistic lifestyles: the basis for new market niches. But the real cause of the materialism of the baby-boomers was the shock of emerging onto housing and labour markets just as the economies of the core countries were experiencing a phase of stagflation compounded by the recessionary effects of the OPEC oil price rise. Wages stood still while consumer prices ballooned. Millions of baby-boomers, raised to take steady improvements in levels of living for granted, found themselves unable to fulfil the American Dream (or the European or Japanese version of it). Their response was to pursue materialism for its own sake. They saved less, borrowed more, deferred parenthood, comforted themselves with affordable luxuries that were marketed as symbols of style and distinctiveness, and generally surrendered to the hedonism of lives infused with extravagant details: gourmet foods, designer clothes, winter vacations, extravagant gadgets and jewellery. The point here is that all of this represented consumer demand not for mass-produced products but for a rapidly changing array of high-quality products whose value as positional goods had a relatively short life. This meant that producers had to be flexible enough to be able, like Benetton, to identify and exploit finely differentiated market niches.

positional goods
Consumer goods that are acquired (in part, at least) in order to denote affluence, social status, and/or style.

195

How Benetton satisfies changing consumer demand

In 1965, the Benetton company began with a single factory near Venice. In 1968, the company acquired a single retail store in the Alpine town of Belluno, marking the beginning of a remarkable sequence of corporate expansion. By 1990 Benetton was a global organization with more than 5000 retail outlets in a total of more than 60 countries, and with its own investment bank and its own financial services organizations. It achieved this growth by exploiting computers, new communications and transportation systems, flexible out-sourcing strategies, and new production-process technologies (such as robotics and CAD/CAM systems) to the fullest possible extent.

In 1997, the corporate payroll was only 2500 employees worldwide, 1300 of whom were located in the company's home base of Treviso, in Italy. From Treviso, Benetton managers coordinated the activities of more than 250 outside suppliers in order to stock its world-wide network of franchised retail outlets. In Treviso, the firm's designers create new shirts and sweaters on CAD terminals; but their designs are produced only for orders in hand, allowing for the coordination of production with the purchase of raw materials – the Japanese *kanban*, or 'just-in-time' system. In factories, rollers linked to a central computer spread and cut layers of cloth in small batches according to the numbers and colours ordered by Benetton stores around the world. Sweaters, gloves and scarves, knitted in volume in white yarn, are dyed in small batches by machines similarly programmed to respond to sales orders. Completed garments are warehoused briefly (by robots) and shipped out directly (via private package delivery firms) to individual stores, to arrive on their shelves within 10 days of being manufactured.

Sensitivity to demand, however, is the foundation of Benetton's success. Niche marketing and product differentiation have been central to this sensitivity, which requires a high degree of flexibility in exploiting new product lines. Key stores patronized by trend-setting consumers (such as the store on the Rue Faubourg St Honoré in Paris) are monitored closely, and many Benetton stores' cash registers operate as point-of-sale terminals, so that immediate marketing data are available to company headquarters daily. Another notable feature of the company's operations is the way that different market niches are exploited with the same basic products. In Italy, Benetton products are sold through seven different retail chains, each with an image and decor calculated to bring in a different sort of customer.

Between core countries, meanwhile, there has been a homogenization of markets. Similar trends in income distribution and consumer tastes have been reinforced by television (especially CNN, MTV and syndicated light entertainment series) and international travel. Together with decreasing relative costs of transport and communications, this has meant that *market niches have merged across national boundaries, thus making it possible for producers to exploit economies of scale in the production of up-market products.* To a lesser degree, the same processes have extended from core countries to the more affluent consumers of *peripheral* and *semi-peripheral* countries, thus allowing the marketing of 'world products' (e.g. German luxury automobiles, British raincoats, Italian sweaters, Swiss watches, French wines, American soft drinks, Japanese consumer electronics) to global market segments. Barnet and Cavanagh (1994, pp. 15–16; 166) describe the emergence of what they call the 'Global Cultural Bazaar' and the 'Global Shopping Mall':

> The Global Cultural Bazaar is the newest of the global webs, and the most nearly universal in its reach. Films, television, radio, music, magazines, T-shirts, games, toys, and theme parks are the media for disseminating global images and spreading global dreams. Rock stars and Hollywood blockbusters are truly global products. All across the planet people are using the same electronic devices to watch or listen to the same commercially produced songs and stories. Thanks to satellite, cable, and tape recorders, even autocratic governments are losing the tight control they once had over the flow of information and their hold on the fantasy lives of their subjects. . . .
>
> The Global Shopping Mall is a planetary supermarket with a dazzling spread of things to eat, drink, wear, and enjoy. . . . [Through] the rise of global advertising, distribution, and marketing. . . . dreams of affluent living are communicated to the farthest reaches of the globe. . . . By 1990 34 per cent of people in developing countries were living in cities where daily exposure to global products through television, radio, and billboards was inescapable. Even in the rural areas of the Philippines any city of over 20,000 will have at least one 'supermarket', usually a one-room affair about the size of an old New Hampshire general store. In the fishing and rice-farming town of Balanga, Bataan, the San José Supermarket offers Philip Morris's Tang and Cheez Whiz, Procter and Gamble's Pringle's potato chips, Hormel's Spam, Hershey's Kisses, RJR Nabisco's Chips Ahoy, Del Monte's tomato juice, Planter's Cheez Curls, and Colgate-Palmolive's toothpaste. Above the cash register is a large poster celebrating 'Sweet Land of Liberty' with a picture of the American flag.

The logic of corporate integration

One of the most important preconditions for the emergence of advanced capitalism and the globalization of business activity has been the restructuring of the corporate world. In response to changing circumstances, private business has had to develop new strategies in order to survive; strategies that have significantly altered the fortunes of different kinds of cities, regions and nations.

Two of the most important outcomes, overall, have been corporate concentration and centralization. **Concentration** involves the elimination of small, weak firms in particular spheres of economic activity: partly through competition, and partly through mergers and takeovers. **Centralization** involves the merging of the resultant large enterprises from different spheres of economic activity to form giant 'conglomerate' companies with a diversified range of activities. Such companies are often *transnational* in their operations, having established overseas subsidiaries, taken over foreign competitors, or bought into profitable foreign businesses.

In every industry, there are limits both to the extent to which productivity can be increased and to which consumers can be induced to purchase more. As competition to maintain profit levels becomes more intense, some firms will be driven out of business while others will be taken over by stronger competitors in a process of

concentration
A trend towards the reduced numbers of firms in any given industry or economic sector. It is partly the result of the elimination of smaller, weaker firms through competition, and partly the result of corporate mergers and takeovers.

centralization
A trend towards corporate economic integration that involves the ownership of private enterprise by progressively fewer corporations. It is the result of corporate mergers and takeovers, and it is characterized by diversified, conglomerate, transnational corporations.

horizontal integration
Corporate mergers involving firms formerly competing in the same market(s) with similar goods or services.

vertical integration
Corporate mergers involving firms that are engaged in different aspects of the same industry or enterprise.

diagonal integration
Corporate mergers involving firms that are engaged in separate and distinct enterprises, producing different goods or services for different markets. The result is a diversification of corporate interests.

horizontal integration. A successful bicycle manufacturer, for example, might buy out other bicycle manufacturers.

Meanwhile, the chances of new firms being successful tend to be diminished, since the larger existing corporations are able to draw on economies of scale in order to edge out smaller competitors by price cutting. But even giant corporations cannot forestall market saturation indefinitely; and they are in any case always vulnerable to unforeseen shifts in demand. A common corporate strategy has therefore been to indulge in **vertical integration** (taking over the firms which provide their inputs and/or those which purchase their output) in an attempt to capture for themselves a greater proportion of the final selling price. The bicycle manufacturer, for example, might take over companies that make specialized components like gears or tyres; and/or companies that distribute or sell bicycles retail. The net result is the *concentration of production*, within most industries, in the hands of a diminishing number of increasingly large companies.

Alternatively – or in addition – **diagonal integration** (taking over firms whose activities are completely unrelated to their own) offers the chance of gaining access to more profitable markets and/or less expensive factors of production. Staying with the same example, the bicycle manufacturer may buy into sportswear companies, real estate companies, or entertainment companies. The net result in this case is the *centralization* of assets, jobs, production and decisions about economic life in the hands of an even smaller number of even larger companies.

The extent of these trends can be illustrated in relation to the US economy during the postwar period. Spearheaded by the large corporations that had established themselves through early flurries of horizontal integration in the 1900s (e.g. US Steel, International Harvester, American Tobacco, General Electric) and vertical integration in the 1920s (e.g. General Foods, B. F. Goodrich and the major petroleum companies), 'Big Business' began to exert an increasing influence on economic life. Although the incidence of horizontal mergers was greatly reduced by anti-trust (anti-monopoly) legislation (e.g. the Celler Kefauver Act, 1950), vertical and diagonal integration proceeded at unprecedented rates, generating around 3000 mergers per year in the peak years of the late 1960s. By the early 1970s, concentration ratios (the percentage of total sales attributable to the four largest firms) had increased significantly across a broad spectrum of industries. Several major industries, including motor vehicles, batteries, telephone equipment, turbines, and cereal breakfast foods, were each almost completely dominated by four (or fewer) firms (Table 6.1).

Meanwhile, giant conglomerates had begun to emerge as a result of diagonal integration. The first of these was Textron Incorporated, established only in 1943 when it sold blankets and other textile products to the US Army. By 1980 it had been involved in buying or selling over 100 different companies in industries as diverse as textiles, aerospace, machinery, watch bracelets and pens (Bluestone and Harrison, 1982). Textron was by no means an isolated example, however. As early as 1955, the majority of mergers taking place in the USA were diagonal, conglomerate mergers. By the early 1990s, nine out of every ten mergers involved conglomerate companies.

As a result of all this merger activity, the world's core economies are increasingly influenced by giant conglomerates. For example, between 1950 and 1980, the 50

Table 6.1: Percentage of sales accounted for by the four largest US producers in selected manufacturing industries, 1947–97

Industry	Percentage of sales		
	1947	**1972**	**1997**
Cereal breakfast foods	79	90	93
Chewing gum	70	87	99
Malt beverages	21	52	87
Weaving mills, cotton	18	31	40
Knitted underwear mills	21	46	94
Mens' and boys' suits and coats	9	19	67
Womens' and misses' suits and coats	31	13	68
Sawmills and planing mills	11	18	73
Greeting card publishing	39	70	88
Cutlery	41	55	96
Turbines and turbine generators	90	93	69
Printing trades machinery	31	42	52
Refrigeration and heating equipment	26	40	54
Household laundry equipment	40	83	100
Electric lamps	92	90	95
Telephone and telegraph apparatus	90	94	21
Electronic components	13	36	41
Primary batteries, dry and wet	76	92	94
Motor vehicles and car bodies	92	93	92
Aircraft engines and engine parts	72	77	78
Photographic equipment	61	74	94
Hard surface floor covering	80	91	100

largest US corporations increased their share of the total value added in *all* manufacturing from less than 20 per cent to nearly 30 per cent; and the largest 200 increased their share from 30 per cent to 50 per cent. Similar trends have occurred in the service sector (where the control of variety stores, department stores, car rental firms, motion picture distribution and data processing had become particularly centralized), and in the agricultural sector (where the largest 10 per cent of all farms, in terms of sales, now account for roughly two-thirds of the total market value of all agricultural commodities).

Because of their size, the larger elements of US conglomerates have also come to exert an increasing influence on the *international* economy. As we saw in Chapter 2 (Fig. 2.9), the annual sales of the very largest business enterprises are of comparable magnitude to the GNP of nation states such as Sweden and Portugal. Perhaps more significant is the fact that *the combined overseas output of US-based multinational companies is now larger than the GNP of every country in the world except the US itself.* As we shall see (Chapter 11), this new dimension of the international economy has come to represent a serious economic threat to small nations, and it has prompted the creation of a variety of supranational organizations.

It should be stressed that the USA has by no means been the only core nation to generate giant conglomerates. Major multinational companies have been bred in Australasia, Canada, Japan and Western Europe in response to the same logic that has applied to the USA; and even some NICs and peripheral countries now have large home-based transnational corporations. Indeed, these companies have been increasing their share of world markets at the expense of US-based companies; and some of them have extended their operations to the USA itself. The Japanese electronics conglomerate NEC, for example (the 45th-biggest corporation in the world in 1995, with a global work-force of 152 719 and sales of $US 45,557 million) had 30 different manufacturing affiliates in 17 different countries in 1995, and 4 of them were in the United States. In addition, 12 of its 47 sales and service affiliates were in the United States. Overall, the number of transnational conglomerate corporations has increased from just over 7000 in the early 1970s to nearly 40 000 in the mid-1990s. Together, these corporations have more than 170 000 foreign affiliates and account for more than US$5.5 trillion in world-wide annual sales.

The evolution of transnational corporate activity

The current importance of transnational corporations in the world economy is the result of an evolutionary process that can be characterized in terms of three distinctive phases. With reference to the experience of US-based transnationals, these phases were as follows:

- *Phase I*: Beginning in the nineteenth century and extending to 1940, this phase was dominated by investment directed at obtaining raw materials – mainly oil and minerals – for domestic manufacturing operations.
- *Phase II*: After the Second World War, some of the leading corporations began to use direct foreign investment in overseas production operations as a means of penetrating foreign consumer markets. Initially, the focus of this investment was Western Europe, where the Marshall Aid programme, NATO rearmament and the US military presence in West Germany provided useful information feedback and points of entry to an expanding consumer market. Meanwhile, the establishment of the US dollar as the world's principal reserve currency at the 1944 Bretton Woods Conference (see Chapter 2), had made it much easier for US companies to buy into foreign industries. It was not long before many US firms began to penetrate the expanding markets of parts of the rest of the world, particularly in Latin America. The resulting mergers and acquisitions often led to the restructuring of corporate production processes:

> Bulova Watch provides a clear example. Bulova now manufactures watch movements in Switzerland and ships them to Pago Pago, in American Samoa, where they are assembled and then shipped to the United States to be sold. Corporation President Harry B. Henshel said about this arrangement: 'We are able to beat the foreign competition because we *are* the foreign competition.'
>
> (Bluestone and Harrison, 1982, p. 114)

Between 1957 and 1967, 20 per cent of all new US machinery plants, 25 per cent of new chemical plants, and over 30 per cent of new transport equipment plants were located abroad. By 1970, almost 75 per cent of US imports were transactions between the domestic and foreign subsidiaries of multinational conglomerates. By the end of the 1970s, overseas profits accounted for a third or more of the overall profits of the hundred largest multinational producers and banks.

- *Phase III*: During the 1970s, the crisis and destabilization associated with the episode of stagflation brought growing competition from goods produced in the NICs with cheap labour. In addition, the collapse of the Bretton Woods agreement in 1971 increased the value of the US dollar, thus making imported goods cheaper and so making it easier for European and Japanese multinationals to penetrate US markets. In response, US multinational companies began to restructure their production processes once again, eliminating the duplication of activities between domestic and foreign-based facilities and reorganizing the division of tasks between them.

  Effectively, this third phase has meant:

1. the further redeployment of capital, bringing peripheral nations into the production space of US companies in order to benefit from lower labour costs (in 1992, the costs of hourly compensation for production workers in manufacturing industries in Austria, Denmark, France and Italy were about 80 per cent of those for US workers; in Australia, Ireland and the United Kingdom, they were about 60 per cent; and in Brazil, Hong Kong, South Korea, Mexico and Taiwan they were between 10 and 15 per cent);
2. withdrawing from locations where unskilled and semiskilled labour is more expensive (i.e. North America and northwestern Europe); and
3. retaining existing facilities which require high inputs of technology and/or skilled labour. Thus, for example, General Electric added 30 000 foreign jobs to its payroll during the 1970s while reducing its US employment by 25 000. Similarly, the RCA Corporation increased its foreign workforce by 19 000 while reducing its US payroll by 14 000 (Bluestone and Harrison, 1982).

A good example of the consequences of this third phase is provided by Barff (1995, p. 59):

> Nike, the athletic footwear marketer, used to own manufacturing plants in the United States and United Kingdom, but presently subcontracts 100 per cent of its production capacity to suppliers in South and East Asia. The geography of Nike's production partnerships has evolved over time, a change powered in large part by changing labor costs in Asia. Initially, production of Nike shoes took place in Japan. Soon, subcontracting arrangements diffused to factories in South Korea and Taiwan. Presently, those partnerships are diminishing in importance as labor costs rise and new networks of subcontractors become established in Indonesia, Malaysia, and China where workers involved in shoe production are paid about one-thirtieth of the wage their counterparts make, working for other companies, in the United States.

PATTERNS AND PROCESSES OF GLOBALIZATION

Most of the world's population now lives in countries that are either integrated into world markets for goods and finance, or rapidly becoming so. As recently as the late 1970s, only a few peripheral countries had opened their borders to flows of trade and investment capital. About a third of the world's labour force lived in countries like the Soviet Union and China with centrally planned economies, and at least another third lived in countries insulated from international markets by prohibitive trade barriers and currency controls. Today, with nearly half the world's labour force among them, three giant population blocs – China, the republics of the former Soviet Union, and India – are entering the global market. Many other countries, from Mexico to Thailand, have already become involved in deep linkages. According to World Bank estimates, fewer than 10 per cent of the world's labour force will remain isolated from the global economy by the year 2000.

Globalization, although incorporating more of the world, more completely, into the capitalist world-system, has intensified the differences between the core and the periphery. According to the United Nations Development Program, the gap between the poorest 20 per cent of the world's population and the wealthiest 20 per cent increased threefold between 1960 and 1990. Some parts of the periphery have almost slid off the economic map. In sub-Saharan Africa, economic output fell by one-third during the 1980s, and now people's standard of living there is, on average, lower than it was in the early 1960s. Indeed, the structural irrelevance of sub-Saharan Africa to the contemporary global economy is probably a much more threatening condition than the dependency of the colonial period.

Meanwhile, globalization has resulted in the consolidation of the core of the world-system. The core is now a close-knit triad of the geographic centres of North America, the European Union of Western Europe, and Japan. Most of the world's flows of goods, capital, and information are within and between these three centres. Between them, they dominate the world's periphery, with each centre having particular influence in its own regional expansion zone: its nearest peripheral region (see Figure 6.3). Around this triangle of wealth, power, and technology, as Castells notes (1996, p. 101), 'the rest of the world becomes organized in a hierarchical and asymmetrically interdependent web, as different countries and regions compete to attract capital, human skills, and technology to their shores'.

Allen Scott (1996) has represented this situation schematically (Fig. 6.4) as a patchwork of global metropolitan regions, 'regional economic motors', each one being the site of dense networks of specialized but complimentary forms of economic activity, together with large and multifaceted labour markets and specialized infrastructures offering powerful agglomeration economies. The central metropolitan area of each regional motor is surrounded by a hinterland occupied by ancillary communities, prosperous agricultural zones, local service centres, and the like. As indicated by Fig. 6.4, the hinterlands of some of these regional motors may coalesce with one another (as in the actual cases of Boston/New York/Philadelphia, Los Angeles/San Diego/Tijuana, Milan/Turin/Genoa and Tokyo/Nagoya/Osaka).

These regional economic motors are linked by intense flows of capital, information, goods and people. Beyond them, there are large residual expanses of the

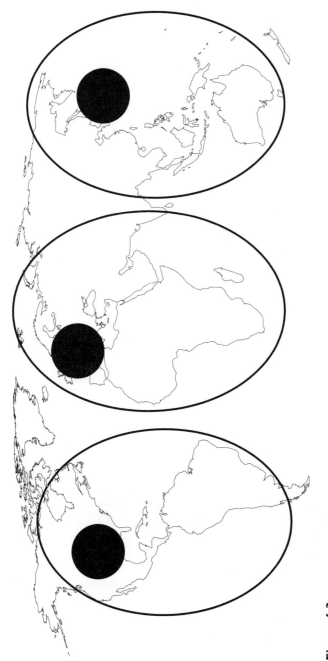

Figure 6.3
The triadic core of the world economy

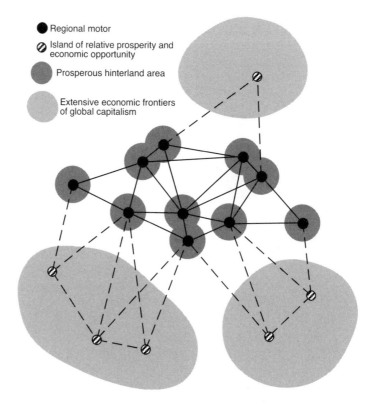

Figure 6.4
A schematic representation of the contemporary geography of global capitalism

Source: Scott (1996), pp. 92–3

contemporary world that lie at the extensive economic frontiers of advanced capitalism (former colonies, ex-socialist states, physically isolated regions, and so on). These are underdeveloped areas that have been unable to build the economic organizations that might provide the basis for sustained growth in a global economy. Scott writes:

> Even so, these areas are occasionally punctuated by islands of relative prosperity and economic opportunity, and some of these may well be on a trajectory that takes them to much higher levels of (agglomerated) development. In the 1960s and 1970s, places like Hong Kong, Singapore, Taiwan, the Seoul region, and central Mexico were all positioned at different stages along this trajectory. Today, a number of Third World metropolitan areas, such as Bangkok, Kuala Lumpur, and São Paulo–Rio de Janeiro, are following on the heels of these pioneers, while parts of Nigeria, the Ivory Coast, India, Indonesia, and possibly Vietnam, seem to be poised at the initiatory phase. (1996, pp. 402–3)

International redeployment and locational hierarchies

The cumulative result of the globalization of economic activity has been the creation of *locational hierarchies* of activities. The aggregate outcome involves:

1. localized concentrations of high-level management in 'world cities' (see Chapter 7);
2. smaller concentrations of mid-level management and administration in large metropolitan areas in core countries and in the capital cities of NICs and some peripheral countries;
3. clusters of Research and Development (R & D) activity in high-tech, innovative milieux – 'technopoles' – (see Chapter 7) within the core countries;
4. regions specializing in advanced, high-tech industrial production, mostly within the core countries; and
5. decentralized pockets of routinized industrial production – branch plants – in (a) the peripheral regions of core countries, and (b) the metropolitan areas of NICs and some peripheral countries.

These tendencies, and the fact that they have been influenced so much by the locational strategies of transnational corporations equipped to undertake a 'global scan' in pursuit of the most profitable redeployment of activities, have contributed to the idea of the emergence of a 'New International Division of Labour' or NIDL (see Chapter 1).

It should be acknowledged that not all firms or industries are equal in their need or their capability to engage in international redeployment of this kind. 'Global

Figure 6.5

Volkswagen's global assembly line

Source: B. Marshall (1991), pp. 92–3

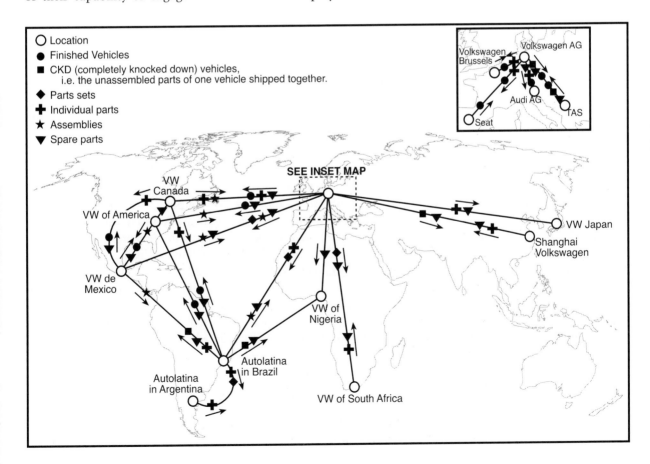

scan' is more of a tendency than accomplished fact. It is the largest companies – the transnational conglomerates – that are in the best position to take advantage of the advances in circulation and production technology. Probably the best-developed example – and the most-researched – is provided by the automobile industry, where the clearly defined national markets of the early postwar period have been almost entirely replaced by production and marketing on a global scale. In 1976, Ford introduced the Fiesta, a vehicle designed to sell in Europe, South America, and the Asian market as well as North America. The Fiesta was assembled in several different locations from components manufactured in an even greater number of locations. The Fiesta became the first of a series of Ford 'world cars,' that now includes the Escort, the Mondeo, and the Contour. The components of the Ford Escort, for example, are made and assembled in 15 countries across three continents. Ford's international subsidiaries, which used to operate independently of the parent company, are now being functionally integrated, using supercomputers and video teleconferences. Meanwhile, other automobile companies have organized their own global assembly lines for their world cars (Fig. 6.5): the Volkswagen Rabbit/Golf, for example, Volkswagen's new Concept 1 vehicle, GM's Corsa, and Fiat's Project 178. Today, most of the 40 million or so vehicles that roll off production lines each year are made by just 10 global corporations (in order of size: General Motors, Ford, Toyota, Volkswagen, Nissan, Fiat, Peugeot–Citroen, Honda, Mitsubishi, and Renault).

Flexible production systems

Concurrent with firms' changing competitive strategies there have been some significant changes in the organization of production systems in many industries. These are often expressed in terms of a transition from Fordism to flexible production systems, or **Neo-Fordism**. In Neo-Fordism, the logic of mass production coupled with mass consumption has been modified by the addition of more flexible production, distribution, and marketing systems.

This flexibility is rooted in forms of production that enable manufacturers to shift quickly and efficiently from one level of output to another and, more importantly, from one process and/or product configuration to another. It must be understood as a change that involves flexibility both *within* firms and *between* them. Within firms, a great deal of the flexibility of Neo-Fordism is attributable to the exploitation of new technologies. Computerized machine tools are capable of producing a variety of new products simply by being reprogrammed, often with very little downtime between production runs for different products. Different stages of the production process (sometimes located in different places) can be integrated and coordinated through computer-aided design (CAD) and computer-aided manufacturing (CAM) systems. Computer-based information systems can be used to monitor retail sales and track wholesale orders, thus allowing producers to reduce the costs of raw materials stockpiles, parts inventories and warehousing through sophisticated small-batch, just-in-time production and distribution systems. The combination of computer-based information systems, CAD/CAM systems and computerized machine tools have also helped firms to be flexible enough

Neo-Fordism
Sometimes referred to as 'post-Fordism' or 'flexible accumulation,' this term identifies the regime of accumulation that has succeeded Fordism within parts of the world's core economies. Rather than being predicated on the mutual reinforcement of mass production and mass consumption, it depends on flexible production systems to exploit specific market segments and/or niches.

Table 6.2: Contrasts in the production process and the labour process: Fordism and flexible accumulation

Fordism	Flexibility
The production process	
Mass production of homogeneous products	Small batch production
Uniformity	Flexible and small batch production of a variety of product types
Large buffer stocks and inventory	No stocks
Testing quality ex-post (rejects and errors detected late)	Quality control part of process (immediate detection of errors)
Rejects are concealed in buffer stocks	Immediate reject of defective parts
Loss of production time because of long set-up times, defective parts, inventory bottlenecks, etc.	Reduction of lost time – diminishing 'the porosity of the working day'
Resource driven	Demand driven
Vertical and (in some cases) horizontal integration	(Quasi-)vertical integration sub-contracting
Cost reductions through wage control	Learning-by-doing integrated in long-term planning
Labour	
Single task performance by worker	Multiple tasks
Payment per rate (based on job design criteria)	Personal payment (detailed bonus system)
High degree of job specialization	Elimination of job demarcation
No or only little on-the-job training	Long on-the-job training
Vertical labour organization	More horizontal labour organization for core
No learning experience	On-the-job learning
Emphasis on diminishing worker's (responsibility disciplining of labour force)	Emphasis on worker's co-responsibility
No job security	High employment security for core workers (lifetime employment). No job security and poor labour conditions for temporary workers/increasing informal activities

Source: Albrechts and Swyngedouw (1989), Fig. 1.

to exploit specialized niches of consumer demand, rendering geographically scattered upscale markets accessible to economies of scale in production. This kind of flexibility depends on new labour practices as well as new technologies, however (Table 6.2). There are two main aspects to this. One is the increasingly flexible use of labour within firms, which requires individual workers to perform a wider variety of tasks. Taken to its extreme, this trend has in some instances substituted 'craftwork' for production-line work. The other is the increasingly flexible size and quality of the labour force required at any one plant. This trend has substituted overtime, part-time and temporary employment for permanent, full-time jobs.

Between firms, the flexibility inherent to Neo-Fordism has been achieved through the *externalization* of certain functions. One way of doing this has been to restructure permanent and hierarchically structured administrative, managerial and technical units within large corporations into flatter, leaner and more flexible forms of organization that can make increased use of outside consultants, specialists and subcontractors. This has led to a degree of vertical disintegration among firms (see Chapter 3). Another route to externalization has been to participate in joint ventures, in the licensing or contracting of technology, and in **strategic alliances** involving design partnerships, collaborative R & D projects, and the like. Strategic alliances have been an important contributor to the intensification of economic globalization. For example, 244 strategic alliances existed in 1996 between the world's 41 largest auto makers, including parts-sharing agreements and joint ventures in research, as well as in manufacturing. Peugeot of France, for instance, had 22 agreements with other car companies, including a partnership with Taiwan's Chinese Automobile Co. to build Citroen C15s, and one with Fiat to produce a commercial van. Other products of strategic alliances include the Geo Prizm, a Toyota Corolla that is made in California and marketed in the United States by General Motors; the Geo Metro, made by Suzuki and Isuzu, and marketed by General Motors; the Jaguar, made in England by a wholly owned Ford subsidiary; and the Mazda Navajo, really a Ford Explorer made in Kentucky.

Such alliances have become an important aspect of global economic geography in the 1990s, as transnational corporations seek to reduce their costs and to minimize the risks involved in their multimillion dollar projects. Strategic alliances serve several functions:

strategic alliances
Commercial agreements between transnational corporations, usually involving shared technologies, marketing networks, market research, or product development.

- allow transnational companies to link up with local 'insiders' in order to tap into new overseas markets;
- provide a quick and inexpensive means of swapping information, on a limited basis, about technologies that help to improve their products and their productivity;
- reduce the costs of product development;
- spread the costs of market research.

The Nestlé food company, for example, has a strategic alliance with the Swiss Ciba-Geigy AG in the area of microbiology. It cooperates with Calgene in researching the regeneration process of soybean plants and the development of substitutes for cocoa butter and other vegetable oils. Nestlé also cooperates with the Coca-Cola company in exchanging technologies and in marketing. Nestlé will, for example, use Coca-Cola's distribution network for products such as Nescafé instant coffee.

The changing geography of the clothing industry

The clothing industry provides a good example of the way that local economic geographies are affected by an industry's response to globalization. In the nineteenth century, the clothing industry developed in the metropolitan areas of core countries, with many small firms using cheap migrant or immigrant labour. In the first half of the twentieth century, the industry, like many others, began to modernize. Larger firms emerged, their success based on the exploitation of mass production techniques for mass markets, and on the exploitation of principles of spatial organization within national markets. In the United States, for example, the clothing industry went through a major locational shift as a great deal of production moved out of the workshops of New York to big, new factories in smaller towns in the South, where labour was not only much cheaper but less unionized.

Then, as the world economy began to globalize, semi-peripheral and peripheral countries became the least-cost locations for mass-produced clothing for global markets. In 1960, less than 7 per cent of all apparel purchased in the United States had been imported; by 1980, more than half was imported. Leisure wear – jeans, shorts, T-shirts, polo shirts, and so on – was an important component of the homogenization of consumer tastes around the world, and it could be produced most profitably by the cheap labour of young women in the peripheral metropolitan areas of the world. Whereas skilled clothing operatives in Toronto's garment district, for example, are paid US$350 a week, plus benefits, for a 33-hour week, their counterparts in Hong Kong are paid US$250 for a 60-hour week, with no benefits (and in Indonesia skilled garment workers can be hired for only US$50 for a 60-hour week). The retail margin on clothing made in workshops in countries like Indonesia and Thailand and sold in Europe and the United States is 200 to 300 per cent, compared to a margin of only 70 per cent or so on domestically made clothing. It is worth noting that studies of the industry have shown that some of the young female workers are as young as 12 years old – girls from rural villages who have been sent to work as sewing machinists in city workshops, sleeping eight to a room, sewing seven days a week from 8 a.m. to 11 p.m.

This globalization of production has resulted in a complex set of commodity chains. Many of the largest clothing companies, such as Liz Claiborne, have most of their products manufactured through arrangements with independent suppliers (over 300 in the case of Claiborne), with no one supplier producing more than 4 or 5 per cent of the company's total output. These manufacturers are scattered throughout the world, making the clothing industry one of the most globalized of all manufacturing activities (Fig. 6.6). Claiborne, for example, has its goods produced simultaneously in as many as 40 countries, including Brazil, China, Costa Rica, Hong Kong, Hungary, Italy, the Philippines, Portugal, South Korea, Taiwan, Thailand, Turkey, and Yugoslavia. The actual geography of commodity chains in the clothing industry is rather volatile, with frequent shifts in production and assembly sites as companies and their suppliers continuously seek out new locations with lower costs.

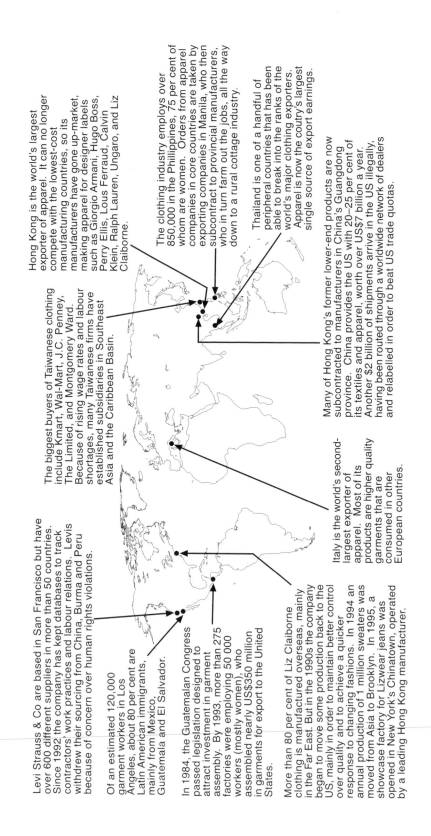

Levi Strauss & Co are based in San Francisco but have over 600 different suppliers in more than 50 countries. Since 1992, the company has kept databases to track contractors' work practices and labour relations. Levis withdrew their sourcing from China, Burma and Peru because of concern over human rights violations.

Of an estimated 120,000 garment workers in Los Angeles, about 80 per cent are Latin American immigrants, mainly from Mexico, Guatemala and El Salvador.

In 1984, the Guatemalan Congress passed legislation designed to attract investment in garment assembly. By 1993, more than 275 factories were employing 50 000 workers (mostly women), who assembled nearly US$350 million in garments for export to the United States.

More than 80 per cent of Liz Claiborne clothing is manufactured overseas, mainly in the Far East. But in the 1990s the company began to move some production back to the US, mainly in order to maintain better control over quality and to achieve a quicker response to changing fashions. In 1994 an annual production of 1 million sweaters was moved from Asia to Brooklyn. In 1995, a showcase factory for Lizwear jeans was opened in New York's Chinatown, operated by a leading Hong Kong manufacturer.

Hong Kong is the world's largest exporter of apparel. It can no longer compete with the lowest-cost manufacturing countries, so its manufacturers have gone up-market, making apparel for designer labels such as Giorgio Armani, Hugo Boss, Perry Ellis, Lous Ferraud, Calvin Klein, Ralph Lauren, Ungaro, and Liz Claiborne.

The biggest buyers of Taiwanese clothing include Kmart, Wal-Mart, J.C. Penney, The Limited, and Montgomery Ward. Because of rising wage rates and labour shortages, many Taiwanese firms have established subsidiaries in Southeast Asia and the Caribbean Basin.

The clothing industry employs over 850,000 in the Phillippines, 75 per cent of whom are women. Orders from apparel companies in core countries are taken by exporting companies in Manila, who then subcontract to provincial manufacturers, who in turn farm out the jobs, all the way down to a rural cottage industry.

Thailand is one of a handful of peripheral countries that has been able to break into the ranks of the world's major clothing exporters. Apparel is now the coutry's largest single source of export earnings.

Many of Hong Kong's former lower-end products are now subcontracted to manufacturers in China's Guangdong province. China provides the US with 20–25 per cent of its textiles and apparel, worth over US$7 billion a year. Another $2 billion of shipments arrive in the US illegally, having been routed through a worldwide network of dealers and relabelled in order to beat US trade quotas.

Italy is the world's second-largest exporter of apparel. Most of its products are higher quality garments that are consumed in other European countries.

Figure 6.6
The changing global distribution of clothing manufacturing

Although cheap leisure wear can be produced most effectively through arrangements with multiple suppliers in peripheral, low-wage regions, higher-end apparel for the global marketplace requires a different geography of production. These products – women's fashion, outerwear, and lingerie, infants' wear, and men's suits – are based on frequent style changes and high-quality finish. This requires short production runs and greater contact between producers and buyers. The most profitable settings for these products are in the metropolitan areas of the core countries – London, Paris, Stuttgart, Milan, New York, and Los Angeles – where, once again, migrant and immigrant labour provides a work-force for 'designer' clothing that can be shipped in small batches to upscale stores and shopping malls around the world.

The result is that commodity chains in the clothing industry are quite distinctive in terms of the origins of products destined for different segments of the market. Fashion-oriented retailers in the United States who sell designer products to up-market customers obtain most of their goods from manufacturers in a small group of high-value-added countries including France, Italy, Japan, the United Kingdom, and the United States. Department stores that emphasize 'private label' products (that is, store brands, such as Nordstrom) and premium national brands will obtain most of their goods from established manufacturers in semi-peripheral East Asian countries. Mass merchandisers who sell lower-priced brands buy primarily from a third tier of lower-cost, mid-quality manufacturers, while large-volume discount stores like Wal-Mart import most of their goods from low-cost suppliers in peripheral countries like China, Bangladesh, and the Dominican Republic. Finally, some importers operate on the outer fringes of the international production frontier, seeking out very cheap but low-quality products from new sources in peripheral countries with no significant experience in clothing manufacture.

Interpretations of corporate flexibility

The increased flexibility in economic organization can be interpreted in two rather different ways. One interpretation, that of *flexible specialization*, sees the trend towards flexible production systems in a permanent and positive light. In short, new technologies have opened up the possibility for the decline of the large integrated firm and for the growth of a production system organized around clusters of small firms (e.g. Piore and Sabel, 1984). Much is made of cases such as the 'Third Italy' (Central and Northeast Italy) where such clusters have emerged over the past 30 years. Alfred Marshall's (1920) model of the industrial district is sometimes used to provide a theoretical argument for the clustering of specialized industries in specific localities: this emphasized specialized labour pools, external economies from proximity accruing to firms in the same industry and the availability of specialized inputs and services. This is, of course, nothing more than a restatement of the main arguments for any kind of agglomeration (localization) economy (Krugman, 1991, pp. 35–67). It has, however, become popular to restate them in terms of the 'new institutional economics' (Williamson, 1985) as 'internalizing transaction costs' within regions rather inside firms. A more radical point of view is that trust, loyalty and partnership between firms are vital to the establishment of the 'new' industrial districts, if not to the old ones that Marshall was interested in. Thus, the social conditions for small-scale production, in combination with (1) a history of artisanal activity, (2) nearby centres of innovation, (3) assistance from local governments, and (4) consensus between labour and management, are all basic requirements for the functioning of industrial districts engaged in flexible production (Harrison, 1992). From this point of view, there is a sociology to the new industrial district that sets limits to its diffusion elsewhere.

On the basis of a series of national case studies (Silicon Valley in the USA, the Ile de France technopole in France, and the Third Italy), Scott (1988, p. 106) has used the phrase 'regime of flexible accumulation' to describe this new wave of economic-geographical organization. The basic proposition is that irrespective of the particular industries involved (e.g. shoes, clothing, machine tools, computers) there is a major drive towards geographical concentration of industries even for manufacturing so-called 'mature products' (those towards the end of a product life-cycle). Rather than specialization on a regional scale of agglomeration, however, a more localized pattern of specialization is now underway. This is because small firms are its major agents and they prosper best as the providers of goods to rapidly changing markets when they are able to share information, labour traditions and inter-industry links.

Responding to criticisms of the industrial district model (such as that it ignores the continuing importance of large firms and exaggerates endogenous or local conditions relative to world markets and the international division of labour), several authors have provided more synthetic accounts. For example, Scott (1992) now argues that large producers can play an important part in inducing and maintaining the growth of (high technology) industrial districts. He suggests that the usual division of production units between flexible and mass producers is insufficient. He identifies a third type, the *systems house*, flexible producers that benefit from economies of scope flowing from research and development or design synergies, with a variegated internal structure of job specialization and batch (as opposed to mass) production of complex products. These systems houses do not stand alone or operate with branch plants. They are usually connected with nearby flexible producers. They are the 'hubs' for high technology industrial districts such as those in Southern California and in Japan.

Scott (1992), for one, does not see this model of industrial districts as incompatible with an internationalized world economy. Indeed, he sees this phenomenon as itself 'the interlinkage of industrial districts across the globe, ... as a mosaic of regions consisting of localized networks of transactions (i.e. industrial districts) embedded in global networks of transactions' (Scott, 1992, p. 274). One study (Henderson, 1989) has attempted to show for one industry, semiconductor (computer chip) production, how its technical and social divisions of labour (broadly, research and development/unskilled, and white male/female immigrant (US) or female rural–urban migrants (East Asia) divisions) and the need for access to some protected markets (Europe) have produced a specific set of locational patterns of production activities. The 'American' semiconductor industry now links *production complexes* (Henderson does not use the language of industrial districts or flexible production) in the United States with others in, for example, Scotland and Hong Kong. Contrary to NIDL models, however, cheap labour is not the sole attraction of any of these locations. For example, there is now an indigenous electronics industry in some countries of East Asia oriented to local rather than American or other external demand.

A second interpretation of the Fordist/Neo-Fordist divide, that of **flexible accumulation** sees the methods of flexible production more as a response to a crisis of capital accumulation in some sectors of industrial production rather than a fully fledged new mode of production (e.g. Harvey, 1990). From this point of view it is the growth and transformation of financial markets in the 1980s and the introduction of flexible production as a means of disciplining the power of labour that attract attention. In the first case it is argued that through the 'explosion in new financial instruments and markets (e.g. junk bonds), coupled with the rise of highly sophisticated systems of financial coordination on a global scale', the financial system has forced the increase in 'the geographical and temporal flexibility of capital accumulation' (Harvey, 1990, p. 194). In the second case the declining rate of profit in the 1970s led firms to a strategy of decentralized production to undermine the power of labour (and reduce wage bills), which had increased under Fordism, yet still maintain centralized control. Many of the often 'idealized' small firms of the Third Italy are indeed sub-contractors for larger firms searching for alternatives to their large unionized labour forces.

It is important to note, however, that several criticisms have been directed at these interpretations, most especially the first one of flexible specialization:

- First, not all the apparently 'flexible' methods of production are in fact that flexible. Simple oppositions such as 'rigid' and 'flexible' and 'Fordist' and 'Neo-Fordist' impose a structure on an industrial history that is not all that simple. In particular, the labour processes in different industries, the market and macroeconomic features of different sectors and the 'organizational cultures' of the firms and areas involved combine to produce a range of industrial geographies along the spectrum between locational fixity and global mobility (Amin and Thrift, 1992, p. 574).
- Second, large firms are now adopting many of the methods that were seen as the exclusive province of small firms clustered in industrial districts (as acknowledged by Scott). Some enter into strategic alliances for the production of certain items even with firms that are their direct competitors (e.g. GM and Toyota). They can do this even with dispersed production facilities.
- Third, the geographical boundaries of industrial districts are not usually carefully defined; partly this is a result of a lack of consensus about the transactions and flows that must be internalized geographically for a district to 'exist'. Some new industrial districts sprawl over large areas and overlap with other districts whereas others are small and exclusive. Does the same logic of production govern both of these types of industrial space?
- Fourth, missing from most discussion of both models is attention to the specificity (in terms of industries, technologies and limited areas) of the 'new spaces' associated with flexible production, prior spatial divisions of labour in

flexible accumulation
A phase of capitalist development (a regime of accumulation, in the terminology of Regulation Theory) characterized by a set of production technologies, labour practices, inter-firm relations and consumption patterns that have evolved in order to allow greater economic and geographic flexibility in economic affairs.

213

the affected areas (and elsewhere where a shift to flexible production is not taking place), and the influence of government policies, especially with respect to technical education, innovation policy, tax incentives, and trade barriers.

- Fifth, little is known of how local conditions interact with global competitive conditions to affect the fortunes of industrial districts. What is known suggests that such districts are not immune to the problems of international competition that afflict 'Fordist' firms (e.g. shoe producers in the Third Italy are in crisis because of Korean and Brazilian competition). Some industrial districts, such as California's Silicon Valley, relied heavily on the defence spending justified by the Cold War. Recent decreases in US defence spending may seriously affect the fortunes of such areas.

- Sixth, the 'shift' to flexible production is emphasized by scholars who live in places experiencing the new forms of economic development. The so-called Los Angeles 'school' (Scott is a leading member) uses examples of new production complexes from Southern California as if they were drawn from a universal sample or provide a window on the future elsewhere. The view from within the Northeast USA would be considerably less sanguine about the the break with mass production and the possible universality of flexible production. Many large firms in the Northeast continue to move production abroad or invest in automation. In many manufacturing industries Fordist principles of production still prevail. (For an excellent review of the various problems with flexible production models, see Gertler, 1992.)

The global office and the informational economy

business services
Services that enhance the productivity or efficiency of other firms' activities or that enable them to maintain their specialized roles. Examples include advertising, personnel training, recruitment, finance, insurance, and marketing.

The globalization of production systems and the growth of transnational corporations have brought about another important change in patterns of local economic development: banking, finance, and **business services** are now no longer locally oriented ancillary activities but important global industries in their own right. The new importance of banking, finance, and business services was initially a result of the globalization of manufacturing, an increase in the volume of world trade, and the emergence of transnational corporate empires. It was helped along by advances in telecommunications and data processing. Satellite communications systems and fibre-optic networks made it possible for firms to operate key financial and business services 24 hours a day, around the globe, handling an enormous volume of transactions. Linked to these communications systems, computers permit the recording and coordination of the data.

As banking, finance, and business services grew into important global activities, however, they were themselves transformed into something quite different from the old, locally oriented ancillary services. The global banking and financial network now handles trillions of dollars every day (estimates in 1996 ranged from 3 to 7 trillion dollars) – no more than 10 per cent of which has anything to do with the traditional world economy of trade in goods and services. International movements of money, bonds, securities, and other financial instruments have now become an end in themselves because they are a potential source of high profits from speculation and manipulation. Several factors have supported this development:

- The institutionalization of savings (through pension funds and so on) has established a large pool of capital managed by professional investors with few local or regional allegiances or ties.
- The deregulation of banking and financial services, as governments in many countries have lifted restrictions and regulations in the hope of capturing more of their growth.
- The quadrupling of crude oil prices in 1973 generated so much capital for oil-rich countries that their banks opened overseas branches in order to find enough borrowers. In many cases, the borrowers were companies and governments in underdeveloped, peripheral countries that had previously been considered poor investment prospects. The internationalization of financial services soon paid off for the big banks. By the mid-1970s, about 70 per cent of Citibank's overall earnings came from its international operations, with Brazil alone accounting for 13 per cent of the bank's earnings in 1976.
- A persistent trade deficit of the United States *vis à vis* the rest of the world (a result of the postwar recovery of Europe and Japan) created a growing pool of dollars outside the United States, known as 'Eurodollars'. This supply in turn created a pool of capital that was beyond the direct control of the US authorities.
- 'Hot' money (undeclared business income, proceeds of securities fraud, trade in illegal drugs, and syndicated crime), easily laundered through international electronic transactions, also found its way into the growing pool of Eurodollars. It is estimated that 100 billion US dollars is laundered each year through the global financial system.
- The initial response of many governments (including the US government) to balance-of-payments problems was to print more money – a short-term solution that eventually contributed to a significant surge in inflation in the world economy. This inflation, because it promoted rapid change and international differentials in financial markets, provided a further boost to speculative international financial transactions of all kinds.

Together, these factors have amounted to a change so important that a deep-seated restructuring of the world economy has resulted. Banks and financial corporations with the size and international reach of Citibank or Nomura or Salomon Brothers are able to influence local patterns and processes of economic development throughout the world, just like the major transnational conglomerates involved in the global assembly line. In addition, key producer or business services (such as market research, accountancy, advertising, banking, corporate insurance, and cor- porate legal services) have proliferated, adding an important new dimension to the world's economic landscapes. In 1995 more than three out of every five jobs in the USA were in the service sectors. In Britain only five out of 459 local authority districts had more than half their jobs in manufacturing employment in 1994.

Perhaps most important of all, the combined effect of all this has been to create an emergent *informational economy*. The informational economy represents a new mode of economic production and management in which productivity and com- petitiveness rely heavily on the generation of new knowledge and on the access to, and processing of, appropriate information. As we saw in Chapter 1, the most important economic sectors in this informational economy are high-technology manufacturing, design-intensive consumer goods, and business and financial services. In the industrial mode of development, the main source of productivity lies in the introduction of new energy sources, and in the ability to decentralize the use of energy throughout the production and circulation processes. 'In the new, informational mode of development the source of productivity lies in the technology of knowledge generation, information processing, and symbol communication' (Castells, 1996, p. 17). A recent report by the OECD estimates that more than half of total GDP in the core economies is now knowledge-based, including industries such as computers, software, pharmaceuticals, education, and television. High-tech industries have almost doubled their share of manufacturing output over the past two decades, to around 25 per cent, and knowledge-intensive services are growing even faster, accounting for eight out of every ten new jobs in the core economies (OECD, 1996).

Business services and metropolitan growth

The fastest-growing service sector in Europe, North America, Japan and in scattered locations elsewhere (e.g. Hong Kong, Singapore) has been that of business services. Most of these activities employ personnel either in managerial or information-processing positions. Demand for these services comes from other firms rather than the general public. Consequently, these services are located close to their main customers, overwhelmingly in and around large cities. Several factors have acted to reinforce this trend. The first is their *internationalization*. Agglomeration economies (access to clients and competitors, proximity to technical services, availability of qualified personnel) are so powerful across most of these services that they are disproportionately located in major cities. But ease of communication has made it possible for firms to operate across different cities rather than restrict themselves to one. The spread of manufacturing and conglomerate multinationals has encouraged successful business service firms to follow suit, establishing multi-city offices to service their multinational accounts. Second, the *deregulation* of national markets (especially in banking and finance) has also strengthened the relationship between certain large cities with well-established institutions (exchanges and commodity markets) and business services. A relatively small number of centres (London, Tokyo, Chicago, New York) has benefited disproportionately from this trend. These *world cities* have become vital control points within a world economy, breaking the bounds previously imposed by national restrictions.

Business services and flexible economies

It can be argued that the processes of sub-contracting and small firm growth associated with flexible production give rise to 'new' service activities and increase the dependence of manufacturers on the purchase of services from independent vendors. Typically, 'intermediary' functions in economic activity, such as wholesaling, have been regarded as internal to large *vertically integrated firms* (all functions carried out within one firm) or ignored because of an assumption that producers trade directly with one another. However, wholesaling has persisted and, recently, expanded. Glasmeier (1990) makes a plausible case for the view that the emergence of 'high tech' industrial districts, such as Silicon Valley and Austin, Texas, depended from the start on the coexistence of manufacturers and merchant wholesalers. In the Austin case, which Glasmeier examines in detail, the wholesalers serve as agents of inter-regional trade, bringing parts and products from outside the local complex into the local economy. Over time, national and regional wholesalers have displaced local ones in importance to the local manufacturers. This suggests both the importance of merchant wholesalers to the development of industrial districts and the role of exogenous (extra-local) agents in local development. The growth of industrial districts cannot be explained just in terms of local social conditions or the nature of manufacturing processes.

Christopherson (1989) argues that the attention given to flexible production in manufacturing has obscured the increasing importance of flexibility in the labour markets of service industries. She points out that in the United States in the 1980s 80 per cent of the new jobs were in retail, health and business services and that perhaps 25 per cent of all service jobs are 'flexible' jobs involving part-time work or independent sub-contracting. Large firms increasingly dominate the growing service industries but to cut costs they make expanded use of sub-contracting and part-time (usually female and minority-group) employees. Indeed, a major feature of the restructuring of labour markets in the USA and Britain in the period 1970–90 has been the 'feminization' of employment in the expanding service sectors at the same time that more highly paid and predominantly male manufacturing jobs have been disappearing (Kosters, 1992; McDowell, 1991). In the retail and health sectors 'worksites' are decentralized and administrative functions separated spatially from the delivery of the services themselves even as large firms become dominant. The services are increasingly standardized from place to place; much like the physical settings such as regional shopping centres, shopping malls, and suburban medical buildings in which they are located.

Spatial homogenization rather than local specialization, therefore, characterizes the emerging spatial pattern of major service industries and the flexible employment upon which they are coming to rely. This flexibility is more difficult to romanticize than that associated with manufacturing. It involves serious reductions in incomes compared to those paid in the 'old' Fordist manufacturing industries. It also reduces the overall 'power' of the work-force through exploiting gender and ethnic divisions (e.g. it is 'natural' to pay women less) and spatially dispersed worksites to restrict employment security and limit labour organizing.

Summary

In this chapter we have seen how the crisis of Fordism, coupled with trends associated with corporate integration, technological advances, and shifting consumer demand have begun to result in the globalization of economic activities that were previously localized within the core economies. Among the salient features of these changes are:

- the emergence of production hierarchies within the large companies that have come to dominate most industries. These hierarchies have tended to result in separate locational settings for (1) high-level corporate control, (2) production requiring high inputs of skilled labour and new technology; intermediate administration and R & D activities; and (3) routine production.

- The organization of the world economy into three broad international regions:
  1. the highly integrated and very diversified industrial cores and control centres of the 'North';
  2. the recently industrialized and newly industrializing semi-peripheries, mostly in the world's 'middle regions' (eastwards from Mexico, through the Mediterranean to South East and East Asia) and in parts of the southern hemisphere;
  3. the relatively thinly industrialized periphery that makes up most of the 'South' and is highly dependent on the 'North'.

- The persistence within each of these broad regions of nested hierarchies of nations and regions at different levels of economic development. Thus the periphery contains core regions and semi-peripheral regions (as, for example, the Lagos/Ibadan region and Abidjan region, respectively, in West Africa), the semi-periphery contains core regions and peripheral regions (e.g. the Calcutta–Hooghly–Howra conurbation and Uttar Pradesh, respectively, in India), and the core contains regions which are, relatively, peripheral and semi-peripheral (e.g. northern Scandinavia and southern Italy, respectively, in Western Europe).

Key Sources and Suggested Reading

Albrechts, L. and Swyngedouw, E. 1989. In L. Albrechts et al. (eds), *Regional Policy at the Crossroads*. London: Jessica Kingsley Publishers.

Amin, A. and Thrift, N. 1992. Neo-Marshallian nodes in global networks. *International Journal of Urban and Regional Research*, 16, 571–87.

Balassa, B. 1979. *The changing international division of labor in manufactured goods*. Washington, DC: The World Bank. Working Paper 329.

Barff, R. 1995. It's gotta be da shoes. *Environment and Planning A*, **27**, 55–79.

Barnet, R, J, and Cavanagh, J. 1994. *Global Dreams: Imperial Corporations and the New World Order*. New York: Simon & Schuster.

Bingham, R. D. and Hill, E. W. (eds) 1997. *Global Perspectives on Economic Development*. New Brunswick, NJ: Center for Urban Policy Research.

Bluestone, B. and Harrison, B. 1982. *The Deindustrialization of America*. New York: Basic Books.

Castells, M. 1996. *The Rise of the Network Society*. Oxford: Blackwell.

Chandler, A. D. 1992. Organizational capabilities and the economic history of the industrial enterprise. *Journal of Economic Perspectives*, **6**, 79–100.

Christopherson, S. 1989. Flexibility in the US service economy and the emerging spatial division of labour. *Transactions of the Institute of British Geographers*, **14**, 131–43.

Dicken, P. 1994. 'Global-local tensions: firms and states in the global space-economy. *Economic Geography*. **70**, 101–27.

Dicken, P. and Thrift, N. 1992. The organization of production and the production of organization: why business enterprises matter in the study of geographical industrialization. *Transactions of the Institute of British Geographers*, **17**, 279–91.

Dunning, J. H. 1979. Explaining changing patterns of international production: in defence of eclectic theory. *Oxford Bulletin of Economics and Statistics*, **41**, 269–96.

Galbraith, J. K. 1977. *The Affluent Society*. Boston: Houghton Mifflin.

Gertler, M. S. 1992. Flexibility revisited: districts, nation-states, and the forces of production. *Transactions of the Institute of British Geographers*, **17**, 259–78.

Glasmeier, A. 1990. The role of merchant wholesalers in industrial agglomeration formation. *Annals of the Association of American Geographers*, **80**, 394–417.

Hamilton, F. E. I. 1984. Industrial restructuring: an international problem. *Geoforum*, **15**, 349–64.

Harrison, B. 1992. Industrial districts: old wine in new bottles? *Regional Studies*, **26**, 469–83.

Harvey, D. 1990. *The Condition of Postmodernity*. Oxford: Blackwell.

Henderson, J. 1989. *The Globalization of High Technology Production: Society, Space and Semiconductors in the Restructuring of the Modern World*. London: Routledge.

Howells, J. and Wood, M. 1993. *The Globalisation of Production and Technology*. London: Belhaven Press.

Kosters, M. H. 1992. *Workers and Their Wages*. Washington, DC: American Enterprise Institute.

Krugman, P. 1991. *Geography and Trade*. Cambridge, MA: MIT Press.

McDowell, L. 1991. Life without father or Ford: the new gender order of post-Fordism. *Transactions of the Institute of British Geographers*, **16**, 400–19.

Marshall, A. 1920. *Principles of Economics*. London: Macmillan.

Marshall, B. (ed.) 1991. *The Real World*. Boston: Houghton Mifflin.

OECD 1996. *The Knowledge-Based Economy*. Paris: OECD.

Piore, M. and Sabel, C. 1984. *The Second Industrial Divide*. New York: Basic Books.

Porter, M. E. (ed.) 1986. *Competition in Global Industries*. Cambridge, MA: Harvard Business School Press.

Quinn, J. B. 1987. The impacts of technology in the service sector. In B. R. Guile and H. Brooks (eds), *Technology and Global Industry: Companies and Nations in the World Economy*. Washington, DC: National Academy Press, 119–59.

Sachar, A. and S. Öberg (eds) 1990. *The World Economy and the Spatial Organization of Power*. Brookfield, VT: Gower.

Sadler, P. 1992. *The Global Region*. Tarrytown, NY: Pergamon.

Scott, A. J. 1988. *New Industrial Spaces: Flexible Production Organization and Regional Development in North America and Western Europe*. London: Pion.

Scott, A. J. 1992. The role of large producers in industrial districts: a case study of high technology systems houses in Southern California. *Regional Studies*, **26**, 265–75.

Scott, A. J. 1996. Regional Motors of the Global Economy, *Futures*, **28**, 391–411.

Taylor, M. J. and Thrift, N. J. 1983. Business organisation, segmentation and location. *Regional Studies*, **17**, 445–65.

Thrift, N. J. 1995. A Hyperactive World. In R. J. Johnston, P. J. Taylor and M. Watts (eds), *Geographies of Global Change: Remapping the World in the Late Twentieth Century*. Oxford: Blackwell 18–35.

Thurow, L. 1993. *Head to Head: The Coming Economic Battle Among Japan, Europe, and America*. New York: Warner Books.

Vernon, R. 1966. International investment and international trade in the product cycle. *Quarterly Journal of Economics*, **80**, 190–207.

Williamson, O. E. 1985. *The Economic Institutions of Capitalism*. New York: Free Press.

CHAPTER 7

THE SPATIAL REORGANIZATION
OF THE CORE ECONOMIES

The evolution of advanced capitalism and the emergence of an informational economy have led to a significant reorganization of the economic geography of places and regions throughout most of the world. We shall examine the nature and implications of these changes for the economic landscapes of the NICs and LDCs in Chapter 9. In this chapter, we focus our attention on urban and regional change in the core nations, emphasizing the overall impact of corporate reorganization in creating new industrial spaces and affecting regional economic well-being.

It is important to note at the outset that the globalization of the economy described in the previous chapter has resulted in a relative *increase* in the importance of cities and regions as agents of economic development:

> . . . in a world economy whose productive infrastructure is made up of information flows, cities and regions are increasingly becoming critical agents of economic development. . . . Precisely because the economy is global, national governments suffer from failing powers to act upon the functional processes that shape their economies and societies. But regions and cities are more flexible in adapting to the changing conditions of markets, technology and culture. True, they have less power than national governments, but they have a greater response capacity to generate targeted development projects, negotiate with multinational firms, foster the growth of small and medium endogenous firms, and create conditions that will attract the new sources of wealth, power, and prestige.

> (Castells and Hall, 1994, p. 7)

As at the international level, the major components of urban and regional change have hinged on the redeployment of routine production capacity from high-cost to low-cost locations and the retention/localization of facilities requiring high inputs of technology and/or skilled labour in key locations with appropriate resources and amenities. As a result, two countervailing trends characterize the 'new' economic geographies of Europe, North America, Australasia and Japan: *decentralization* and *consolidation*. Decentralization has led to an attenuation of regional and inter-urban gradients in economic well-being; consolidation has contributed to an increased spatial differentiation in terms of the conditions of production and exchange and the hierarchical structure of control.

At the same time, we have to consider the effects on core countries of wider changes in the world economy and in the regime of capitalist accumulation. We begin, therefore, with a brief outline of the main outcomes of these secular changes. In the broadest of terms, three key changes can be identified: a transformation of the relationship between capital and labour; the creation of new regional divisions of labour; and the development of new roles for the state. Together, they amount to the beginning of a distinctive new context for economic development in core countries.

A NEW CONTEXT FOR URBAN AND REGIONAL CHANGE

Neo-Fordism has emerged as a new *regime of accumulation* as companies throughout the developed world have exploited new technologies and new strategies in order to remain competitive in a globalizing economy. In the process, the relationship between capital and labour has been transformed, with capital recapturing the initiative over wage rates and conditions that had been established under 'organized' industrial capitalism. New technologies have played a major role in this transformation. The introduction of robotics in factories and information-processing technologies in offices, for example, has made for dramatic increases in productivity but has also created a long-term threat: that of substituting machines for workers, thus placing labour in a weak bargaining position.

Neo-Fordism has also created a new international and inter-regional division of labour, as large corporations have pursued flexible strategies in order to deal with, and exploit, the 'time-space compression' introduced by new transport and tele-communications technologies such as long-distance fibre-optic systems, regional telecommunications systems, satellite teleports, 'smart' buildings, telefax, and microwave communications. Paradoxically, this 'annihilation of space' and 'electronic colonialism' has *heightened the importance of geography*. The reduction of spatial barriers has had the effect of greatly magnifying the significance of what local spaces contain because the new flexibility of the business world enable relatively small difference between places to be quickly, if temporarily, exploited to good effect. As a result, there has been an acceleration of shifts in the patterning of uneven development on the basis of particular local mixes of skills and resources: *a continuously variable geometry of labour, capital, production, markets and management*.

Last, but not least, Neo-Fordism has required the development of new roles for the state and the public sector: reduced government intervention in the economy and a decreased emphasis on collective consumption (school, hospitals, community services, etc.). The dilemma facing most governments was that the deindustrialization recession accentuated the vulnerability of more and more people while making it increasingly difficult – politically as well as economically – to finance existing programmes. As a result, a 'new conservativism' in the orientation of central and local governments emerged. This new conservatism was associated with an ideological stance based on the assertion that the welfare state had not only generated unreasonably high levels of taxation, budget deficits, disincentives to work and save, and a bloated class of unproductive workers, but also that it may have fostered 'soft' attitudes towards 'problem' groups in society. The consequent restructuring of the welfare state was most pronounced in the United Kingdom and the United States, where the Thatcher and Reagan administrations respectively embarked on programmes of privatization in health, housing and education, accompanied by cuts (some absolute, some relative) in higher education, in programmes for the unemployed, the disabled and the elderly, and in regional policy budgets. In the United Kingdom, closer controls on local government expenditure by the central government led to corresponding cuts at the local level, particularly in depressed towns and cities where the incidence of need for welfare services is high but local fiscal resources are low.

This last point is central to the emergence of a new *mode of regulation* that, in turn, is tied in to the emergence of the new Neo-Fordist or 'flexible' regime of accumulation. Following Jessop (1992), we can characterize this emerging mode of regulation in general terms as involving a commitment to supply-side innovation in flexibility, with specific manifestations in several areas:

- A change in the regulation and conduct of labour markets, involving (1) a shift away from centralized collective bargaining towards company- or plant-level negotiations, and (2) an increasing tolerance of insecurity and marginality in the wage relations and employment conditions of unskilled workers.
- Flatter, leaner and more flexible forms of corporate organization that are suited not only to externalization but also to joint ventures and to public–private cooperation.
- More flexible forms of credit: the result of deregulation of financial markets and the internationalization of finance and financial services, which has effectively reduced the degree of control that can be exerted by individual national governments.
- The displacement of Keynesian welfare states by 'workfare states':

> The emerging state form will no longer be concerned mainly with securing full employment within a national economy but with guiding and promoting the structural competitiveness of the national economy by intervening on the supply-side to encourage innovation; and it will no longer be concerned to generalise norms of mass consumption but to articulate policies to the need to promote greater flexibility.
>
> (Jessop, 1992, p. 32)

Table 7.1 illustrates some of the main contrasts between the characteristics of states under the two systems.
- The 'hollowing out' of the nation state as a result of (1) the displacement of national power upwards through pan-regional agreements and transnational organizations (see Chapter 12) and downwards to regional and local governments (see Chapter 13), and (2) increasing cooperation among local and regional governments in key fields such as R & D and technology transfer in ways that bypass their respective national states.

Table 7.1: Contrasts in the characteristics of states: Fordism and flexible accumulation

Fordism	Flexibility
Regulation	Deregulation/regulation
Rigidity	Flexibility
Collective bargaining	Division/individualization
Socialization of welfare (the welfare state)	Privation of collective needs and social security
	Soup kitchen-state for the underprivileged
International stability through multilateral agreements	International destabilization
Centralization	Decentralization and sharpened interregional/intercity conflicts
The 'subsidy' state/city	The 'entrepreneurial' state/city marketing
Indirect intervention in markets through income and price policies	Direct state intervention in markets through procurement
Firm-financed R & D	State-financed R & D
Industry-led innovation	State-led innovation

Source: Albrechts and Swyngedouw (1989), Fig. 1.

It has been suggested that this emerging mode of regulation, together with the imprint of Neo-Fordism and the globalization of economic activity from the core (as described in Chapter 7) mark the beginning of a new phase of capitalism. Lash and Urry (1987) describe this phase as one of *disorganized capitalism*, to distinguish it from the previous phase that was dominated by a closely regulated and highly organized relationship between labour, capital and government at the level of nation states. Disorganized capitalism, in contrast, is characterized by:

1. a deconcentration of capital within national markets, a growing separation of finance from industry, and the decline of cartels (as a result of the growth of a world market, the increasing scale of industrial commercial, and banking enterprises, and the general decline of tariffs);
2. a decline in the absolute and relative size of the core working class and the expansion of a service class of professional, white-collar workers in core economies as they are deindustrialized;
3. a decline in the importance and effectiveness of national-level collective bargaining and a growth in company and plant-level bargaining (as companies exert their new leverage in order to impose more flexible forms of organization);
4. an increasing independence of large monopolies from direct control and regulation by individual nation states;
5. a decline in average plant size because of shifts in industrial structure, substantial labour-saving capital investment, the hiving off of various sub-contracted activities, and the export of labour-intensive activities to underdeveloped countries and to peripheral regions within core nations;

Table 7.2: Contrasts in ideology: Fordism and flexible accumulation

Fordism	Flexibility
Mass consumption of consumer durables: the consumption society	Individualized consumption: 'Yuppie'-culture
Modernism	'Post'-modernism
Totality/structural reform	Specificity/adaptation
Rationalism	Deconstructionism
Socialization	Individualization
	The 'spectacle' society

Source: Albrechts and Swyngedouw (1989), Fig. 1.

6. The decline of metropolitan dominance within core nations: the loss of jobs and population from inner-city areas and an increase of jobs and population in smaller towns and some rural areas;

7. a weakening of the degree to which industries are concentrated in specific nations and regions as a result of the new, variable geometry of the division of labour: 'each nation or locality develops its own kind of industrialization process even as it may depend on, and partake of, global investment flows and multinational industrial production systems' (Storper, 1987, p. 591); and

8. a decline in the salience and class character of political parties, an increase in cultural fragmentation and pluralism, the emergence of a 'global' culture and an international consciousness, and the ascendance of a 'postmodern' cultural-ideological configuration (Table 7.2) (Lash and Urry, 1987, pp. 5–7).

SPATIAL REORGANIZATION OF THE CORE ECONOMIES

In this section, we examine two important trends in the economic geography of the core economies, both of which have occurred within the context of the globalization of economic activity and the secular shifts within core countries from manufacturing to informational economies. The first of these trends is the regional, inter-metropolitan and metropolitan *decentralization* of certain categories of both manufacturing and service employment. The second is the regional and inter-metropolitan *consolidation* of other kinds of activities.

Spatial decentralization and external control

Decentralization has operated at regional, metropolitan and inter-metropolitan scales in response to a variety of complex and often cross-cutting processes of reorganization and adjustment.

Regional decentralization

Regional decentralization is a product of the migration of some firms and 'births' and 'deaths' of others, together with the transfer of productive capacity by plant shutdowns in core, metropolitan regions and the opening of new branch plants (or the expansion of existing ones) in declining or peripheral cities and regions. The result has been the creation of what have been called 'branch-plant economies' and 'module production places' in the peripheral regions of most core countries.

A useful distinction can be made between **diffuse industrialization** and **branch-plant industrialization** (Hudson, 1983). The former has been directed towards the reserves of unskilled labour in peripheral rural regions, while the latter has been directed towards the skilled manual-labour reserves of declining industrial regions. Central and northeastern Italy provide classic examples of *diffuse* industrialization, much of it resulting from the decentralization of companies from the Milan–Turin area in response to the increasing shortage, cost and militancy of labour there. Typically, diffuse industrialization involves activities in which labour costs are an important part of overall production costs *and* in which there has been little scope for reducing labour costs through technological change: it can thus been seen as an expression of the product-cycle model of industrial location. Empirical studies have shown that the main attractions of rural locations for such activities have been:

- the availability of relatively low-cost labour
- inexpensive supplies of easily developed land
- lower levels of taxation
- low levels of unionization.

Branch-plant industrialization proper, on the other hand, typically involves activities which require significant inputs of technology and of skilled (or at least experienced) labour and which also require a certain degree of centrality in order to assemble and distribute raw materials and finished products. Good examples are provided by many former textile cities – Dundee and Rochdale in the United Kingdom, for example, and Amiens in France – where branch plants in a variety of 'light' industries (from car batteries, cash registers and cameras to tyres, watches and light engineering) have moved in to take advantage of 'surplus' labour, cheap factory space and an established infrastructure. It is not only manufacturing activities that are being decentralized, however. While many places have developed branch-plant economies on the basis of assembly-line activities, some have attracted white-collar information-processing or wholesaling functions (Hepworth, 1992). Omaha, Nebraska, for example, has become the '800' telephone exchange centre of the United States; Sioux Falls, South Dakota, is now Citicorp's credit operations centre; and Roanoke, Virginia, has become the centre for a number of mail order and TV shopping operations.

The twin processes of diffuse and branch-plant industrialization, combined with the process of mergers and acquisitions, have meant that regional decentralization has come to be characterized by increasing levels of **external control**. In the Northern region of England, for example, 78 per cent of manufacturing employment in 1973 was controlled by companies with headquarters outside the region,

diffuse and branch-plant industrialization The growth of manufacturing employment in peripheral rural regions of core economies – a result of the availability of reserves of relatively cheap and less militant unskilled and semi-skilled labour suitable for employment in branch factories of manufacturing enterprises based elsewhere.

external control This term refers to situations where employment opportunities and decisions about investment and production in a given plant or locality are controlled by corporate managers based in other cities, regions, or countries.

226

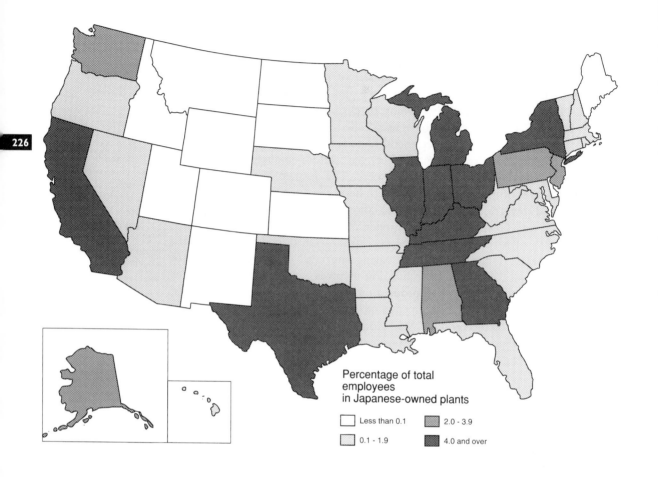

Percentage of total
employees
in Japanese-owned plants

| | Less than 0.1 | | 2.0 - 3.9 |
| | 0.1 - 1.9 | | 4.0 and over |

Figure 7.1

Employment in
Japanese-owned
manufacturing plants in
the United States,
1990

Source: Chang (1989), Fig. 1,
p.320

compared with 57 per cent in 1963 (Smith, 1979). In the southern states of the USA, fewer than one third of the new jobs created in manufacturing plants between 1969 and 1976 belonged to southern-based corporations (Birch, 1979).

As at the international level, it has been the large transnational conglomerates that have been particularly important in influencing the extent and spatial pattern of external control. In the United States, the total number of jobs in foreign-owned firms jumped from 2.0 million in 1980 to 3.2 million in 1986. Most of these were in manufacturing, and most were controlled by either British, German, Canadian, Japanese, Dutch, French, or Swiss companies. In some states, Japanese-owned companies alone account for more than 4 per cent of all manufacturing employees (Fig. 7.1). In high-tech, high-growth industries, more than 25 per cent of all employment in the United States – and as much as 50 per cent in several states – is controlled by foreign firms (Fig. 7.2).

Because of the degree of external control involved in regional economic decentralization, it has become a moot point as to how much long-term benefit will accrue to the regions involved. On the *positive* side, it can be argued that branch-plant economies and module production places benefit by having access to the financial resources and technological and administrative innovations of the parent firm. On

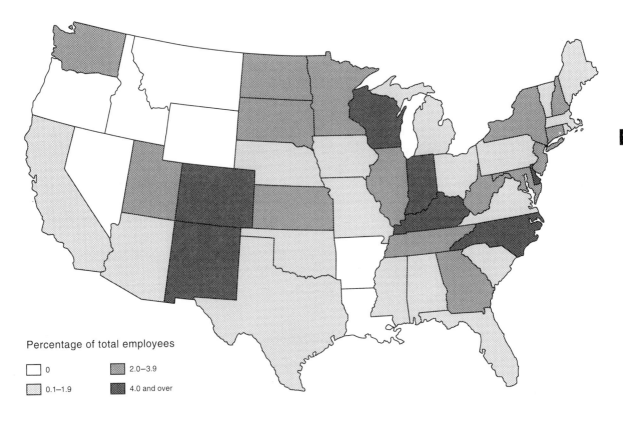

Percentage of total employees

☐ 0
☐ 0.1–1.9
▨ 2.0–3.9
■ 4.0 and over

the *negative* side, it has been suggested that the absence of 'higher-order' corporate functions:

1. limits the profile of local employment opportunities, leading to a *deskilling* of the local workforce, to the suppression of entrepreneurial drive and enthusiasm, and to the retardation of technological innovation;
2. results in a very open regional economy, so that international economic fluctuations are transmitted into the region relatively quickly. The corollary of this is that because externally controlled plants are poorly integrated with the local economy, their own potential multiplier effects are limited; and
3. increases the vulnerability of branch-plant economies to the further redeployment of capital: branch-plant economies in the core nations are placed in direct competition with those of the NICs, which typically have much lower factor costs.

Figure 7.2
Foreign-owned high-technology employment as a percentage of all employment in high-technology industries conterminous United States, 1990.
Source: Warf (1990) Fig. 1, p. 427

Metropolitan decentralization

Metropolitan decentralization (the exodus of industry and employment from inner-city areas to suburbs) can in fact be traced to the 1930s; but since the early 1970s the process has begun to dominate patterns of urban development in a number of countries. Historically, the major impetus for metropolitan decentralization has been employers' desire to sidestep the increasing militancy of labour in inner-city neighbourhoods. Suburban locations have also been attractive to many industries because of the availability of larger tracts of relatively cheap land. Given this basic attractiveness, successive improvements in transport and communications have greatly accelerated the process of decentralization.

Residential suburbanization, meanwhile, has provided labour supplies – including cheap, non-unionized, female labour – that have encouraged the suburbanization of more firms. A mutually reinforcing process was thus set in motion. At the same time, the intensification of some of the locational disadvantages of inner-city areas – higher taxes, congestion, restricted sites, and so on – began to push some firms out. The stagflation episode of the 1970s threw the cumulative effect of all these factors into focus for many firms, as profits were sharply squeezed. What was most pronounced was the 'shake-out' of routine and labour-intensive inner-city areas – some of it destined for relocation in the suburbs, but much more destined for relocation in rural areas, peripheral regions, peripheral countries, or the bankruptcy courts. *The overall result was a sharp acceleration in the relative rate of growth of employment in the suburbs and a sudden intensification of the 'inner city problem'.*

Inter-metropolitan decentralization

At the inter-metropolitan level, the most striking aspect of decentralization has involved service industries, particularly business and professional services. Corporate reorganization, facilitated by advances in telecommunications, has resulted in a general decentralization of routine business and professional services down the urban hierarchies of core countries. O hUallacháin and Reid (1991), who have analysed such changes in the United States, showed that the process has been geographically selective, with a relatively small number of metropolitan areas (including Anaheim, San José, Boston, Denver, Newark, Detroit, Phoenix, Fort Lauderdale, Dallas, Tampa, Minneapolis, San Antonio, Sacramento, Indianapolis and Riverside) experiencing a disproportionately high rate of growth in employment in business and professional services. This phenomenon is surprising to some observers, who had expected that new communications technologies would allow for the dispersion of 'electronic offices' and, with it, the decentralization of an important catalyst for local economic development. A good deal of geographic decentralization of offices has occurred, in fact, but it has mainly involved 'back office' functions that have been relocated from metropolitan and business-district locations to small-town and suburban locations (Warf, 1995).

back-office functions
Record-keeping and analytical functions that do not require frequent personal contact with clients or business associates.

Back-office functions are record-keeping and analytical functions that do not require frequent personal contact with clients or business associates. The accountants and financial technicians of main street banks, for example, are back-office workers. Developments in computing technologies, database access, electronic data interchanges, and telephone-call-routing technologies are enabling a larger share of back-office work to be relocated to specialized office space in cheaper settings, freeing space in the high-rent locations occupied by the bank's front office. For example, the US Postal Service is using Optical Character Readers (OCRs) to read addresses on mail, which is then bar-coded and automatically sorted to its appropriate substation. Addresses that the OCRS cannot read are digitally photographed and transmitted to a computer screen, where a person manually types the address into a terminal. In Washington, DC, OCR sorting takes place at the central mail facility, but the manual address entry is done in Greensboro, North Carolina, where wage rates are lower. Workers in Greensboro view images of letters as they are sorted in Washington and enter correct addresses, which are in turn electronically transmitted back to be bar-coded on the piece of mail.

Among the more prominent examples of back-office decentralization from US metropolitan areas have been the relocation of back-office jobs in American Express from New York to Salt Lake City, Fort Lauderdale, and Phoenix; the relocation of Metropolitan Life's back offices to Greenville, South Carolina, Scranton, Pennsylvania, and Wichita, Kansas; the relocation of Hertz's data-entry division to Oklahoma City, Dean Witter's to Dallas, and Avis's to Tulsa; and the relocation of Citibank's MasterCard and Visa divisions to Tampa and Sioux Falls. Some places have actually become specialized back-office locations as a result of such decentralization. Omaha and San Antonio, for example, are centres for a large number of telemarketing firms, while Roanoke, Virginia, has become something of a mail-order centre.

Internationally, this trend has taken the form of offshore back-offices. By decentralizing back-office functions to offshore locations, companies save even more in labour costs. While Zürich-based Swissair, for example, now does its accounting in Mumbai (Bombay), India, several New York-based life-insurance companies have established back-office facilities in Ireland. Situated conveniently near Ireland's main international airport, Shannon Airport, they ship insurance claim documents from New York via Federal Express, process them, and beam the

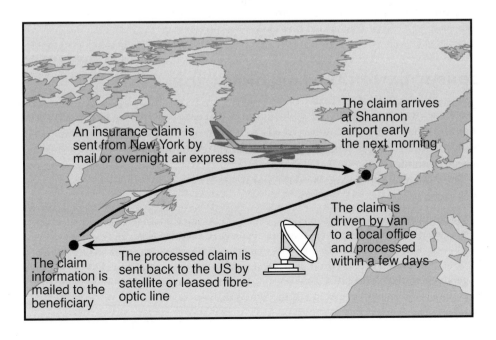

Figure 7.3

International back offices

Source: Warf (1995) p. 374

229

results back to New York via satellite or the TAT-8 transatlantic fibre-optics line (Fig. 7.3).

The Sunbelt

Regional, metropolitan and inter-metropolitan decentralization are individual components in what is ultimately a multi-dimensional dynamic of spatial change. The growth of the US Sunbelt provides a good example. The Sunbelt phenomenon can be interpreted as the *combined* product of diffuse industrialization, inter-metropolitan decentralization and metropolitan decentralization. Such an interpretation is supported by the types of employment growth which characterize the rise of the Sunbelt: (1) production jobs in branch plants in industries such as textiles, clothing and electronics; (2) production jobs in branch plants and in locally based firms in high-growth industries – mainly in computer hardware, scientific instruments, aerospace, and chemicals and plastics; and (3) service jobs catering both to these industries and to the increased population attracted to the retirement and leisure communities.

In very general terms, it appears that peripheral and semi-peripheral Sunbelt states like Arizona, California, the Carolinas, and Texas have been able to benefit from relative advantages in terms of labour costs, labour unionization, land costs, energy costs, local taxation, local government boosterism and federal expenditure patterns. In addition, Sunbelt cities have proved attractive to industries because they did not have a legacy of inefficient layout and infrastructure. As Gordon put it (1979, p. 78):

> They could be constructed from scratch to fit the needs of a new period of accumulation in which factory plant and equipment were themselves increasingly predicated upon a decentralized model. . . . There was consequently no identifiable downtown factory district. . . . Automobiles and trucks provided the connecting links, threading together the separate pieces. The corporate city became . . . The Fragmented Metropolis.

CONSOLIDATION AND AGGLOMERATION

The structural and functional consolidation of certain activities under advanced capitalism has made for counter-trends that have strengthened the economic well-being of many of the largest and most central components of the core nations' space-economies.

The fundamental reason for the consolidation of certain economic activities in such settings is that

> Large towns offer larger local markets, with the associated internal economies of scale, plus greater external economies than are available in smaller places, and together these allow production costs which are often significantly lower than those in smaller towns: once transport costs began to fall substantially, so that they were less than the production cost differential between the large-town and the small-town firm, the former could begin the invasion of the latter's market.

(Johnston, 1980, pp. 110–11)

The sectoral shifts and manufacturing specializations of advanced capitalism have also worked in favour of many large cities and metropolitan regions (Amin and Thrift, 1994). Manufacturers of many sophisticated new high value-added products have been drawn to such locations. The reasons for this are several:

- the complex links that these new products have with established industries;
- their dependence on risk capital in the early stages of development;
- their need for access to a large, affluent and sophisticated market during the early stages of marketing.

231

Similarly, large parts of the rapidly expanding service sector have been drawn towards metropolitan locations because of the kind of environment and work-force required by information-processing, coordinating, controlling and marketing activities.

Corporate restructuring and new competitive strategies have added to the agglomerative and recentralizing trends of certain economic activities in metropol-itan settings. The flexibility of Neo-Fordism requires a new social division of labour with access to a large and fluid labour pool (containing part-time and temporary workers as well as highly skilled workers – attributes that are most easily found in metropolitan settings). Equally important, metropolitan settings are essential to the *externalization* of certain functions and the more extensive use of outside consult-ants, subcontracting, joint ventures, strategic alliances and collaborative R & D that characterize flexible production systems.

Finally, the national and international redeployment of activities by large con-glomerate companies has also contributed to the consolidation of certain activities in the central regions and metropolitan areas of the developed nations. In particular, there has been *a marked localization of two key functions: headquarters offices and R & D establishments*. Indeed, the distribution of these two functions has come to represent an important dimension of the 'new' economic geography of advanced capitalism.

Consolidation in rural regions: agribusiness

Direct corporate involvement in agriculture – agribusiness – has been an inevitable outcome of the logic of specialization and economies of scale. With greater specialization, farms become less autonomous and self-contained as productive units, making for the penetration of an integrated, corporate system of food processing and distribution:

> Agriculture has become increasingly drawn into a food-producing complex whose limits lie well beyond farming itself, a complex of agro-chemical, engineering, processing, marketing and distribution industries which are involved both in the supply of farming inputs and in the forward marketing of farm produce.
>
> (Newby, 1980, p. 61)

It is in the actions of food-processing conglomerates like Associated British Foods, Nestlé and Rank-Hovis-Macdougall, Newby suggests, 'that the shape of agriculture and ultimately of rural society in virtually all advanced industrial societies is decided' (p. 62). The most common form of corporate involvement in agriculture has to do with the forward contracting of produce at a fixed price. This not only weakens the independence of farmers, but also tends to transfer income from farmers and rural communities to the processing industry. Forward-contracting arrangements also reinforce the overall structural changes affecting agriculture:

> They encourage both fewer, larger holdings and increased specialization so that the size of individual enterprises can be enlarged to fully achieve the prevailing scale economies. This trend . . . is likely to lead to both a reduction in the numbers employed in agriculture, and a decline in the managerial role of those farmers remaining . . . leaving them caretaker functions.
>
> (Metcalf, 1969, p. 104)

Rural landscapes have also been affected as the logic of industrial production and centralization has been applied to agriculture. In northwestern Europe, for example, field systems have been rationalized, hedgerows and dykes removed, and mechanization has virtually eliminated the need for gang labour, leaving the fields of most farms devoid of human life for most of the year. Factory farming has brought poultry and pigs indoors permanently, while many cattle spend their winter months indoors, and there are now 'zero grazing' techniques which may see them inside year round. Only sheep steadfastly refuse to acknowledge the laws of industrial production, stubbornly refusing to prosper in regimented and sanitized conditions.

Corporate control centres and world cities

The United States provides a good example of the changing geography of corporate headquarters. Historically, the most striking feature of the geography of corporate headquarters in the USA has been the dominance of the Manufacturing Belt in general and of New York and Chicago in particular. Elsewhere, the pattern of headquarters offices has tended to reflect the geography of urbanization, so that the more important 'control centres', in terms of business corporations, have been the major entrepôts and central places which developed under earlier phases of economic development, as points of optimal accessibility to regional economies.

With the arrival of advanced capitalism the relative importance of the control centres of the Manufacturing Belt has decreased somewhat, with cities in the Midwest, the South and the West increasing their share of major company headquarters offices. Atlanta, Denver, Houston, Minneapolis and Seattle have been the major beneficiaries of this shift, though no new control centres have emerged to counter the dominance of New York, Chicago and the other major cities of the northeast. One interpretation of this shift is that it is simply a reflection of changes in the urban system: high-order urban areas tend to be higher-order business control centres because of their reserves of entrepreneurial talent, the array of

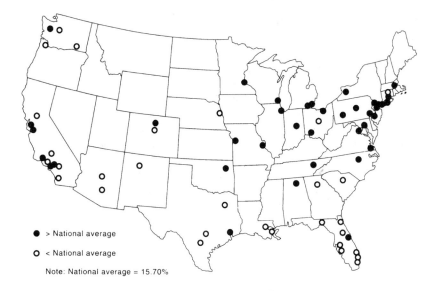

Figure 7.4

US metropolitan areas above and below the national average share of total job growth in business and professional services, 1976–86

Source: O hUallacháin and Reid (1991) Fig. 1, p. 266

• > National average

○ < National average

Note: National average = 15.70%

support services that they can offer, their accessibility in both a regional and a national context.

In overall terms, 'there has been a process of *cumulative and mutual reinforcement* between relatively accessible locations and relatively effective entrepreneurship' (Borchert, 1978, p. 230, emphasis added; see also Dunning and Norman, 1987). This has made for a high degree of inertia in the geography of economic control centres and this, in turn, has consolidated the economic position of the metropolitan areas of the northeast through the *multiplier effects* of concentrations of corporate headquarters, whereby the vitality of the corporate administrative sector contributes to the growth and circulation of specialized information concerning business activity, thus generating further employment in a relatively well-paid sector and sustaining the area's attractiveness for headquarters offices.

The concentration of corporate headquarters offices has contributed to the emergence of a few places within the international urban system as 'world cities', dominant centres and sub-centres of transnational business, international finance, and international business services – what Friedmann (1986) called the 'basing points' for global capital. These 'world cities', it should be stressed, are not necessarily the biggest within the international system of cities in terms of population, employment or output. Rather, they are the 'control centres' of the world economy: places that are critical to the articulation of production and marketing under the contemporary phase of world economic development (Knox, 1995). Because these properties are difficult to quantify, it is not possible to establish a definitive list or hierarchy of world cities. It is possible to identify world cities on the basis of their role in articulating the functions of the global economy associated with financial markets, major corporate headquarters, international institutions, communications nodes, and concentrations of business services (Fig. 7.4). On this basis, all but three of the dominant and major world cities – São Paulo, Seoul and Singapore – are located in core countries. The relative importance of secondary world cities is very much a function of the strength and vitality of the national economies that they articulate. In overall terms, notes Friedmann,

> The complete spatial distribution suggests a distinctively linear character of the world city system which connects, along an East–West axis, three distinct sub-systems: an Asian sub-system centred on the Tokyo–Singapore axis, with Singapore playing a subsidiary role as regional metropolis in Southeast Asia; an American sub-system based on the three primary core cities of New York, Chicago and Los Angeles, linked to Toronto in the North and Caracas in the South, thus bringing Canada, Central America and the small Caribbean nations into the American orbit; and a West European sub-system focused on London, Paris and the Rhine Valley axis from Randstad and [sic] Holland to Zurich. The southern hemisphere is linked into this system via Johannesburg and São Paulo.
>
> (Friedmann, 1986, pp. 72–3)

Centres of innovation

As Castells notes (1996, p. 56), the development of the information technology revolution has contributed to the formation of milieux of innovation where important commercial discoveries interact and are tested in a recurrent process of trial and error. These milieux require 'spatial concentration of research centers, higher education institutions, advanced technology companies, a network of ancillary suppliers of goods and services, and business networks of venture capital to finance start-ups'.

The geography of these milieux has important implications for urban and regional development. Malecki, who has examined the geography of R & D activity in the USA in detail, suggests that the overall pattern can be interpreted in terms of:

1. the availability of highly qualified personnel, and
2. corporate organization.

In relation to the former, he suggests that amenity-rich locations (cities with a wide range of cultural facilities, well-established universities and pleasant environments) which are attractive to highly qualified personnel tend to be favoured as locations for R & D activity. Malecki also notes that existing concentrations of R & D activity tend to be attractive because of the potential for 'raiding' other firms (Malecki, 1991).

In relation to corporate organization, it seems that corporate-level or long-range R & D is best performed in or near headquarters complexes in a central laboratory where intra-organizational interaction can be fostered. In firms with independent divisions producing quite different product lines, however, R & D activity tends to be located in separate divisional laboratories. Such a pattern is particularly common for conglomerates which have acquired firms with active R & D programmes in existing laboratories. Finally, some industries, whatever the organizational structure of the firms involved, require R & D laboratories to have close links with production facilities, resulting in a relatively dispersed locational pattern corresponding to the pattern of plant location.

The net result of these locational forces is in fact *a marked agglomeration of R & D laboratories in major control centres and manufacturing regions*. It is the metropolitan areas of the Manufacturing Belt which dominate the geography of corporate R & D activity. As Malecki points out, most of these are either major control centres with a significant element of headquarters office activity. Elsewhere, R & D tends to be concentrated in 'innovation centres' – university cities with diversified economies, some high-technology activity and a strong federal scientific presence (e.g. Austin, Texas; Huntsville, Alabama; Lincoln, Nebraska).

In terms of locational *trends*, Malecki has shown that:

> Although industrial R&D appears to be evolving away from a dependence on some large city regions, especially New York, it remains, at the same time, a *very markedly large-city activity*. . . . The comparative advantage of city size, particularly in centres of corporate headquarters location, manufacturing activity and university and government research, shows little sign of reversing.
>
> (Malecki, 1979, p. 321, emphasis added)

In short, R & D laboratories, like headquarters offices, exhibit a strong tendency for consolidation, accompanied by a certain amount of decentralization. This pattern has important implications for regional economic development, for the urban areas in which concentrations of R & D activity exist will in future be able to consolidate their comparative advantage over other areas in the generation of new products and new businesses. They will also benefit from the short-term multiplier effects of employment generation in a particularly well-paid sector. Conversely, cities and regions with little R & D activity will be at a disadvantage in keeping up with the new economic content of advanced capitalism.

OLD INDUSTRIAL SPACES

One of the most striking overall changes within core economies has been the decline in the traditional industrial manufacturing employment base. Initially, this took the form of a relative decline: growth in the postwar boom period was much greater in the service sector of most economies than it was in the manufacturing sector. With the globalization of economic activity, however, there has been an *absolute* decline in core manufacturing employment (see, for example, Figs 7.5 and 7.6). Whereas in 1960 manufacturing in the 15 most industrialized OECD countries generated between 25 and 42 per cent of the GDP and accounted for similar proportions of their employment, the comparable figures for 1990 were in the range of 15 to 25 per cent.

The decline has been most pronounced in the early industrializers of north western Europe. In the United Kingdom, for example, more than 1 million manufacturing jobs disappeared, in *net* terms, between 1966 and 1976: a fall of 13 per cent. This decline affected almost every sector of manufacturing: not just the traditional pillars of the manufacturing sector – shipbuilding (–9.7 per cent), metal manufacture (–21.3 per cent), mechanical engineering (–14.5 per cent) and textiles (–26.6 per cent) – but also its former growth sectors and the bases of the fourth Kondratiev cycle – motor vehicles (–10.1 per cent) and electrical engineering (–10.5

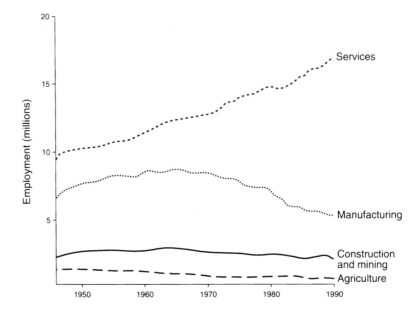

Figure 7.5
Employment by sector
in the United
Kingdom, 1946–91

per cent). In the West Midlands – widely regarded as a 'leading' region within Britain – a net loss of 151 117 manufacturing jobs between 1978 and 1981 helped to redefine the region as part of Britain's 'rust belt'. It is within the peripheral regions of the United Kingdom that the problem has been most acute, however. In

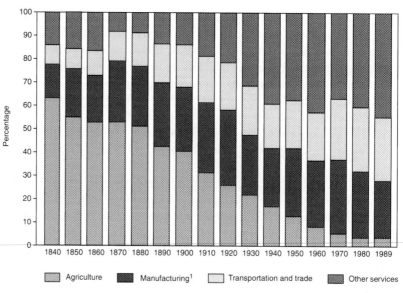

Figure 7.6
Employment shares, by
economic sector, USA,
1840–1989

Source: Council of Economic
Advisors (1991), Chart 4–1,
p. 114

Table 7.3: Jobs created and destroyed as a result of openings, closings, relocations, expansions and contractions of private business establishments in the United States 1969–76 (in thousands of jobs)

Region	Number of jobs in 1969	Jobs created		Jobs destroyed		Net job change	
		By openings and immigrations	Expansion	By closures and outmigrations	Contractions	Number	Per cent
US as a whole	57936.1	25281.3	19056.1	22302.3	13183.2	8851.9	15.2
Frostbelt	32701.2	11321.5	9470.4	11351.7	7212.1	2228.1	6.8
Northeast	15824.6	4940.4	4347.5	5881.5	3589.0	−182.6	−1.2
New England	3905.3	1251.2	1131.0	1437.2	952.1	−7.1	−2.6
Mid-Atlantic	11919.3	3689.2	3216.5	4444.3	2636.9	−175.5	−1.5
Midwest	16876.6	6381.1	5123.0	5470.2	3623.2	2410.7	14.3
East North Central	12563.6	4670.6	3581.8	3962.6	1651.7	1638.1	13.0
West North Central	4313.0	1710.6	1541.2	1507.6	971.5	772.7	17.9
Sunbelt	25234.9	13959.8	9585.7	10950.5	5971.0	6624.0	26.2
South	16044.5	8934.2	5964.6	6824.3	3803.3	4272.2	26.6
South Atlantic	8204.1	4651.2	2013.0	3547.9	2014.2	2002.1	24.4
East South Central	3065.2	1518.2	1089.9	1211.0	631.9	765.2	24.9
West South Central	4775.2	1764.8	1916.7	2065.4	1157.2	2503.9	31.4
West	9190.4	5025.6	3621.1	4126.2	2167.8	2352.7	25.6
Mountain	1914.9	1226.1	953.6	977.9	481.0	720.8	483.
Pacific	7248.5	3799.6	2667.6	3148.3	1686.8	1632.1	22.5

Source: Based on Bluestone and Harrison (1982), Table 2.1, p. 30.

Lancashire, for example, the textile industry alone shed over half a million jobs (Fothergill and Guy, 1992).

In short, the secular decline of the traditional industrial base has been most pronounced in the regions that had come to be most specialized in Fordist industrial manufacturing. For some communities in these regions, the consequences of plant shutdowns have been disastrous. In Youngstown, which became the symbol of American industrial decline, the closure of the Campbell Steel Works in 1977 eliminated over 10 000 jobs at a stroke. In overall terms, it has been estimated that the Mid-Atlantic region (New Jersey, New York, Pennsylvania) experienced a net loss of over 175 000 jobs during the period 1969–76 whereas the South Atlantic Region (Delaware, DC, Florida, Georgia, Maryland, North Carolina, South Carolina, Virginia, West Virginia) experienced a net gain of over 2 million jobs in the same period (Bluestone and Harrison, 1982). This represents a job loss of 1.5 per cent in the Mid-Atlantic region and a gain of 24.4 per cent in the South Atlantic region, compared with a net gain of some 15 per cent in the USA as a whole (Table 7.3).

Deindustrialization on this scale brought with it a number of *downward spiralling multiplier effects*, including the substantial contraction of major segments of intra-regional, vertically integrated production chains (e.g. ore-mining, coalmining, steel production, marine engineering and shipbuilding), and the disappearance of inefficient, more labour-intensive firms and sections of production chains (e.g. in textiles), leaving the region only finishing, specialized and high-quality product lines (e.g. in clothing) which now become dependent on supply linkages that are often 'stretched' overseas.

It would be misleading, however, to place too much emphasis on the demise of old industrial regions. Many cities within such regions have been successful in making the transition to an advanced economy. Within the Ruhrgebiet, for example, an economic renaissance is in evidence as the sites of old iron and steel works have been bulldozed to make way for Europe's largest shopping centre: a US$1.3 billion mega-mall with 230 stores, an 11 000 seat arena, a 1500-seat fast-food court, a 30-restaurant 'gastronomy' annex anchored by Planet Hollywood, two hotels, a tennis complex, a huge aquarium, and an artificial lake with an 'adventure island' for children. The British developers of Centro, as the project is called, expect it to draw up to 30 million visitors a year, and local officials expect the complex to create 10 000 new jobs. Such developments show that the regional economic decline associated with deindustrialization is not irreversible. Nevertheless, it is important to note that, in comparison with the jobs lost in traditional industries, employment in retailing, food and leisure provides a much less desirable base: typically, jobs are less well paid, with less security and fewer benefits.

We should also note that the process of deindustrialization (i.e. the *relative* decline of manufacturing jobs), while localized within old industrial regions has affected every region within core countries (Table 7.3). Even California, the archetypal Sunbelt state, was seriously affected by shutdowns. In Los Angeles alone, almost 18 000 manufacturing jobs were lost between 1978 and 1982, many of them the result of plant closures by large corporations like Ford, Pabst Brewing, Max Factor, Uniroyal and US Steel. In the state as a whole in the single year of 1980, more than 150 large plants closed down, displacing more than 37 000 workers (Bluestone and Harrison, 1982). In short, the overall losses of the Manufacturing Belt conceal a complex and uneven pattern of ebbs and flows.

NEW INDUSTRIAL SPACES

Advanced capitalism has not only seen the evolution and alignment of the 'old' economy, it has also seen the emergence of *new* industries based on entirely new technologies: semiconductors and computer software, for example; and more recently, biotechnology, photovolatics and robotics. These were the precursors of the fifth Kondratiev upswing (p. 9). The possibility thus emerges of an entirely new dimension to the economic landscapes of the developed nations, with concentrations of high-tech ('sunrise') industries initiating new patterns of urban and regional growth through new 'production ensembles' with new multipliers of cumulative causation (Storper and Scott, 1992).

Studies of high-tech industries in the United States confirm that job creation has been significant and is likely to continue to expand. By 1990, over 6.5 million (about 6 per cent of total US employment) were employed in high-tech industries, including nearly 2 million in service industries such as data processing and commercial testing laboratories.

The growth of some of these industries has certainly been explosive. Employment in computer software in the United States, for example, doubled to 250 000 in the 1970s, and grew to 879 000 by 1990; while employment in robotics rose from around 10 000 in 1980 to around 100 000 in 1990 (though a much greater number of jobs in other industries will of course have been *displaced* by the application of robotics). Whether such growth will be sufficient to cancel out the effects of continued deindustrialization and international redeployment is by no means certain, however. By the mid-1990s, many US high-tech companies had begun to experience acute problems as a result of the combination of the sluggishness of the overall economy, overproduction, and the persistence of Japanese non-tariff trade barriers. Some companies – including Texas Instruments, National Semiconductor, Intel, and Micron Technology – laid off workers; others went out of business altogether. Nevertheless, it should be recalled that, in the context of economic long waves, the effects of high-tech innovations are not likely to be fully developed until at least a decade or more after the beginning, around 1990, of the fifth Kondratiev upswing.

It should also be noted that in terms of occupational structure the expansion of high-tech employment is a microcosm of the trends which have dominated advanced capitalism. Studies in California, for example, 'suggest that the occupational, ethnic and gender composition of new jobs in high-tech sectors will tend to worsen the current trend toward the "disappearing middle", that is toward a labour force bifurcated between high-paid professionals and low-paid service workers' (Markusen, 1983, p. 19). In relation to corporate structure, high-tech industry is distinctive for its tendency towards the proliferation of small breakaway companies set up by key employees; but at the same time the larger and more established firms have soon been drawn into the process of mergers and acquisitions, either as the dominant element (in horizontal and vertical integration) or as a subsidiary element (in diagonal integration).

Technopoles

technopole
A planned development, within a concentrated area, for technologically innovative, industrial-related production. Technopoles include science parks, science cities, and other high-tech industrial complexes

The locational impact of these expanding high-tech activities is already emerging, however. The phenomenon that has received most attention has been the emergence of technology-oriented complexes, or **'technopoles'** (Castells and Hall, 1994), and the archetype has been in Santa Clara county – 'Silicon Valley' – in California. In the 1950s, Santa Clara was a quiet agricultural county with a population of about 300 000. By 1980, it had been transformed into the world's most intensive complex of high-tech activity, with a population of 1.25 million. During the 1970s, the high-tech industries of Santa Clara county generated over 40 000 jobs a year, each new job creating at least two or three additional jobs in other sectors – an extremely high multiplier in comparison with the figure of about one new job created for every new manufacturing job in buoyant (by national standards) metropolitan economies such as San Francisco's.

The development of Silicon Valley in the first instance is generally attributed to the work of Frederick Terman, a professor (and, later, Vice President) of Stanford University at Palo Alto, in the northwestern corner of Santa Clara county. As early as the 1930s, Terman began to encourage his graduates in electrical engineering to stay in the area and establish their own companies (one of the first was founded by William Hewlett and David Packard in a garage near the campus; it is now one of the world's largest electronics firms). By the end of the 1950s, Terman had persuaded Stanford University to develop a special industrial park for such fledgling high-tech firms, creating a hothouse of innovation and generating significant external economies – including a specialized workforce and a specialized array of producer services – which have not only sustained the continued agglomeration of high-tech electronics enterprises but also attracted other high-tech industries. Nearly a third of all employment in biotechnology, for example, is located in California, and of this over 90 per cent is located in the San Francisco Bay area. Stanford University, meanwhile, found itself in receipt of an increasing flood of donations from grateful companies. In 1955, these amounted to around US$500 000 annually; by 1965 they exceeded around US$2 million, and in 1976 they had reached US$6.9 million.

This kind of linkage between university research and high-tech activity is seen by many to be the key to the emerging geography of the fifth Kondratiev cycle. Not only do the new industries thrive on a symbiotic relationship with one another and university research departments, but key workers tend to favour technology complexes associated with top-flight universities since they provide abundant social and cultural activities and a job market that allows individuals (and spouses) to switch jobs without relocating. Such areas soon acquire a reputation as 'the right place to be', and this often counts for more than cost-of-living or quality-of-life factors. Where, as in Silicon Valley, the 'right place to be' happens to offer the

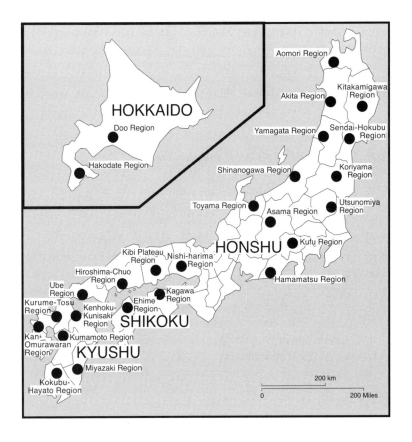

Figure 7.7
Technopolis cities in Japan.
Source: Castells and Hall (1994), Fig. 6.1, p. 118

additional bonus of an attractive environment and climate, the result is explosive growth. An important point in this context, as Hall (1981) observed, is that 'university systems, even in a country as dynamic as the United States, have a great deal of built-in inertia' (p. 536). Large, top-drawer universities like Harvard, MIT, Berkeley and Stanford are secure in their status, but few other institutions seem destined to join them. The result is that, outside these potential areas, there are few places in the USA where a high-tech industrial base is likely to be developed – apart, perhaps from the Research Triangle (Raleigh – Durham – Chapel Hill) that has already been established in North Carolina around Duke University and the University of North Carolina.

Similarly, there are few environments in other developed countries that are likely to attract a critical mass of high-tech activity, despite the proliferation of 'technology parks' – or, to be more accurate, *designated* technology parks. One exception is Japan, where an ambitious 'Technopolis' programme has designated 26 specialized new *towns* (Fig. 7.7), with special tax breaks and subsidies designed to attract not only high-tech industries such as electronics, electromechanical engineering, biotechnology and ceramics but also to develop a supportive infrastructure of universities, R & D labs, cultural facilities and civic centres.

Nevertheless, not all technopoles need be dependent on proximity to large, first-class universities. In fact, technopoles come in a variety of formats:

Most notably, it is clear that in most countries, with the important exception of the United States and, to some extent, Germany, the leading technopoles are in fact contained in the leading metropolitan areas: Tokyo, Paris-Sud, London-M4 Corridor, Milan, Seoul-Inchon, Moscow-Zelenograd, and at a considerable distance Nice-Sophia Antipolis, Taipei-Hsinchu, Singapore, Shanghai, São Paulo, Barcelona, and so on.

(Castells, 1996, p. 390)

The decentralization of high-tech employment

Almost all of the existing high-technology complexes are very much a suburban phenomenon. As Markusen (1983, p. 26) noted in relation to the early development of the Silicon Valley and Route 128 complexes, they are 'newly developed, auto-based, suburban areas whose jobs and tax base do not overlay the inner-city poor nor the central city jurisdiction'. But, because high-tech firms have tended to be very self-conscious about their 'address', these suburban complexes have become crowded and expensive. The outcome has been the familiar combination of corporate functional and spatial reorganization. More routine production tasks and downstream marketing and service functions are beginning to be dispersed, while managerial and developmental activities are retained in order to maximize the external economies of the 'right address'.

Research has suggested that American computer firms have kept their R & D and administrative activities in places like California and Massachusetts while moving their production facilities to southeastern states to take advantage of lower labour costs. Furthermore, some of the larger corporations in the computer and semiconductor fields have already begun to redeploy at the international scale, partly to acquire foreign technology and expertise and partly in search of cheaper labour, both highly qualified and semiskilled. There is now an emergent international division of labour in the US integrated circuit industry, for example, with skilled production functions decentralized to settings such as Central Scotland, assembly and testing in the likes of Hong Kong and Singapore, and assembly operations in metropolitan areas of peripheral countries like the Philippines, Malaysia and Indonesia – leaving only R & D functions to the innovation centres of the USA itself. Because of this sort of international redeployment, the central belt of Scotland – 'Silicon Glen' – now accounts for about 80 per cent of all British and about 20 per cent of all European integrated-circuit ('silicon chip') production, almost all of which came from factories owned by Hughes Aircraft Corporation, Motorola, National Semiconductor, Nippon Electric (NEC) and General Instrument. As with the local branch-plant economies generated by the decentralization of traditional manufacturing industries, these regional concentrations of decentralized high-tech industry do not seem to generate many local linkages or multiplier effects.

Flexible production regions

While the imprint of the new, high-tech industries of advanced capitalism cannot be said to amount to an entirely new dimension of the economic landscapes of the core nations, their new industrial spaces have clearly contributed an additional component to existing landscapes. Meanwhile, however, other industries have been changing, leaving their imprint on the economic geography of the core. The crisis of Fordism, combined with the opportunities afforded by new production-process and circulation technologies, by changing patterns of consumer demand, and by corporate restructuring and new competitive strategies, has led to the emergence of a phenomenon that *can* be described as a new dimension of the economic landscapes of the core nations. 'Flexible production regions' have emerged in many core countries as a result of the interplay of flexible production strategies, existing labour markets, and the fixed capital of older industrial spaces.

Flexible production regions, which may contain elements of *branch-plant industrialization* (see p. 225) along with a mixture of other new functions and activities, are seen as the product of Neo-Fordism, in which the emphasis on flexibility results in the externalization of certain functions and the vertical disintegration of organizational structures – which in turn lead to locational convergence and spatial agglomeration. Allen Scott, who has contributed most to this interpretation, sums up the central tendency as follows: 'vertical disintegration encourages agglomeration, and agglomeration encourages vertical disintegration' (1986, p. 224). The result is a series of regions or production complexes whose dynamics 'revolve for the most part around the social division of labour, the formation of external economies, the dissolution of labour rigidities, and the reagglomeration of production' (1988a, p. 181).

The archetypal flexible production region is the so-called Third Italy (Emilia-Romagna, Tuscany, the Marches, the Abruzzi, and Venetia), where branch-plant industrialization has combined with highly skilled local labour markets, well-developed infrastructure, and economies of scale and scope arising from the division of labour between specialized firms to create a regional network of innovative, flexible, and high-quality manufacturers whose products include textiles, knitwear, jewellery, shoes, ceramics, machinery, machine tools, and furniture. Other examples of flexible production regions based on a similar mixture of design- and labour-intensive industries include Jutland (Denmark), the Swiss Jura, and Southern Germany (Table 7.4).

Networks of manufacturers in high-technology industries form the basis of a second group of flexible production regions. Here, the agglomerating tendencies of new industries have resulted in localized growth in relatively new metropolitan settings: Orange County and Silicon Valley in California, Scientific City, Grenoble, and Montpelier in France, and the M4 Corridor in Britain, for example (Table 7.4).

These examples support Scott's observation that flexible production regions 'are almost always some distance – socially or geographically – from the major foci of Fordist industrialization' (1988b, p. 14). The argument, as we have seen, is that the interests of flexibility are best served by avoiding the rigidities (from outdated infrastructure to outdated institutions and labour relations) of Fordist settings. Yet we must recognize that it is quite possible for flexible, Neo-Fordist manufacturers to establish successful enclaves *within* older industrial regions and metropolitan areas. Examples include the design- and labour-intensive clothing industry in New York, clothing, high-tech electronics and furniture in Milan, and clothing and

Table 7.4: Propulsive industries and new industrial spaces

Propulsive sector	Typical features	Cited examples
Craft industries		
(a) labour-intensive craft industries, e.g. clothing, furniture	Exploitation of 'sweatshop' labour, often high level of immigrants. Subcontracting and out-working	New York, USA Los Angeles, USA Paris, France
(b) design-intensive craft industries, e.g. jewellery	High-quality products. Extreme social division of labour (but class polarization subdued in some examples)	Jura, Switzerland Southern Germany Emilia-Romagna, Italy Central Portugal Jutland, Denmark
High-technology industries	Segmented local labour markets with (1) skilled managerial cadres and (2) disorganized and malleable fractions of the labour force	Route 128, Boston, USA Orange County, CA, USA Silicon Valley, CA, USA M4 Corridor, UK Scientific City, France Austin, TX, USA Boulder, CO, USA Cambridge, UK Grenoble, France Montpellier, France Sophia Antipolis, France
Office and business services	Preferentially based on white-collar labour – including low-wage female labour. Very diversified and prone to agglomeration	London, UK New York, USA Tokyo, Japan

Source: Tickell and Peck (1992), p. 199.

motion pictures in Los Angeles. Equally, it is legitimate to interpret the localized networks of business and professional services within world cities as a specialized form of flexible production region. We are forced to conclude, therefore, that although flexible production regions represent a new dimension within the economic landscapes of the core, they do not represent an absolute or fundamental break from the old. As with previous transitions, the old order of things does not, and cannot, simply disappear. Thus,

As far as the geography of change is concerned, it is necessary to grasp the coexistence and combination of localising and globalising, centripetal and centrifugal, forces. The current restructuring process is a matter of a whole repertoire of spatial strategies, dependent on situated contexts and upon balances of power.

(Amin and Robins, 1990, p. 28)

REGIONAL INEQUALITY IN CORE ECONOMIES

We have seen that the evolution of advanced capitalism and the emergence of an informational economy have led to a significant reorganization of the economic geography of places and regions throughout the developed world. This reorganization has modified many of the core–periphery patterns of regional development associated with industrial capitalism. Yet core–periphery patterns and regional economic disparities have by no means disappeared. Rather, they have been reconfigured and intensified. Core regions within developed countries have typically been centred on urban-industrial heartlands, but have recently been modified by the geography of service activities, particularly business services, and by the imprint of the new spatial divisions of labour associated with globalization, decentralization, and agglomeration.

Europe provides a good example of the kind of regional differentiation that is characteristic of contemporary core economies. The overall core–periphery contrasts that were the legacy of industrial capitalism (Fig. 5.2) have carried over into substantial disparities. Fig. 7.8 shows the range and variability of regional economic well-being within the European Union in 1993, as measured by per capita GDP based on purchasing power parity (PPP: see Chapter 2). The 10 least-developed regions, located mainly in Greece and Portugal, had average incomes per capita that were less than one-third of the average of the ten most affluent regions of the continental core of the European Union.

Within this broad pattern of regional inequality there exist sharp local variations. Fig. 7.9 shows the geography of 'economic performance' of local labour market areas in Britain, as measured by an index based on rates of unemployment and average duration of unemployment in 1989, levels of employment in traditional industries and in business services in 1987, and average house prices in 1989 (Champion and Green, 1992). In general terms, the whole of England to the southeast of a line drawn between the Severn Estuary and Lincolnshire can be considered to be Britain's economic core. At the heart of this region is a prosperous hub centred on Greater London and extending outward for a radius of about 100 kilometres, encompassing the M4 motorway corridor to the west of London, the M3 belt to the southwest, and the M11 corridor to the north. In general, high economic performance in Britain is associated with a southeastern location, with planned expansion and proximity to the motorway system, and with relatively high levels of employment in finance, banking, insurance, and related producer services.

Britain's traditional industrial heartland (labour markets in the Northwest, Yorkshire and Humberside, and the northern Midlands), it will be noted, now fares no better than the long-recognized periphery of northern England, Wales and

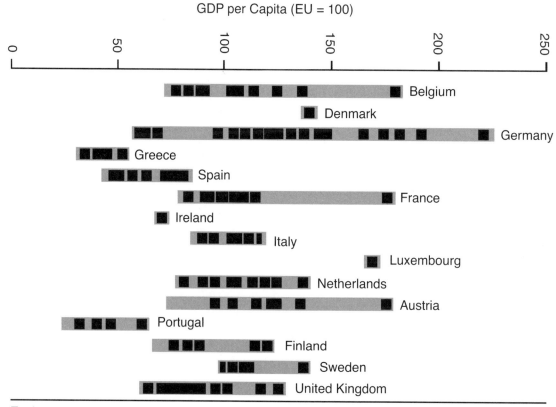

Each square represents one region

Figure 7.8
Regional inequality in the European Union, 1993

Scotland. Within this broad periphery, the worst-performing labour market areas are widely scattered around the coalmining and heavy industrial districts of Central Scotland, South Wales, Tyneside, Teesside, South Yorkshire and Merseyside. The fundamental cleavage reflected in this core–periphery pattern is echoed by a broad spectrum of social and economic data, to the point where it has become common to refer to Britain's political economy in terms of the 'Two Nations' of North and South (Green, 1988; Martin, 1988).

National economic development and regional inequality

The relationship between overall levels of development and the intensity of regional disparities is central to theory in economic geography. Much of the conventional wisdom on the subject is derived from a major study by Williamson (1965), who examined inter-regional income disparities in a sample of 24 countries. The results of this analysis suggested that the greatest regional inequalities were associated with countries at an intermediate level of development, with much smaller differences within both the most- and the least-developed countries (Fig. 7.10).

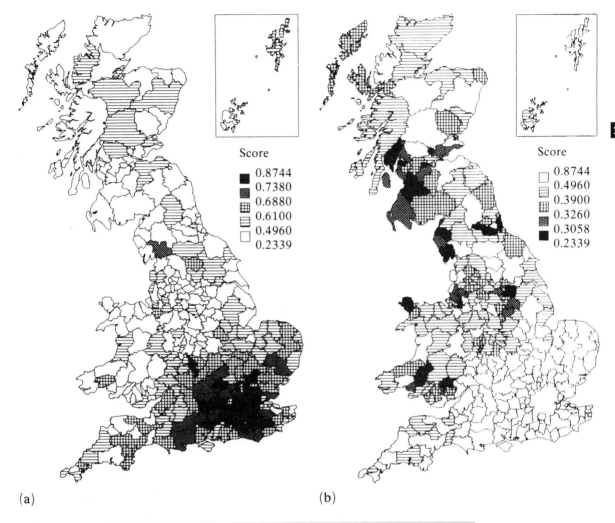

(a)　　　　　　　　　　　　　　　　(b)

Figure 7.9

An index of local economic performance in Britain, by local labour market areas. (A) above median, (B) below median

Source: Champion and Green (1992), Fig. 3

Williamson interpreted these cross-national results as a consequence of the dynamics of economic development, suggesting that the onset of industrial development precipitates sharp increases in regional inequality that are subsequently reduced as the economy matures.

Both the results and the interpretation of Williamson's work have been questioned on several grounds, however. The reliability of data and the appropriateness of measurement techniques are crucial: depending on the researcher's selections, anything can be proved.

Reviewing the large number of empirical studies that have followed Williamson's work, Krebs (1982) showed that, even allowing for different measurement techniques, the idea of divergence followed by convergence in regional disparities does not meet with strong support. In terms of the spatial concentration of both per capita incomes (see, e.g., Fig. 7.11) and of economic performance (see, e.g., Fig. 7.12), the evidence often points to spatial polarization or, at best, no change.

Figure 7.10

The relationship between inter-regional inequality and levels of economic development, as posited by Williamson (1965)

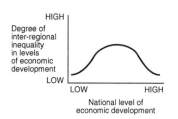

248

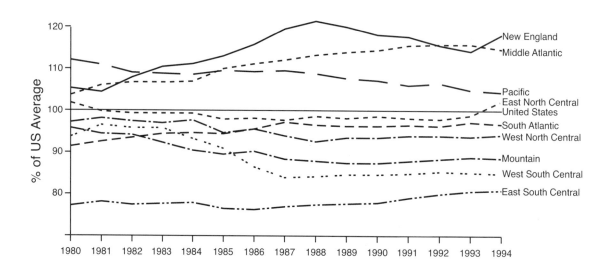

Figure 7.11

Regional trends in per capita incomes in the United States, 1980–94

Finally, we must remember to set these trends within the context of the long-wave rhythms of the world economy. Although the relationships between economic long waves and regional inequality are an under-researched topic, it is clear from our discussion in Chapter 1 that a fundamental consequence of the succession of Kondratiev cycles is that each new phase of economic development, based on new technology systems and requiring new resources and new markets, initiates a round of 'creative destruction' that leaves an indelible imprint on economic landscapes and, therefore, on the pattern and intensity of regional inequality. Furthermore, there are dynamics to the political economy that stem from the rhythm of economic long waves and that feed back to influence the whole question of inequality within countries. Thus, for example, periods of economic growth and low inflation tend to generate widespread satisfaction and a lack of enthusiasm for altering the status quo. During such periods, conservative political parties tend to be in the ascendant, while labour unions tend to become moderate and weak. At the same time, these are the periods during which voters are most likely to feel able to

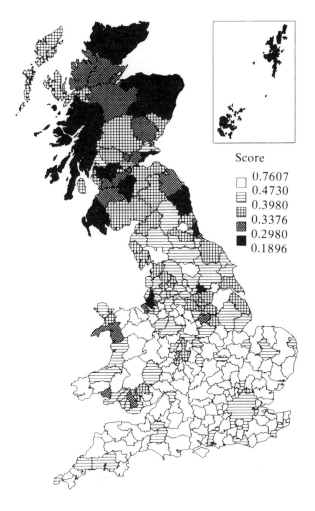

Score
☐ 0.7607
▤ 0.4730
▦ 0.3980
▥ 0.3376
▨ 0.2980
■ 0.1896

Figure 7.12
An index of change
in levels of economic
performance in Britain,
by local labour market
areas, 1985–9; the
lower the index score,
the worse the
performance
Source: Champion and Green
(1992), Fig. 4

afford to pay for redistributive policies. While we have no adequate single theory or explanation to account for these relationships, it is clear that we must consider regional inequality within the broader context of economic change. Viewed in this way, regional inequality never disappears: it is a perennial consequence of uneven development, of the see-sawing of capital from one set of opportunities to another. We may see regional inequality diminish, over the long term, in one part of the world; but elsewhere, and at different spatial scales, inequality will persist or intensify.

Interpretations of regional economic inequality

The most widely known explanation of regional economic inequality is that of Myrdal (1957). This is based on the contention that in a market economy changes in the location of economic activities produce cumulative advantages for one region

cumulative causation
A spiral build-up of advantages that occurs in specific geographic settings as a result of the development of external economies, agglomeration effects, and localization economies.

agglomeration effects
Cost advantages that accrue to individual firms because of their location among functionally related activities.

localization economies
Cost savings that accrue to particular industries as a result of clustering together at a specific location.

rather than a straightforward equalization of growth across all regions. **Cumulative causation** refers to the spiral build-up of advantages that occurs in specific geographic settings as a result of the development of agglomeration effects, external economies, and localization economies. **Agglomeration effects** are the cost advantages that accrue to individual firms because of their location within such a cluster. These advantages are sometimes known as external economies. External economies are cost savings that result from advantages that are derived from circumstances beyond a firm's own organization and methods of production. Where external economies and local economic linkages are limited to firms involved in one particular industry, they are known as **localization economies** These economies are cost savings that accrue to particular industries as a result of clustering together at a specific location.

Myrdal pointed out that the spiral of local growth involved in cumulative causation would tend to attract people – enterprising young people, usually – and investment funds from other areas. In some cases, this loss of entrepreneurial talent, labour, and investment capital is sufficient to trigger a cumulative negative spiral of economic disadvantage. With less capital, less innovative energy, and depleted pools of labour, industrial growth in peripheral regions tends to be significantly slower and less innovative than in regions with an initial advantage and an established process of cumulative causation. This, in turn, tends to limit the size of the local tax base, so that local governments find it hard to furnish a competitive infrastructure of roads, schools, and recreational amenities. These disadvantages were called backwash effects by Myrdal. Backwash effects are the negative impacts on a region (or regions) of the economic growth of some other region. These negative impacts take the form, for example, of outmigration, outflows of investment capital, and the shrinkage of local tax bases. They are important because they help us to explain why regional economic development is so uneven and why core–periphery contrasts in economic development are so common.

Myrdal recognized that peripheral regions do sometimes emerge as new growth regions, and he provided a partial explanation of them in what he called spread (or trickle-down) effects. Spread effects are the positive impacts on a region of the economic growth of some other region. This growth creates levels of demand for food, consumer products, and other manufactures that are so high that local producers cannot satisfy them. This demand provides the opportunity for investors in peripheral regions to establish a local capacity to meet the demand. Entrepreneurs who attempt this are also able to exploit the advantages of cheaper land and labour in peripheral regions. If these effects are strong enough, they can enable peripheral regions to develop their own spiral of cumulative causation, thus changing the inter-regional geography of economic patterns and flows (Fig. 7.13).

Myrdal's influential model was followed by others based on similar logic. Hirschman's (1958) model assumes 'polarization' and 'trickling-down effects' but sees early polarization producing a countervailing trend (and, thus, inter-regional equilibrium) rather than the cumulative intensification of initial advantage. Hicks's (1959) model was more like Myrdal's in its emphasis on cumulative causation but it gives greater attention to flows of labour and capital from growing to lagging regions rather than vice versa.

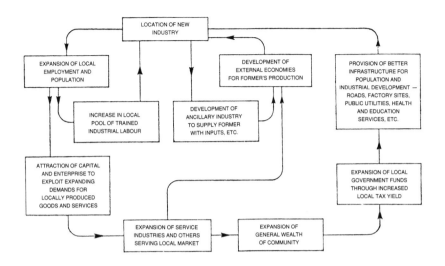

Figure 7.13
Myrdal's model of regional cumulative causation

Not all industries are equal in the extent to which they stimulate growth. Perroux (1955, 1961) argued that the locations of 'propulsive industries' (those that attract other industries and stimulate new ones in the vicinity) serve as the distinguishing characteristic of regions that achieve high rates of economic growth. As a propulsive industry grows, it attracts other linked industries and creates a set of agglomeration economies. A growth pole is formed and a growth centre develops. In the 1920s, shipbuilding was a propulsive industry. In the 1950s and 1960s, automobile manufacturing was a propulsive industry, and in today's informational economy high-tech manufacturing, design-intensive consumer goods, and business and financial services are propulsive industries.

Krugman (1991) suggests a more complex and more formal model to account for a centre–periphery pattern of economic development and for the possibility of its transformation. He argues that the relationship between demand and production established during an early phase of industrialization locks into place an inter-regional imbalance through the conjoint operation of increasing returns to scale in plant operations, transportation costs, and demand. As he puts it verbally, prior to embarking on a formal mathematical presentation,

> Given sufficiently strong economies of scale, each manufacturer wants to serve the national market from a single location. To minimize transportation costs, she chooses a location with large local demand. But local demand will be large precisely where the majority of manufacturers choose to locate. Thus there is a circularity that tends to keep a manufacturing belt in existence once it is established.

(Krugman, 1991, p. 15)

However, once the population of a peripheral region reaches a critical mass it may serve to stimulate production facilities there. A dramatic shift in regional fortunes may follow. Local 'boosterism', faith in a locality's future possibilities and policies that reflect this can also prove decisive in reversing regional imbalance. 'Nothing', ... therefore, ... 'is forever' (Krugman, 1991, p. 26). This is an argument for a rapid reversal in regional fortunes, rather than for slow regional balancing. Centres become peripheries and vice versa. Krugman argues that increasing returns, imperfect competition, and historical accident conspire to produce geographical concentration. But the identity of the favoured region is not set for all time. A new pattern of concentration can break the historic mould.

Summary

In this chapter we have seen how the crisis of Fordism and the sectoral shifts and changing business structures of the emerging era of advanced capitalism have begun to reshape the economic landscapes of the industrial core regions. Several aspects of this transition are of special importance:

- The spatial reorganization of the core economies is the product of several interdependent processes, including the globalization of core-area economic activity, the secular transition away from manufacturing towards service industries, the development of Neo-Fordist production processes and competitive strategies, and the beginnings of a new mode of regulation.

- The imprint of the secular transition to economies dominated by service employment has had two main dimensions: deindustrialization and economic decline in regions of traditional heavy industries, and rapid growth – a new bout of cumulative causation – in larger metropolitan settings that have attracted higher-order business and professional services.

- This secular transition has overlapped with a phase of stagflation and with the globalization of core-area economic activity, with the result that some aspects of change have been intensified and others have been introduced. The globalization of economic activity, for example, has intensified the effects of deindustrialization; while the pressures of stagflation contributed to radical changes in the role of the public economy.

- Corporate reorganization and redeployment have resulted in an increase in the external control of regional economies, with the consequence that the industrial systems of established industrial regions have become fragmented while those of newly industrializing regions have become segmented or truncated.

- The net effects of change have resulted in simultaneous spatial trends involving both decentralization and agglomeration, each highly selective in terms of the regions and economic activities involved.

- The diverse mixes of industry, workers and infrastructure inherited from the industrial era have mediated the broader processes of structural change and reorganization, so that different kinds of regions have evolved in different ways. *There have been four broad trajectories of change*. In the first, restructuring in response to a legacy of declining industry has been the dominant process. Examples include the old manufacturing heartlands of northern England, South Wales, central Scotland and the US manufacturing belt. In some rural areas and peripheral regions, on the other hand, the dominant process has been one of decentralization of footloose, labour-intensive industries from the metropolitan areas and core regions. Examples include parts of the southern USA such as the Carolinas. A third trajectory is characterized by regions whose industry has developed at or just above the national average, sustained by a consistent supply of new investment. Examples include most of Southeast England and the Boston–New York–Washington, DC–Richmond corridor. Finally, there are some regions whose attributes have made them attractive to new industries and/or to new investments aimed at exploiting new competitive strategies and new production processes. These encompass 'new industrial spaces' such as the high-tech concentrations of Orange County and Silicon Valley and 'flexible production regions' such as north-central Italy.

253

- Overlying these categories there has been a general accentuation of the importance and prosperity of large metropolitan areas, while many smaller towns and, in particular, many of the specialized new towns spawned by industrial capitalism (mining towns, heavy manufacturing towns, and so on) have declined.

Key Sources and Suggested Reading

Albrechts, L. and Swyngedouw, E. 1989. The challenges for regional policy under a flexible regime of accumulation. In L. Albrechts *et al.*, *Regional Policy at the Crossroads*. London: Jessica Kingsley Publishers.

Amin, A. (ed.) 1994. *Post-Fordism: A Reader*. Cambridge, MA: Blackwell.

Amin, A. and Robins, K. 1990. The re-emergence of regional economies? The mythical geography of flexible accumulation. *Society and Space*, **8**, 7–34.

Amin, A. and Thrift, N. (eds), 1994. *Globalization, Institutions, and Regional Development in Europe*. Oxford: Oxford University Press.

Birch, D. L. 1979. *The Job Generation Process*. Cambridge, MA: MIT Program on Neighborhood and Regional Change.

Bluestone, B. and Harrison, B. 1982. *The Deindustrialization of America*. New York: Basic Books.

Borchert, J. R. 1978. Major control points in American economic geography. *Annals, Association of American Geographers*, **68**, 214–32.

Castells, M. 1988. High technology and urban dynamics in the United States. In M. Dogan and J. Kasarda (eds), *The Metropolis Era. Vol. 1: A World of Giant Cities*. Newbury Park, CA: Sage, 85–110.

Castells, M. 1996. *The Information Age: Economy, Society and Culture*. Volume 1, *The Rise of the Network Society*. Oxford: Blackwell.

Castells, M. and Hall, P. 1994. *Technopoles of the World: The Making of 21st Century Industrial Complexes*. London: Routledge.

Champion, A. and Green, A. E. 1992. Local economic performance in Britain during the 1980s: results of the Booming Towns study. *Environment & Planning A*, **24**, 243–72.

Chang, K. 1989. Japan's direct manufacturing investment in the US. *Professional Geographer*, **41**, 314–28.

Council of Economic Advisers 1991. *Economic Report of the President*. Washington, DC: US Government Printing Office.

Cox, K. R. (ed.) 1997. *Spaces of Globalization: Reasserting the Power of the Local*. New York: Guilford Press.

254

Dunford, M. and Kafkalas, G. 1992. The global–local interplay, corporate geographies and spatial development strategies in Europe. In M. Dunford and G. Kafkalas (eds). *Cities and Regions in the New Europe*. London: Belhaven Press, 3–67.

Dunning, J. H. and Norman, G. 1987. The location choice of offices of international companies. *Environment & Planning A*, **19**, 613–31.

Fothergill, S. and Guy, J. 1992. *Retreat from the Regions: Corporate Change and the Closure of Factories*. London: Regional Studies Association.

Friedmann, J. 1986. The world city hypothesis, *Development and Change*, **17**, 69–84.

Green, A. E. 1988. The North–South divide in Great Britain: An examination of the evidence. *Transactions, Institute of British Geographers*, **13**, 178–98.

Gordon, D. M. 1979. *The Working Poor: Toward a State Agenda*. Washington, DC: Council of State Planning Agencies.

Hall, P. 1981. The geography of the fifth Kondratieff cycle. *New Society*, **55**, 535–36.

Hepworth, M. 1992. *Geography of the Information Economy*. London: Belhaven.

Hicks, J. R. 1959. *Essays in World Economics*. Oxford: Oxford University Press.

Hirschman, A. O. 1958. *The Strategy of Economic Development*. New Haven: Yale University Press.

Hudson, R. 1983. Regional labour reserves and industrialization in the EEC. *Area*, **15**, 223–30.

Jessop, B. 1992. Post-Fordism and Flexible Specialization. Incommensurable, contradictory, or just plain different perspectives? In H. Ernste and V. Meier (eds), *Regional Development and Contemporary Industrial Response*. London: Belhaven, 25–43.

Johnston, R. J. 1980. *City and Society*. Harmondsworth: Penguin.

Krebs, G. 1982. Regional inequalities during the process of national economic development: a critical approach. *Geoforum*, **13**, 71–81.

Krugman, P. 1991. *Geography and Trade*. Cambridge, MA: MIT Press.

Knox, P. L. 1995. World cities in a world-system. In P. L. Knox and P. J. Taylor (eds), *World Cities in a World-System*. Cambridge: Cambridge University Press, 3–20.

Lash, S. and Urry, J. 1987. *The End Of Organized Capitalism*. Cambridge: Polity Press.

Malecki, E. 1979. Locational trends in R&D by large US corporations, 1965–1976. *Economic Geography*, **55**, 308–23.

Malecki, E. 1991. *Technology and Economic Development: The Dynamics of Local, Regional and National Change*. London: Longman.

Markusen, A. 1983. High tech jobs, markets, and economic development prospects. *Built Environment*, **9**, 18–27.

Martin, R. 1988. The political economy of Britain's North–South divide. *Transactions, Institute of British Geographers*, **13**, 389–418.

Metcalf, D. 1969. *The Economics of Agriculture*. Harmondsworth: Penguin.

Myrdal, G. 1957. *Economic Theory and Underdeveloped Regions*, London: Duckworth.

Newby, H. 1980. Rural Sociology. *Current Sociology*, **28**, 1–141.

O hUallacháin, B. and Reid, N. 1991. The location and growth of business and professional services in American metropolitan areas, 1976–1986. *Annals, Association of American Geographers*, **81**, 254–70.

Perroux, F. 1955. Note sur la notion de pole de croissance. In I. Livingstone (ed.) 1979, *Development Economics and Policy: Selected Readings*. London: Allen & Unwin.

Perroux, F. 1961. La firme motrice dans la region et la region motrice. *Théorie et Politique de la Expansion Regionale*. Liège: Université de Liège.

Scott, A. J. 1986. Industrial organization and location. *Economic Geography,* **62,** 215–31.

Scott, A. J. 1988a. Flexible production systems and regional development. The rise of new industrial spaces in Europe and North America. *International Journal of Urban and Regional Studies.* **12,** 171–86.

Scott, A. J. 1988b. *New Industrial Spaces.* London: Pion.

Sirbu, M. A. Jr., *et al.* 1976. *The Formation of a Technology-Oriented Complex.* Cambridge, MA: MIT Center for Policy Alternatives.

Smith, I. 1979. The effects of external takeover and manufacturing employment change in the Northern region, 1963–1973. *Regional Studies,* **13,** 421–36.

Storper, M. 1987. The new industrial geography, *Urban Geography,* 8, 585–98.

Storper, M. and Scott, A. J. (eds) 1992. *Pathways to Industrialization and Regional Development.* London: Routledge.

Tickell, A. and Peck, J.A. 1992. Accumulation, regulation and the geographies of post-Fordism. *Progress in Human Geography,* **16,** 199.

Warf, B. 1990. U.S. employment in foreign-owned high-technology firms. *Professional Geographer,* **42,** 421–32.

Warf, B. 1995. Telecommunications and the changing geographies of knowledge transmission in the late 20th century, *Urban Studies,* **32,** 374.

Williamson, J.G. 1965. Regional inequality and the process of national development. *Economic Development & Cultural Change,* **13,** 3–45.

255

PART 3

SPATIAL TRANSFORMATION OF THE PERIPHERY

In the next three chapters we examine the world outside the core of the world economy, paying special attention to the changing historical relationships between core and periphery outlined in Chapters 2 and 3. In Chapter 8 we examine the spatial transformations that have occurred as a consequence of both an older colonialism and a more recent interdependent global capitalism. However, attention is also paid to how the consequences have varied depending on local reactions and institutional responses. In Chapters 9 and 10 two major economic activities, agriculture and manufacturing industry, are examined both with respect to their roles in economic development and their changing geographical patterns. The emphasis throughout this section is upon the impacts of and responses to the evolving modern world economy in the 'dependent' world.

Picture credit: World Bank

Picture credit: Paul Knox

CHAPTER 8

The Dynamics of Interdependence: Transformation of the Periphery

If local isolation was rarely complete and development was rarely totally independent, the coming of the modern world economy led to greater and greater interaction between different parts of the world. In this chapter we focus on the cumulative consequences of this increased interdependence for those regions incorporated into the world economy on terms initially and decisively disadvantageous to them. This is not to say that the terms of interdependence have always remained absolutely disadvantageous, although this is true, for example, in the case of Central America and large parts of sub-Saharan Africa. Particularly since the late 1960s, the major oil-producing countries (e.g. Saudi Arabia, Iran, Venezuela, Nigeria, Indonesia) and the Newly Industrializing Countries (NICs) (such as Taiwan and S. Korea) have challenged the static picture of a 'fixed' industrial core and a 'fixed' non-industrial periphery. The world economy now has a vibrant semi-periphery of NICs and some other economies. This chapter begins with a discussion of how existing economies were transformed into colonial ones. A second section identifies the major ways in which these colonial economies were enmeshed and maintained within the world economy. A third section identifies the importance of frameworks of administration introduced by Europeans. A fourth section discusses the cultural mechanisms that facilitated integration into the world economy. The final two sections explore the contexts of change in the nature of interdependence since the 1960s, respectively the global context (New International Division of Labour, decolonization, and the Cold War) and several national political-economic strategies or 'models' of development that have challenged the dominant ones.

COLONIAL ECONOMIES AND THE TRANSFORMATION OF SPACE

The modern world economy began with the global expansion of trade and conquest by European merchants, adventurers and statesmen (Fig. 8.1). But a distinction

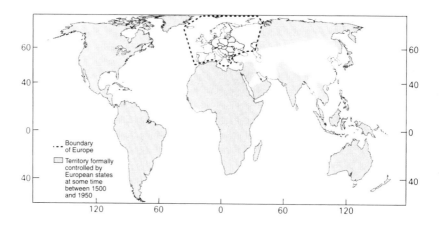

Figure 8.1

The geographical extent of European political control

Source: Taylor (1985), Fig. 5, p. 67

should be drawn between pre-capitalist colonial rule, notably that of Spain and Portugal in Latin America, and the new colonialism that was associated with the growth and global expansion of West European capitalism, beginning in the sixteenth century and itself undergoing successive shifts in development. The major purpose of pre-capitalist colonialism was the extraction of tribute from subject peoples and its major mechanisms involved political-territorial control. In contrast, the 'new colonialism' was associated primarily with economic objectives and mechanisms. Direct political-territorial control, though often advantageous, was not essential. The emphasis initially was upon the exploitation of raw materials. After the industrial revolution in Britain, however, markets for manufactured goods became an equally important objective. Realizing both of these objectives required a restructuring of the economic landscapes of the colonized societies.

Territorial conquest, with or without the elimination of indigenous peoples, and the planting of either settler enclaves or slave plantations and mining enterprises were the major features of European expansion through the eighteenth century. For much of the nineteenth century, however, many societies that remained or became formally independent were under the economic domination of European and, increasingly, American capitalists. With the 'German challenge' to British hegemony in the 1870s, there was a new 'scramble' for territorial conquest as rival colonial powers attempted to pre-empt one another, especially in Africa. This coincided with the emergence of capital export as a major stimulus to intervention and domination, as profit rates in the periphery exceeded those in the core.

In both territorial (colonial) and interactional (commercial) forms, capitalist expansion entailed a forcible transformation of pre-capitalist societies whereby their economies were internally disarticulated and integrated externally with the world economy. They were no longer locally oriented but had now to focus on the production of raw materials and foodstuffs for the 'core' economies. Often they became extremely dependent on monoculture in order to confer the blessings of 'comparative advantage' (the purported benefits of specializing in goods that a country can produce at a lower relative cost while importing goods for which its own production costs are relatively higher) on the developed world. At the same time they also provided markets for the manufactured goods exchanged in return. For many parts of the world this relationship still holds true today.

The dramatic transformation of the existing geography of production that the reorientation towards the core entailed is not sufficiently noted. Before the industrial revolution and European capitalist expansion, Asia and other parts of what we now conventionally call 'the Third World' contained a far larger share of world manufacturing output than did Europe. Bairoch's (1982) calculations of industrial output by world region during the course of the nineteenth century, for example, reveal a picture not simply of a higher rate of industrialization in core countries but also of deindustrialization in the periphery as the cheaper European products forced traditional producers out of business (see Chapter 3, p. 71).

The division between 'developed' and 'underdeveloped' economies, therefore, which had been moderate before now grew enormously. This was the main geographical consequence of the coming of the world economy. The conditions for the development of 'national economies' did not exist nor were permitted to exist in the 'dependent world' that came into existence during the course of the expansion of the European world economy (see Chapter 3). The unequal core–periphery structure to the world economy has been in place since the Europeans first ventured out into the world in the sixteenth century. From this point of view, underdevelopment is not an original condition, equivalent to 'traditionalism' or 'backwardness.' To the contrary, it is a condition created by integration into the world economy. At the same time, however, the die is not permanently cast in confining some places and peoples to an underdeveloped condition. The examples of the United States, Japan and the NICs suggest that upper mobility is possible for some states/regions initially in an underdeveloped state. What is equally clear is that for large parts of the world such mobility is either difficult or next to impossible. To understand the geography of the world economy we need to understand such cases as well as the 'successful' ones.

During the early years of incorporation into the world economy the periphery tended to specialize in the extraction of raw materials (from gold bullion to furs and spices) and production of plantation crops (such as tobacco, cotton and sugar). As time wore on, plantations and extractive industries were sometimes supplemented by labour-intensive manufacturing that took advantage of cheap colonial labour. By the mid-twentieth century, Latin America, Asia and Africa were organically linked to and financially dependent upon Western Europe and the United States. The emergence of the United States as a dominant force and the growth of the Soviet bloc, however, undermined the monopoly of political control exercised by the European powers over large parts of the world. A process of decolonization began with the independence of the South Asian countries in 1947. This brought the possibility, however constrained (as the experience of Latin America, 'independent' since 1820, shows), of more autonomous development.

The imposition of regional specializations

The late nineteenth century was an especially critical period in the creation of colonial economies (Fig. 8.2). Whole regions became specialized in the production

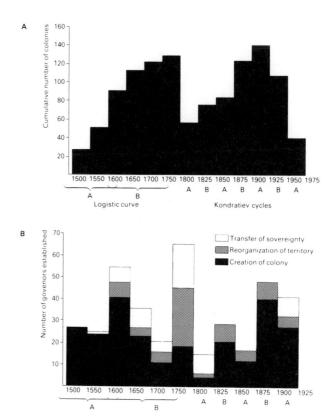

Figure 8.2

Long waves and colonization. (A) The two long waves of colonial expansion and contraction; (B) establishment of colonies, 1500–1925

Source: Taylor (1985), Figs 6 and 7, p. 76

of a specific raw material, food crop or 'stimulant'. Many of these had a prior history, such as the sugar-producing areas of the Caribbean or the cotton-growing regions of the United States, India, and Egypt. But the Long Depression (1883–96), a major downturn in the world economy (the IIB phase of Kondratiev's long-wave cycle) – due, among other things, to decreased profitability in manufacturing – ushered in a major spurt in the global expansion of capitalism and intensified regional specialization. During this time period more and more resources and labour were drawn into an increasingly differentiated world economy (Wolf, 1982).

Adam Smith and David Ricardo, writing well before this period, had envisaged a global division of labour in which each country would freely choose the commodities it was most suited to produce and freely exchange its optimal commodity for the optimal commodities of others. Unfortunately, this economic vision of comparative advantage, basic to most modern theories of international trade, ignores both the historical-political conditions under which commodities were selected and the costs faced by a specialized commodity economy in terms of vulnerability to the vagaries of 'world' demand. Competitive advantage, established through market dominance and political power, makes more sense as the significant determinant of the global map of production.

In the late nineteenth century, 'choice' of commodity was often imposed by force or through market domination. Moreover, once embroiled in the global system of regional specialization, an economy had to organize its factors of production in order to foster capital growth or fall by the wayside. At the same time, other regions, without some initial advantage in raw material, climate, social organization or accessibility, became providers of labour power to the new outposts of global capitalism. Three examples, out of a host of possibilities, illustrate how externally oriented colonial economies were created on the basis of regional specialization: bananas in Central America, rubber in Malaya and tea in Sri Lanka (see Wolf, 1982, Chapter 11 for other examples).

Bananas are hardly a major food staple, but the creation of banana plantations in the late nineteenth century affected many areas, especially in Central America. Introduced into the Americas by the Spaniards, the banana became a staple crop among the lowland populations of Central America. In the 1870s it became a plantation crop as an American entrepreneur engaged in railroad construction in Costa Rica experimented with commercial banana production to increase the profitability of his railroad. As a result of this initiative was developed the United Fruit Company, incorporated in 1889 as a corporation engaged in the marketing of bananas from Central America in the United States. Over the years, the Company produced bananas on plantation-estates in Costa Rica, Panama, Honduras, Colombia and Ecuador. Geographic dispersal had a number of advantages; it

> enabled the Company to offset political pressures in any one host country. Dispersal also allowed it to take advantage of suitable environments in different locations, thus reducing the chance that floods, hurricanes, soil depletion and plant diseases could bring production to a halt in any one of them. To further reduce these risks, the Company acquired a great deal more land than it could use at any one time, to hold as a reserve against the future. In some areas it formed relationships with local cultivators who grew bananas and then sold them to the Company.
>
> (Wolf, 1982, p. 324)

Much of the labour on the plantations was recruited locally, especially in Colombia and Ecuador, but in parts of Central America workers were brought from the English-speaking islands of the Caribbean. This resulted because of the difficulty the Company faced in obtaining labourers from the populated highlands to work in the lowlands and the firm's preference for a work-force that could be socially isolated and made wholly dependent upon the Company. The role of these foreign workers gradually decreased as host governments limited immigration and encouraged their native populations to engage in wage labour on the plantations. Bananas are still an important export crop, especially for Panama and Costa Rica.

Wild rubber from Brazil dominated the world market for most of the nineteenth century. In 1876, however, Amazonian rubber seeds were smuggled from Brazil to England where they were prepared for planting in Malaya. Malayan rubber plantations grew from 5000 acres in 1900 to 1 250 000 acres in 1913. During this expansion, an original planter class was supplanted by a class of managers for companies operating from London. Labourers were initially imported from southern India but over time many plantations came to employ local Malays. Although plantation production remained dominant, many Malay cultivators tapped their own rubber trees as a source of cash income. Rubber increasingly replaced irrigated rice, a food staple, as the major commodity produced by small-scale proprietors. It remains so today.

Finally, among the range of commodities destined for consumption in the industrial world, some were neither foodstuffs (such as bananas) nor industrial crops (such as rubber). Such commodities as sugar, tea, coffee, cocoa, tobacco and opium were of fundamental importance in the global expansion of the world economy. Explaining the popularity of these 'stimulants' is not easy. Some accounts suggest that the work behaviour required under industrialization favoured the sale of these stimulants (except opium, a special case, as its main initial market was China) because they provided 'quick energy' and prolonged work activity. Others suggest that some (e.g. sugar and cocoa) provided low-cost substitutes for the traditional and increasingly costly diet of pre-industrial Europe. Whatever the basis to demand, by the late nineteenth century the stimulants were of great and increasing importance in world trade (see Mintz, 1985, for a splendid discussion of sugar and its role in world trade). Tea had become 'the drink' of English court circles in the late seventeenth century. It came entirely from China. Demand was so great that in the early eighteenth century tea replaced silk as the main item carried by British ships in the Chinese coastal trade. At the time of the American War of Independence, as the 'Boston Tea Party' reminds us, tea was the third largest import, after textiles and iron goods, of the American colonies. Some tea plantations were established in Assam (Northeast India) in the 1840s, but until the opening of the Suez Canal Indian tea could not compete with the Chinese tea carried by the famous clipper ships around the Cape of Good Hope. With the opening of the canal and decreasing cost of steamship transportation, Indian 'black' teas became commercially competitive with the green teas of China. In the 1870s tea plantations spread with great speed throughout the uplands of Ceylon (Sri Lanka). This was done by confiscating peasant land through the device of 'royal condemnation' and then selling it to planters. By 1903 over 400 000 acres were planted in tea shrubs.

Tea cultivation is extremely labour-intensive. To obtain the necessary labour, the Ceylon tea planters imported Tamil labourers from southern India. These Tamils, not to be confused with the long-resident Sri Lankan Tamils of northern and eastern Sri Lanka, now number over 1 million people, in a region in which the upland or Kandyan Sinhalese are about 2 million. As a consequence an ethnic conflict has been imposed on top of an economic conflict between Sinhalese cultivators and Tamil plantation workers. Tea remains an important export crop in the contemporary Sri Lankan economy.

Many other examples could be added to these three to demonstrate the degree to which regional specialization was the classic motif of the colonial economies as they developed at an accelerating rate in the late nineteenth century. The Long Depression of that time in Europe and the United States stimulated an unprecedented expansion of the world economy into all parts of the globe as European and American capitalists sought to maintain capital accumulation in the face of declines in industrial production. Commodity production for a world market was not new but in its late nineteenth-century 'explosion' it 'incorporated pre-existing networks of exchange and created new itineraries between continents; it fostered regional specialization and initiated worldwide movements of commodities' (Wolf, 1982, p. 352).

ECONOMIC MECHANISMS OF ENMESHMENT AND MAINTENANCE IN THE COLONIAL WORLD ECONOMY

How was it that the regional specialization that began in the late nineteenth century was possible? And how did it evolve over time? Answering these questions requires us to focus on the means by which an integrated world economy was created: flows of capital investment, networks of communication and marketing, movements of commodities and people, and transportation – urban networks as channels of diffusion and concentration.

Trade and investment

The period 1860–70 inherited from the earlier centuries of colonial expansion two major systems of economic interaction, an Atlantic system built upon the 'Triangular Trade' between Europe, Africa and the Americas (Fig. 8.3), and a Eurasian system built upon trade with India and the Far East. In the mid-nineteenth century

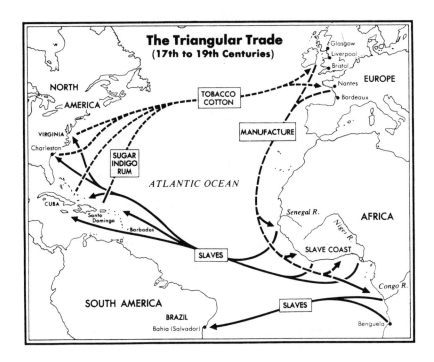

Figure 8.3

The Atlantic system, 1650–1850

Source: Duignan and Gann (1985), Map 1, p.12

265

Table 8.1: Stock of foreign capital investment held by Europe 1825–1915 (US$ billion)

	1825	1855	1885	1915
UK	0.5	2.3	7.8	19.5
France	0.1	1.0	3.3	8.6
Germany	0.1	1.0	1.9	6.7
Others	–	–	–	11.4
Total	0.7	4.3	13.0	46.2

Source: Warren (1980), p. 62.

the Atlantic system in its classic form collapsed. It was replaced by a system based on a mix of competitive colonialism, regional specialization, and investment in infrastructure (especially railways).

Between 1830 and 1876 there was a vast increase in the number of colonies and the number of people under colonial rule. There was also a tremendous expansion of foreign investment by European states and capitalists in the late nineteenth and early twentieth centuries (Table 8.1). Moreover, there was an important shift in the geographical distribution of both investment and exports. British trade and investment, to use the most important example, shifted away from India, Europe and the United States, especially with respect to investment in the first two and with respect to exports to the third, from the 1870s on. South America and the British Dominions (Australia, New Zealand, Canada and South Africa) became more important, especially with respect to investment. However, the pattern was to fluctuate considerably over the years as some regions/states increased and others decreased in attractiveness to investors (Fig. 8.4).

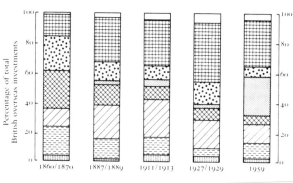

Figure 8.4

The geographical distribution of British foreign investment, 1860–1959

Source: Hobsbawm (1968) p. 303

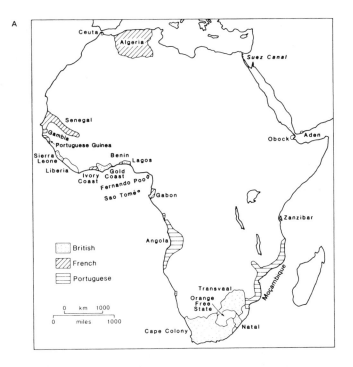

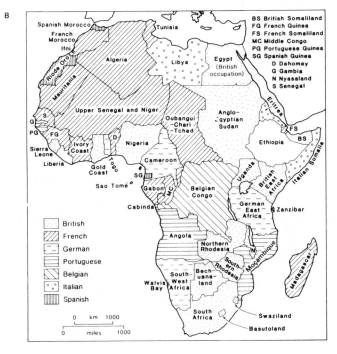

Figure 8.5
The colonization of
Africa: (A) 1850;
(B) 1914

Source: Christopher (1984)
Figs 2.1 and 2.2, pp. 28–9

The geographical switching of investment, however, was not always obviously
economic in motivation. In particular, European incorporation of Africa into the
world economy was based largely on competitive colonialism. Local settlers, as in

South Africa, sometimes developed their own local 'imperialisms' and when challenged by other settlers or hostile natives called in the Motherland. The relative weakness of many African polities also invited direct intervention. Once one European state was involved, others were tempted to engage in pre-emptive strikes to limit the damage to their 'interests'. Between 1880 and 1914 Africa was divided into a patchwork of European colonies and protectorates (Fig. 8.5).

Transport routes

At the global level the colonial system was bound together by a network of steamship and communication (postal, telegraph and, later, telephone) routes.

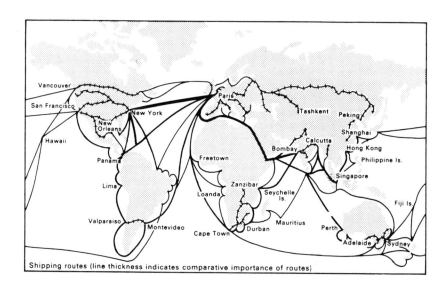

Figure 8.6

World steamship routes, by volume of trade, 1913

Source: Latham (1978), Map 2, p. 33

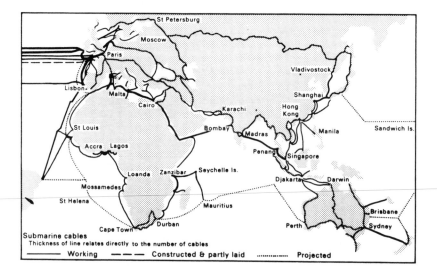

Figure 8.7

The telegraph system in Asia and Africa, 1897

Source: Latham (1978), Map 3, p. 36

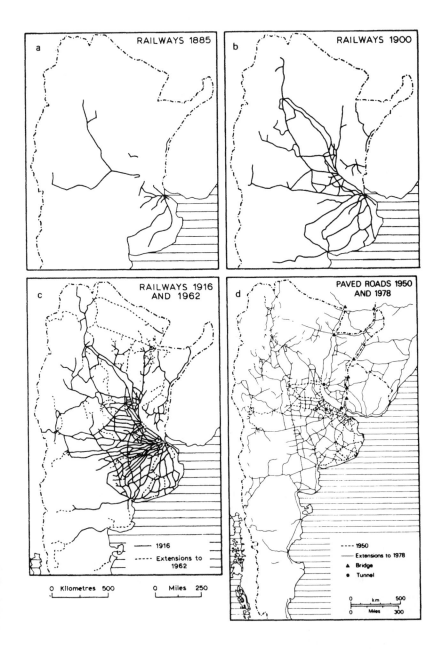

Figure 8.8
The development of
roads and railways in
the River Plate region
of South America,
1885–1978
Source: Crossley (1983),
Fig 9.3, p. 401

These became progressively more dense and interconnected from the 1860s
onwards. The Suez and Panama Canals were important in providing shorter and
less hazardous routes between 'home ports' in Europe and North America and
colonial destinations. By 1913 the world economy was effectively integrated by a
system of regularly scheduled steamship routes (Fig. 8.6). A world telegraph system
enabled orders to be placed and shipments to be embarked for a large number of
ports around the world (Fig. 8.7).

Within colonies, railway building was the major mechanism of spatial transformation. In most of Africa and Asia, with the important exceptions of India,

South Africa and north China, railways were not mechanisms for creating integrated colonial economies but, rather, means for moving a basic export commodity for shipment to Europe or North America. However, railways were often an important investment in their own right rather than a burdensome state responsibility. This was especially the case in South America and China, if much less so in India and Africa. In Argentina, for example, though the road and railway networks were oriented towards the River Plate estuary and the capital city of Buenos Aires, they provided a relatively dense grid for the rich commercial agriculture of the Argentinian pampas (Fig. 8.8). This produced a transport system considerably more interconnected and integrated than the simple linear systems prevalent in Central America, Southeast Asia and most of Africa. In short, spatial integration into the colonial world economy did not take the same form everywhere.

Whatever the precise nature of the railway networks, there was a tendency for all networks to focus on one or, at most, several coastal ports. These became 'privileged' locations, often assuming the role of administrative as well as economic centre for the entire colony. Specialization in the export of raw materials and concentration of administrative functions thus had the effect of stimulating the disproportionate growth of these 'links' to the world economy. This was especially marked in India (with Bombay, Calcutta and Madras) and Africa (for example, Lagos, Dakar, Cape Town).

Settlement systems

But the character of the colonial system also put limits on the growth of the dominant or 'primate' settlements. There was only a limited stimulus to the growth of a distinctive urban economy. The orientation of urban networks was towards exploitation of hinterlands rather than an industry- and service-based urban economy. It was only with political independence that a new dynamic for urban growth occurred as the primate cities shifted from being mechanisms for colonial control to their contemporary role 'as the corporate representative of the people of the former colony' (Fiala and Kamens, 1986, 28). As Rondinelli (1983, p. 49) points out, 'colonial activities often stimulated the growth of secondary cities. In some cases they were encouraged to grow as colonial administrative posts or as transfer and processing centres for the exploitation of mineral and agricultural resources in the interior of a country.'

Regions without a history of urbanization before colonialism were not surprisingly the most easily and strongly reoriented to the colonial world economy. In Malaysia, for example, cities grew up in the interior where crops were grown for export or where other exportable commodities (tin, especially) were exploited. These cities were connected by railway and road to port cities which grew as processing and transfer points. In western Nigeria, a region with a long pre-colonial history of urbanization, roads and railways were often built to bypass traditional centres of trade such as Ife, Benin City and Sokoto. New, more effectively colonial cities grew up at nodes in the transportation network. As one study notes 'fortune rode the trains. [Towns] that received terminals grew, but those that did not stagnated or declined, as did many river ports' (Gugler and Flanagan, 1978, pp. 27–8).

The realignment of trade and production

The outcome of the new extension of the world economy and the intensification of trade within regions already incorporated was a substantial increase in world trade. The nature of the system of trade has led to its naming as a **crossover system of trade**. How important was the contribution of Asia, Africa and Latin America? The answer is: of great importance (Tables 8.2 and 8.3). By 1913 Asia and Africa provided more exports to the world economy than either the USA and Canada or the UK and Ireland. In 1913, Asia also had a share of world imports almost as large as the USA and Canada combined. What happened was that the industrializing countries of Europe and North America bought increasing amounts of raw materials and foodstuffs from the undeveloped economies and ran up large trade deficits with these regions. Britain, however, as a result of its free trade policy, ran up substantial deficits as a result of importing manufactured goods and investing heavily in the industrializing countries (especially the United States and Germany). In turn, Britain financed its deficits through the export of manufactured goods to the undeveloped world. Thus the circle of international trade and dependence was closed.

India and China were particularly important to this world pattern of trade and payments. It was Britain's trade with India and China that compensated for a negative balance of payments with the United States, industrial Europe, Canada, South Africa and New Zealand. Without the 'Asian surplus' Britain would not have been able to subsidize the growth of these other economies. So, far from being 'peripheral' to the growth of the world economy, the undeveloped world, especially India and China, was vital.

Between 1918 and 1939 this system of multilateral trade suffered a number of setbacks. One was the overall decline in trade as the world experienced a major

271

crossover system of trade
A system of multilateral trade established in the late nineteenth century, which lingered into the 1930s. Europe and North America bought raw materials from less-developed countries (LDCs). In return, Britain imported manufactures from and exported capital to Europe and North America. Britain's assets were boosted by the return on foreign investment and the export of manufactured goods to the LDCs (both colonies and independent states).

Table 8.2: World exports by geographical region, 1876–1937, per cent of total

	1876–80	1896–1900	1913	1928	1937
US and Canada	11.7	14.5	14.8	19.8	17.1
UK and Ireland	16.3	14.2	13.1	11.5	10.6
Northwest Europe	31.9	34.4	33.4	25.1	25.8
Other Europe	16.0	15.2	12.4	11.4	10.6
Oceania	} 24.1	} 21.7	2.5 } 26.3	2.9	3.5
Latin America			8.3	9.8	10.2
Africa			3.7	4.0	5.3
Asia			11.8	15.5	16.9

Source: Yates (1959). Table 6, p. 32.

Table 8.3: World imports by geographical region, 1876–1937, per cent of total

	1876–80	1896–1900	1913	1928	1937
US and Canada	7.4	8.9	11.5	15.2	13.9
UK and Ireland	22.5	20.5	15.2	15.8	17.8
Northwest Europe	31.9	36.5	36.5	27.9	27.8
Other Europe	11.9	11.0	13.4	12.5	10.2
Oceania	} 26.3	} 23.0	2.4 } 23.4	2.6	2.8
Latin America			7.0	7.6	7.2
Africa			3.6	4.6	6.2
Asia			10.4	13.8	14.1

Source: Yates (1959). Table 7, p. 33.

depression. But the worst was the decline in Britain's relative position as the linchpin of the colonial world economy. This reflected both successful industrialization in India and China displacing British products (especially cotton textiles) and increased competition from Japan in Britain's 'traditional' colonial markets. But another problem was the overproduction of the main export crops and raw materials. As a consequence, commodity prices fell and so did demand for manufactured goods. The successful expansion of plantations and mines, therefore, ultimately undermined the system of capital circulation and accumulation that their introduction had brought into existence in the nineteenth century.

The onset of the Great Depression of the 1930s effectively ended the expansionist regime of international trade established in the late nineteenth century. The major industrial states reacted to the Depression by raising tariffs and devaluing their currencies. These shifts in economic policy were premised on the assumption that Britain would remain 'open' as the linchpin of the colonial world economy. But, as Stein (1984, p. 375) puts it: 'Depression left Britain unable and unwilling to accept an increasingly asymmetric bargain.' Not until the 1970s would world trade return to the relative levels that it had achieved in the early 1900s (see Chapter 3).

The keys to economic recovery in Western Europe and the United States in the 1940s and 1950s were provided by military spending and massive increases in domestic consumption of domestic manufactures. Although this did lead to increased demand for many of the industrial raw materials and foodstuffs produced in the 'periphery', there was no longer the crossover system of trading linkages. If anything, the European colonial states and, above all, the United States now came to have direct links to specific sites of exploitation in the periphery without the necessity of the infrastructure and administrative investments that had limited short-run payoffs. This approach favoured direct investment and the creation of subsidiaries by multinational firms rather than portfolio investment and conventional trade. Advantages hitherto specific to the United States – the cost-effectiveness of large plants, economies of process, product and market integration – had become the proprietary rights of large firms. The world was now their oyster, rather than that of the colonial states. 'American governments could preach against colonialism while large American [and other] firms colonized the world' (Agnew, 1987, p. 62).

In the early part of the twentieth century the major share of accumulated foreign direct investment (FDI) was in the 'dependent world', 62.8 per cent in 1914. Total FDI came overwhelmingly from Britain (45.5 per cent) and the USA (18.5 per cent). Since the Second World War, however, most foreign direct investment has been between the industrialized economies. In 1971 only 30.9 per cent of FDI was directed to the 'less developed countries' (LDCs). Though, after Western Europe, Latin America was the major recipient region. The USA was now the major source of total FDI: 48.1 per cent in 1971. This has decreased since with the growth of Japanese, European and some NIC FDI (Dunning, 1983). The dramatic post-Second World War expansion in FDI, therefore, has not involved all parts of the world on equal terms. In terms of flows of FDI, the LDCs on the whole have become less central to the world economy than they were previously. From this point of view at least, the end of colonialism was something of a mixed blessing.

THE INFLUENCE OF COLONIAL ADMINISTRATION ON INTERDEPENDENCE

Many of those who colonized the world from Europe, in both the sixteenth and the nineteenth centuries, saw their activities as part of a historic 'mission' of Western civilization: to bring progress to backward and barbarian peoples. Lord Lugard, the famous British colonial administrator, maintained that Britain stood in a kind of apostolic succession of empire: 'as Roman imperialism . . . led the wild barbarians of these islands [the British Isles] along the path of progress, so in Africa today we are re-paying the debt, and bringing to the dark places of the earth . . . the torch of culture and progress' (quoted in Ranger, 1976, pp. 115–16). At best the political ideas of the European imperialists were that 'political power tended constantly to deposit itself in the hands of a natural aristocracy, that power so deposited was morally valid, and that it was not to be tamely surrendered before the claims of abstract democratic ideals, but was to be asserted and exercised with justice and mercy' (Stokes, 1959, p. 69).

The chief problem was to understand and pacify the indigenous colonized. The Nigerian novelist Chinua Achebe (1975, p. 5) puts this as follows:

> To the colonialist mind it was always of the utmost importance to be able to say: I know my natives, a claim which implied two things at once: (a) that the native was really quite simple and (b) that understanding him and controlling him went hand in hand – understanding being a precondition for control and control constituting adequate proof of understanding.

This approach provided the ideology for what Hopkins (1973, p. 189) referring to the British in Africa, has called the 'art of light administration', administration without too much long-run investment or explicit (and expensive) violence.

The colonial regimes themselves never amounted to more than a thin veneer of European officials and soldiers on top of complex networks of local collaborators.

In India in the 1930s, for example, 4 000 British civil servants, 60 000 soldiers and 90 000 civilians ruled a country of 300 million people. The British were able to do this

> by constructing a delicately balanced network through which they gained the support of certain favoured economic groups (the Zamindars acting as landed tax collectors in areas such as Bengal, for example), different traditional power holders (especially after the Great Mutiny of 1857, the native princes), warrior tribes (such as the Sikhs of the Punjab), and aroused minority groups such as the Muslims.

(Smith, 1981, p. 52)

This kind of brokerage system was to be found in every colonial territory without a large European settler population. Sometimes a foreign economic presence was crucial (the Chinese in Southeast Asia; the Lebanese in West Africa; European settlers in Algeria and Kenya). Often there were alliances with new or traditional ruling groups (the Princely States in Malaya; the Ottoman bureaucracies in Tunisia and Morocco; the Hashemite family in Mesopotamia and Syria). Above all, local rivalries were exploited to advantage, as in Madagascar, India and China. Even in the face of nominal local political independence, as in China or Latin America, colonial imperatives and administrative models had considerable influence through imported school curricula and business practices (e.g. British influence was strong in Argentina and Venezuela; German influence was strong in Chile and Brazil).

Alliances and administrative forms were far from static and differed from colony to colony and between colonial powers. But one change was permanent. The new colonies, often vastly bigger than the territorial units they superseded, created markets of unprecedented size. Internal tolls and other restraints on trade disappeared. Sumptuary laws that prevented people of low status from acquiring luxury goods were abolished. All forms of servitude that interfered with the wage economy were outlawed. The great tribal migrations of eastern and southern Africa were brought to a close. New judicial methods were introduced and old ones were eliminated. Schools and hierarchical systems of local administration were established.

The colonial powers operated in different ways. In Africa, for example, the British administration was more civilian and decentralized than the French and Belgian administrations. Its officials 'prided themselves on being gentlemen and amateurs, rather than on being military, legal or administrative specialists. The British pioneers set up an administrative hybrid based partly on British metropolitan models and partly on models derived from colonial India and Ireland' (Gann and Duignan, 1978, p. 355). In particular, there was a dispersal of administrative power.

Any description of the particularities of administration in the various territories would require much more space than is available here (see Gann and Duignan, 1978; and Gifford and Louis, 1971 for some of the details). One example must suffice. In Nigeria, Britain's most populous colony in Africa, the coastal (Lagos) and northern (Kaduna) regions were administered in completely different ways. The coastal region had a long-standing commercial base and export trade, tied to Liverpool and England's northern industries. Consequently, 'Lagos governors . . . tried to please north country British businessmen by emphasizing the needs of trade, communications and public health, by avoiding wars and punitive expeditions, by their reluctance to impose direct taxation, and by their determination to maintain a policy aimed at "peaceful penetration" and commercial development' (Gann and Duignan, 1978, p. 209). The northern region was a borderland and its international trade was limited. 'In this region the tone of administration was military; the British ruling group was linked to London and the Home Counties [the Southeast of England] rather than to Lancashire. . . . The northern administrative ethos was shaped by Lord Lugard, whose administrative gospel blended muscular Christianity with a military puritanism that exalted the virtues of physical fitness, self-denial and "character"'. Government emphasized prestige instead of profit, hierarchy in place of diversity' (Gann and Duignan, 1978, p. 209).

In all colonies, however, priority was given to communication, transport and medical care. Railways were built both to promote agricultural and mineral exports and to facilitate the movement of police and army detachments. Post offices, telegraphs and telephones gradually tied together the local administrative units. Indicative of the centrality of transportation networks to colonial administration was the fact that public works departments were often the first government units established in a territory. As Gann and Duignan (1978, p. 271) emphasize: 'By 1914 all the British African dependencies possessed a basic infrastructure of specialized services, the most important of which was the creation of a modern transportation network.'

An important cultural import into the colonies, therefore, was the assumption that the state should both encourage development and provide social services – education, agricultural instruction, etc. 'The very notion of the state as a territorial entity independent of ethnic or kinship ties, operating through impersonal rules, was one of the most revolutionary concepts bequeathed by colonialism to post-colonial precedent, . . . All of them have taken over, in some form or other, both the boundaries and the administrative institutions of their erstwhile Western overlords' (Gann and Duignan, 1978, p. 347).

MECHANISMS OF CULTURAL INTEGRATION

The imposition of colonial rule, and, more generally, Western penetration of societies outside Europe, involved a great deal of violence and war. But, once established, 'law and order' involved the imposition of Western values as much as terminating local conflicts and suppressing practices (witchcraft, infanticide, bride burning) that Europeans regarded as 'barbaric'.

The social effects of European values were paradoxical. On the one hand, old values were destroyed, as missionaries and schoolteachers attacked animistic creeds, polygamy and other customs. Families often broke up as some members 'converted' and others did not. On the other hand, the new ways were used by some people to establish new bases to authority. In particular, Western-educated natives became indispensable to European rule and influence. Interpreters, clerks, foremen and police sergeants were cultural pioneers; they represented the new order and profited from it.

The economic effects were also double-edged. As restraints on trade disappeared, commercial agriculture and trade spread in extent and intensity. Yields increased as agricultural techniques improved, and trade proved more profitable as new communications linked previously isolated interiors with coastal entrepôts. Yet, as a consequence, the certainties and rhythms of local life broke down and traditional skills were devalued. Above all, new types of consumption, while adding to the comforts of life of those with sufficient disposable income, led to the destruction of many local industries and the growth of dependence on manufactured imports from the colonial 'motherland'. In this context, obligations to community and chief began to weaken. Money became the major metric for assessing social status. This was in part because money could help purchase an education. 'Education, in turn, brought power and influence. These new opportunities profoundly affected life in the village, and the village ceased to be an almost self-contained unit, absorbing all the interests of its people. Instead, cash-cropping and wage labour for limited periods gradually came to occupy a much more central position in the cultivator's life' (Gann and Duignan, 1978, p. 367).

The growth of 'free' labour was a process that was 'always uneven and idiosyncratic' (Marks and Rathbone, 1982, p. 13). It depended upon spatial variation in the extent of competition for labour; conflicts of interest between firms and the colonial states; and the availability of alternatives to wage labour. But once colony-wide labour and other markets were effectively created, the prospering of commercial enterprises (both foreign and indigenous), such as mines and plantations, depended upon an increasingly efficient and productive labour force to operate new equipment and machinery. This required measures to both increase labour-force stability (housing, minimum wages) and attempts to upgrade the health, literacy and skills of employees. Of course, employee organizations also played a role in pressuring for these changes. Consumption demands, and hence demands for higher incomes, tended to increase in concert with the increase in permanent wage employment.

It is evident that in many colonies there were dramatic improvements in education and health. In Africa in the period 1910–60 the number of children attending school grew much faster than had school enrolments in Europe in the boom years 1840–80. There were also significant increases in life expectancy (e.g. in Ghana in 1921 it was 28 years; in 1980 it was 44.8), and reductions in infant mortality rates. To the British colonial authorities education was an important means of inculcating both 'modern' work habits and a commitment to the class structure of colonialism. At the centre of colonial education, was

> the idea of work – taught to those who lacked property – emphasizing regularity, the organization of time and human energy around the work routine, and the necessity of discipline. It was a moral and cultural concept . . . Prohibitions against drinking and dancing . . . were as much a part of changing concepts of labour as forced recruitment, vagrancy laws, and the insistence that workers put in regular hours.
>
> (Cooper, 1980, pp. 69–70)

But, above all, it was necessary 'to get workers to internalize cultural values and behaviour patterns that would define their role in the economy and society' (Cooper, 1980, p. 70).

This conception of education was particularly characteristic of the British colonies. Other colonies either failed to develop the capitalist labour markets to which it was a reaction (e.g. sections of the French colonies on the southern margins of the Sahara Desert) or were severely underfunded for public activities (e.g. the Belgian Congo and Portuguese colonies). But British policy brought a price. It was precisely the educated élite that 'formed the vanguard of the nationalist movements, and the more "disciplined" African workers became, the more effective were their trade union organizations in pursuing not only economic, but political anti-colonial objectives' (Sender and Smith, 1986, p. 66). As taught in colonial schools, Western concepts of 'democracy' and 'justice' served to undermine the legitimacy of the Western empires that had introduced them.

The growth of wage labour incorporated women as well as men into the colonial world economy, although at significantly lower levels of participation. More importantly, however, the spread of wage employment disrupted existing **sexual divisions of labour,** often to the detriment of women. Women remained powerfully constrained by family, marriage, religion, and so-called 'domestic duties', even while engaged in wage employment. But norms, power and traditional patterns of authority based on gender, as well as age, caste and lineage, were subject to radical challenge because of wage labour, education, migration, and urbanization. Lonsdale (1985, p. 730) quotes from one Zulu chief in southern Africa who in 1905 expressed his opposition to the cultural changes of the time as follows:

> Our sons elbow us away from the boiled mealies in the pot when we reach for a handful to eat, saying, 'we bought these, father', and when remonstrated with, our wives dare to raise their eyes and glare at us. It used not to be thus. If we chide or beat our wives and children for misconduct, they run off to the police and the magistrate fines us.

Cultural change was not always one way, however. Some pre-capitalist and pre-colonial social and religious institutions were strengthened. More accurately, perhaps, 'new' syncretic traditions were invented out of elements of past traditions. Mission-educated élites often invented mythic histories of ancient empires and new nationalist traditions both to legitimize their quests for political power and to protect themselves in the new labour markets. Adaptation was at least as common as straightforward assimilation in many colonial settings.

Ultimately, however, cultural incorporation, whether by adaptation or assimilation, undermined the socio-political relation of dominance/subordination that colonialism existed to reproduce. Even when the idiom of dominance was *improvement* rather than *order*, native groups were not always persuaded of the natural superiority of colonial ways. Langley's (1983, p. 223) conclusion concerning American colonial adventures in the Caribbean is a fitting epitaph to the cultural contradictions of the entire colonial enterprise:

> Striving to teach by example, they found it necessary to denigrate the cultural values of those whom they had come to save. . . . Their presence, even when it meant a peaceful society and material advancement, stripped Caribbean peoples of their dignity and constituted an unspoken American judgement of Caribbean inferiority. Little wonder, then, that the occupied were so 'ungrateful' for what Americans considered years of

277

sexual divisions of labour
The division, specialization and different rewards between occupations predominantly occupied by either men or women.

benign tutelage. But, then, Americans do not have in their epigrammatic repertory that old Spanish proverb that Mexicans long ago adopted: 'The wine is bitter, but it's our wine.'

THE CHANGING GLOBAL CONTEXT OF INTERDEPENDENCE

The colonial world economy began to disintegrate after the Second World War. The crossover trading system effectively ended in the 1930s. The Second World War's build-up of industrial capacity for military purposes in the United States and the Soviet Union produced two superpowers without overseas colonial empires and European decolonization further undermined the colonial world order. In the 1950s and 1960s the so-called Third World (of politically independent but often politically non-aligned and always less prosperous nations) was born. Large parts of Asia and Africa now joined Latin America as a largely non-industrial and ex-colonial but still 'dependent world' (Fig. 8.9). The initial tendency was to attempt to achieve economic self-reliance through national strategies of industrialization and diversification of trading partners. This was the first change in relation to the global context of interdependence.

Figure 8.9
Decolonization after the Second World War
Source: Edwards (1985), p. 210

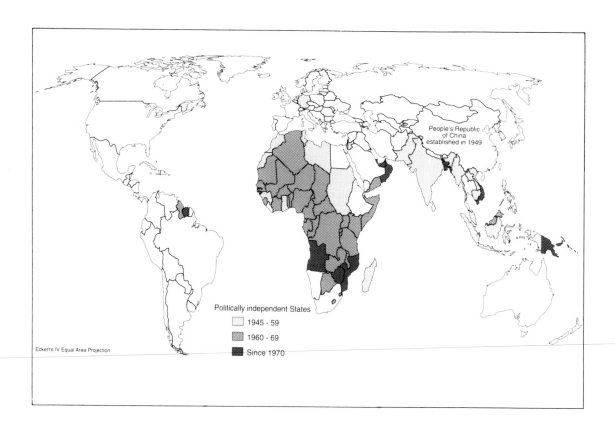

The end of colonialism and national strategies of development

As new states came into existence, so did attempts to stimulate industrial develop-ment and economic growth. From the Depression of the 1930s on, and especially during the Second World War, stagnation and shipping blockades had encouraged some import substitution in the colonies. Increasingly traders and merchants in the richer peripheral countries looked to manufacturing industry as a source of capital accumulation. Leading politicians also saw in industrial development both national and personal advantage (see Chapter 10 for more details on the preference for industrialization).

Most importantly, however, the barter terms of trade (the ratio between the prices of exports and the prices of imports) for many of the basic commodities exported to the 'core' appeared to have deteriorated, as had the general barter terms of trade between richer and poorer countries (Table 8.4). Basic commodity prices (largely those for raw materials) declined relative to those for manufactures. One consequence was a widening gap between what peripheral countries received for their exports and what they had to pay for their imports. This was widely viewed as inevitable, given the low-income elasticity of demand for 'less developed countries'' (LDCs) primary products (as incomes rose in developed countries there was not a parallel increase in the consumption of goods from LDCs), the persisting low wages in the primary production sector in the LDCs (lowering local consumption power), and the high protective barriers for competing primary production (mainly temper-ate agricultural products, such as sugar beet and cereals, and largely erected in the 1930s) in the 'developed countries' (DCs).

In fact, the evidence for deteriorating terms of trade is much more tenuous than it once seemed. Though real prices for almost all primary commodities have declined since the early twentieth century, the rate and stability of decline vary considerably across primary commodities and relative to manufactures (Diako-savvas and Scandizzo, 1991). Some research suggests that the barter terms of trade is subject to sharp jumps along a downward path rather than a simple secular decline (Powell, 1991). What is important, however, is that a *perception* of a deteriorating barter terms of trade inspired the first post-independence leaders to embark on policies that led away from specialization in primary commodity production. Recent years have not offered much encouragement. During the period 1984–93, the average annual rate of change in monthly prices for the principal non-fuel primary commodities in world trade has been between −3.6 and −5.6 in constant US dollars. This indicates for this period at least a net downward trend in most basic commodity prices (UNCTAD 1994).

Pessimism about the future prospects for primary products was reinforced by the view that the primary sector was inherently backward compared to the manufactur-ing sector: because of the latter's multiplier effects and, allegedly, greater economies of scale. Consequently, the pursuit of industrialization could be justified 'theoret-ically' as well as materially (see Chapter 10).

The promotion of industry came from either protection – from reserving domestic markets for domestic industry – or, later, establishing export enclaves and attracting foreign and local capital to finance branch plants. Once established, and/ or protected through their early, vulnerable years, new industries would be able to

Table 8.4: A hundred years of the barter terms of trade

	LDCs against DCs		Primaries against manufactures	
	all products	all products (excluding fuels)	LDCs' primaries and DCs' mfrs	World primaries and world mfrs
1880	126	n.a.	130	114
1900	114	n.a.	103	91
1913	177	n.a.	185	106
1928	172	n.a.	178	98
1937	150	n.a.	152	78
1952	115	122	109	118
1962	100	101 (1960)	100	100
1968	101	102 (1965)	97	93
1970	101	105	96	97
1975	143	78	198	156
1980	177	76	309	211

Source: Edwards (1985), Table 4.2, p. 62.

compete globally. A predilection for protection, in many cases, reflected both a positive interpretation of the past practices of such countries as the United States, Japan and Germany, which had in their day protected 'infant industries' and the conception of an 'activist' state common to many ex-colonial territories.

During the 1960s exports of manufactures from LDCs grew quickly, from around US$3 billion in 1960 to over US$9 billion in 1970 (UNIDO 1981). As a percentage of total world trade in manufactures this was an increase of from under 4 per cent to 5 per cent. In the 1970s growth was even more rapid. By 1980 LDC manufactured exports were more than US$80 billion, or over 9 per cent of the world total. This growth is part of the New International Division of Labour (NIDL) (see also pp. 40–8). One of the most notable features of the period 1960–80, however, and perhaps even more notable since then, has been the polarization of performance and prospects between different regional groupings of LDCs. More than 80 per cent of the total of LDC manufactured exports comes from ten NICs (Brazil, China, Hong Kong, India, Malaysia, Mexico, Pakistan, Singapore, South Korea and Taiwan). In most of these cases, especially Hong Kong, Malaysia, Mexico, Singapore, South Korea and Taiwan, industrial growth has been export-led (export-enclave) rather than import substitution (protection). Since 1980 this pattern has become institutionalized, with the bulk of the foreign direct investment, bank lending and trade relating to manufacturing production outside of the DCs concentrated in East Asia, parts of Latin America and Eastern Europe (Cook and Kirkpatrick, 1997).

More than 60 per cent of all LDC manufactured exports are now sold to DCs. This market is limited by both high levels of protection in the DCs and the risk of increased protection in the future. Since the early 1970s the level and uncertainty of

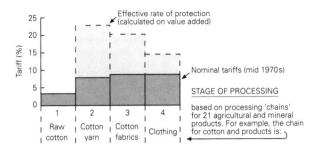

Figure 8.10

How protectionism in developing countries varies by stage of processing: the case of cotton and cotton products

Source: Edwards (1985), Fig. 8.5, p. 220

281

protective barriers (tariffs, quotas, etc.) to LDC manufactured exports have increased tremendously, particularly so-called hard-core nontariff barriers such as quotas, voluntary export restraints, and the Multifibre Agreement (MFA) (covers textiles and clothing). This is especially the case for relatively more finished products (Fig. 8.10 shows how average tariffs vary for cotton at different stages of processing). Between 1966 and 1986, the share of imports affected by all non-tariff measures increased by more than 20 per cent for the USA, around 40 per cent for Japan and 160 per cent for the EC. By 1986 21 per cent of LDC exports to the industrialized world were covered by these barriers even as average tariff rates and coverage declined (World Bank, 1991).

Manufactured goods can be classified into two main product groups: capital goods (machinery and equipment, including transport equipment) and consumer and intermediate goods, of which textiles and clothing are the largest single category in international trade. Capital goods account for about half of all manufactured goods traded in the world economy (46 per cent in 1990). But different world regions account for different shares of the two product categories. DCs supply 92 per cent of world exports of capital goods and 52 per cent of textiles and clothing. They also supply 81 per cent of other miscellaneous manufactures. LDCs are important only as suppliers of textiles and clothing and certain light industrial products, mainly consumer goods (with the important exception of electronics components and automobile components, in the case of Brazil and Mexico). The specialization of trade flows between LDCs and DCs, therefore, extends today beyond the distinction between primary commodities and other goods to apply *within* the category of manufactured goods (World Bank, 1991).

The role of transnational corporations

The growth of trade in manufactures in the 1970s, after a 40-year period in which manufacturing production was intensively concentrated in the DCs and there was more limited DC–DC as well as DC–LDC trade in manufactures, was influenced by the growing significance of multinational corporations and of contractual cooperation between firms in different countries (De Vroey, 1984). Transnational corporations (TNCs) have long been active in manufacturing in LDCs. As we saw in Chapter 6 and discuss later in Chapter 10, they tended at first to duplicate plants around the world in order to gain access to protected markets, or to make use of local raw materials. The production by TNCs of cars (e.g. in Brazil), agricultural engineering products (e.g. in South Africa and Mexico), and pharmaceuticals (e.g. in India) across a range of LDCs are examples. This kind of manufacturing production still exists, especially in countries with large internal markets. In Brazil, for example, in the mid-1970s to take an extreme case, almost 50 per cent of industrial output was produced by TNCs and more than 90 per cent of TNC production was sold locally (Joekes, 1987).

Since the 1960s, however, much TNC involvement in LDCs has also involved what is known as **global sourcing**. As a result of technical change, especially reductions in transportation costs, and the appeal of cheap (often female) labour in certain countries, production activities that once were adjacent spatially can now be dispersed widely. Many so-called light industrial processes are especially suited to the separation of various stages of production. In particular, labour-intensive stages (as in the product life-cycle model) can be located to take advantage of both the enormous international spread in wage levels and the exchange-rate fluctuations between currencies that have been a feature of the world economy since the early 1970s. With respect to wage levels, the shoe industry faces wage costs of US$6 per hour in the United States but only US$0.85 to US$1.39 in East Asia; Chinese textile workers earn US$0.37 per hour compared to an American rate of $10.02.

One industry that has engaged in global sourcing on a massive scale (perhaps, and for this reason, somewhat exceptional) is the consumer electronics components and products industry. This industry has two characteristics that have encouraged the shift to global sourcing: discrete production segments, of which some are extremely labour-intensive and require 'flexible response' because of short product-cycles that make automation uneconomic (Eisold, 1984), and compact products (parts and components) that can be shipped relatively cheaply. East Asian locations with cheap, reliable, literate and tractable (largely female) labour forces have been especially attractive to this industry (and some others such as textiles and clothing). Governments have often facilitated the process of establishing component and assembly plants through the provision of export-processing zones, subsidies, and tax advantages and the enforcement of the 'political stability' highly valued by TNCs and their local sub-contractors.

Changes in markets for primary commodities

For most LDCs, however, there is still a heavy dependence on trade in primary commodities (see Fig. 2.12: Index of Commodity Concentration of Exports). But the primary commodity sector has become extremely heterogeneous with respect to trading conditions since the Second World War. Four major categories stand out in this regard: fuels (mainly petroleum), non-fuel minerals, grains, and other agricultural products. These product groups have experienced very different price movements and, to some extent, quantity fluctuations over the past 30 years. The non-food commodity prices have been especially volatile. Generally, manufactures have increased in price to the disadvantage of primary commodity exporters. But there have been two periods, 1949–52 and 1973–80, when demand for primary non-food commodities was extremely strong and commodity prices surged. In particular, in the years 1973–80, inflation and uncertain economic conditions in the DCs boosted the prices of agricultural raw materials. Since 1980, however, the

global sourcing
The use of multiple sources in different countries for the components of a particular product that is assembled elsewhere.

relative price strength in the fuels and agricultural products groups has largely disappeared and the value of commodity export earnings in these sectors has sunk precipitously in relation to the prices of manufactures and minerals. Perhaps the most negative price movement from the perspective of most LDCs has been in the price of grains. There has been a long-term decline in world grain prices. This reflects tremendous increases in production the world over, but especially in the United States and other DCs. Normal yields per hectare are now twice what they were in 1950. The real price of wheat, however, is now about half what it was 100 years ago.

Across all primary commodities, commodity agreements between producing and consuming countries (cocoa, tin, sugar and natural rubber) and producer cartels (most famously, OPEC for petroleum) have failed to reduce volatility and raise the prices of primary commodities relative to those of manufactures because of fundamental differences of interest between producers and consumers and among producers. Even OPEC, after successfully raising the price of oil from 1973 to 1979, has been riven by conflict and the failure to attract some major oil producers (such as Mexico, Britain, Norway) to its ranks. This failure has encouraged further attempts at industrialization as the major strategy of economic development. However, price volatility acts to reduce the industrial potential of 'mineral economies' (Auty, 1991). This is one of the 'Catch-22s' of the contemporary world economy.

Some countries, especially those in sub-Saharan Africa, could probably benefit from increased attention to primary commodities. Sender and Smith (1986) show that the macroeconomic policies of many African governments have worked to undermine their region's shares of world export markets across a range of primary commodities (Table 8.5). Since the early 1970s real export earnings have remained stagnant or declined significantly in 25 out of 33 countries in sub-Saharan Africa for which data are available. In 1988, the whole of this region, with over 400 million people, had export revenues less than those of Singapore, a city state of 2.5 million people (Svedberg, 1991).

Inelasticity of demand in the DCs (expressed in a deteriorating barter terms of trade) cannot explain the magnitude of these declines. Neither can the absence of commodity diversification, since those countries with a relatively diversified structure of agricultural exports – such as Tanzania – have not experienced more favourable trends in export earnings than more specialized ones. Sender and Smith (1986, p. 127) explain the absolute decline of sub-Saharan Africa's contributions to world commodity markets in terms of 'the continued dominance of anti-trade ideologies and export pessimism' that are 'probably explained by the political hegemony of nationalism. It remains expedient for the national bourgeoisie, or for those determining the form and nature of state intervention, to deflect criticism by resort to anti-imperialist rhetoric and to blame foreign scapegoats for economic failure.'

However, this is probably too narrow a perspective. Political agendas and social problems of a more general nature also have played important roles. In the immediate post-independence period, considerable political energy was expended in diversifying import and export markets rather than building larger ones. This was a direct result of trying to slay 'the colonial dragon' as the newly independent

Table 8.5: Sub-Saharan African exports as a percentage of total world exports of selected primary commodities, 1961–92

	1961–62	1969–71	1984–85	1991–92
Coffee	25.6	29.3	21.8	13.7
Tea	8.7	14.4	15.7	12.0
Groundnut oil	53.8	57.6	30.6	33.1
Groundnuts	85.5	69.1	7.2	3.4
Palm kernel oil	55.2	54.8	7.5	4.5
Palm oil	55.0	16.4	1.8	2.5
Bananas	10.9	6.5	2.7	3.2
Cotton	10.8	15.5	11.9	12.6
Rubber	6.8	6.8	5.3	5.4
Tobacco	12.1	8.2	9.6	14.0

Source: UNCTAD (1987, 1994), various tables.

countries tried to become less dependent on their former colonial powers. Governments have also been faced with major ethnic divisions and rivalries, fragile political institutions, and 'superpower' infiltration and manipulation (Jackson, 1986). The Nigerian Civil War, frequent military *coups d'état*, and American or Soviet covert operations in most African countries are symptomatic examples of the diversions from economic policymaking that have faced political élites in sub-Saharan Africa (and to a lesser extent also in Latin America and Asia) since the 1940s (see Grant and Agnew, 1996).

Disparities within the periphery

Disparities between LDCs have increased substantially since the 1970s. Several groups have emerged and can be distinguished. First, there are the NICs which have grown rapidly and are important exporters of manufactured goods. These include the 'old' NICs such as South Korea, Taiwan and Hong Kong, and newer NICs, predominantly in Asia (e.g. Malaysia, Thailand) but including Turkey and Mexico. In 1960 the Asian NICs only accounted for 5 per cent of total LDC exports and in 1980 for 10 per cent but by 1989 this had risen to 32.1 per cent, reflecting both their growth and the relative stagnation in the 1980s of other NIC and 'near-NIC' economies.

Then there is a group of countries that experienced reasonable growth until the late 1970s but because of high debt loads stagnated in the 1980s. Examples would include Argentina, Brazil (often thought of as a NIC), the Philippines and Morocco. A couple of countries in this group with less serious debt problems, Costa Rica and Colombia, managed to continue their economic diversification away from primary commodities. A third group remains very dependent on raw material exports but export demand has held up reasonably well. Examples would include oil exporters such as Nigeria, Iraq, Ecuador and Cameroon as well as exporters of other commodities such as Bolivia and Jamaica.

Finally, there are two groups of low-income countries mainly in sub-Saharan Africa and Asia. The Asian group, China, Bangladesh, India, Pakistan, Sri Lanka, Nepal, and Afghanistan, is populous and until recently they isolated themselves from the world economy through protectionist policies. China has been the most aggressive in 'opening up' to trade and foreign investment and the impact of this is now apparent in China's coastal areas, especially around Hong Kong and in the vicinity of Shanghai. In many respects China is a NIC in the making. Sub-Saharan Africa has had the poorest record of economic growth over the past 20 years. Most countries in the region have experienced declining or stagnant export earnings in the 1980s. They are heavily dependent on foreign aid and investment by multi-lateral institutions such as the World Bank (but on the dangers of over-aggregating the African case, see Grant and Agnew, 1996).

The effects of the Cold War

The end of colonialism did not usher in an era of equivalently 'sustainable' national development everywhere in the former colonial world; that much should be clear from the preceding discussion. The factor initially most responsible for this was the Cold War between the United States and the Soviet Union, that, while encouraging 'aid' programmes of one kind or other, also encouraged militarization and political instability. After the Second World War the world was effectively divided into two spheres of influence with large parts of the new 'Third World' of former colonies as a zone of superpower competition (Figure 8.11). In certain cases, such as, for example, South Korea and Taiwan, superpower aid (American in these cases) contributed to economic growth. In many African countries aid has helped achieve major improvements in physical and social infrastructure, although much of the most productive aid has not come from the superpowers, who have specialized in military aid and technical assistance (intelligence gathering) rather than direct economic assistance. International agencies (the UN, World Bank, etc.), and some

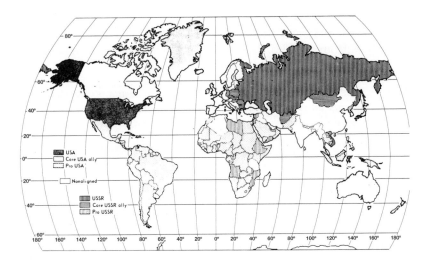

Figure 8.11
The Soviet and US spheres of influence, 1982
Source: O'Loughlin (1986), Fig 10.1, p. 242

European countries (particularly the Scandinavian countries and the Netherlands) have provided much of the more economically 'useful' aid (on aid see Chapter 2).

In other cases, 'models' of development were imported from either the United States ('free enterprise') or the Soviet Union (central planning) and then supported/ undermined from outside by each of the superpowers. This often led to increased militarization both of governments and national budgets as internal opponents were repressed and external patrons satisfied. Between 1960 and 1982, more than 10 700 000 people were killed in 65 major wars (those with more than 1 000 deaths attributable to them). Most of these wars were fought in the LDCs.

The globalization of capital

A second change in the global context of interdependence since the demise of colonialism and the rise of national development strategies, and increasingly important since the 1970s, has been the increased pace and internationalization of the world economy as noted in Chapters 3 and 6. Capital has become much more mobile, both in time and space. For example, before 1973 currency exchange-rates changed once every four years on average, interest rates moved twice a year, and companies made price and investment decisions no more than once or twice a year (see Fig. 3.2). This has all changed. There is now an almost constant review of prices and investment decisions, a constant instability and disorder. This places an even greater premium on the ability of firms and governments in LDCs to react within some coherent national framework to changes in the global political-economic environment.

Nowhere is this clearer than with respect to the world financial system. As we have seen, the growing integration of the world economy and the loosening of government control over exchange rates when the major industrial governments, led by the United States, abandoned the Bretton Woods system of semi-fixed currency exchange rates in 1973, stimulated the growth of a massive *private* international monetary system. This system was organized around Eurodollars, a term that originally meant US dollar deposits in banks in Europe but now refers to dollars that circulate outside the United States, which are used for world trade, and are not regulated by the US government. This global currency mushroomed between 1974 and 1981 as a result of the enormous dollar surpluses earned by the OPEC countries from oil sales. The large banks which received these funds sought borrowers who

> could be charged enough interest to enable the banks to earn a profit. The banks first moved Eurodollars into Third World countries, saddling them with a US$1 trillion debt burden by the end of 1986, compared to less than US$100 million in 1973. Instead of supporting new productive investment, however, a large portion of this debt went into luxury consumption. Over US$200 billion disappeared through capital flight from the Third World back to the industrial countries.
>
> (Wachtel, 1987, p. 786)

This phenomenon was particularly deleterious for poor oil-importers (who had to finance their oil imports) and countries with ambitious development plans (into which outside capital could be pumped). Many of them are still locked into debt repayment schedules that require them to devote the lion's share of their export earnings to servicing their debts. In Latin America, for example, since 1981 there has been a persistent net outflow of capital 'reflecting a "scissors movement" of declining grant aid and mounting debt repayments' (Cook and Kirkpatrick, 1997, p. 63).

Since 1982, when Mexico and certain other major Latin American borrowers effectively defaulted on loan repayments, the Eurodollars have shifted into funding the US budget deficit, into firm mergers and acquisitions in the United States and Europe, and into the world's stock markets (Lessard and Williamson, 1987). But the burden of debt, in both the form of repayment and the inability to borrow fresh capital with outstanding debt, has emerged as a major new barrier to economic development (see Fig. 2.15). In the 1980s net financial flows to LDCs initially decreased as commercial bank lending dried up. Since 1986, government and multilateral agency (World Bank, etc.) aid and foreign direct investment by TNCs have increased substantially. But only in 1989 did total flows equal the 1982 figure and, within this, aid increased from 38 to 55 per cent of the total (*The Economist*, 14 September 1992, p. 120). Since then private investment has increased somewhat, including both portfolio and foreign direct investment, but debt service still soaks up significant proportions of all capital inflows. This is borrowing to pay back what was previously borrowed rather than productive investment in economic activities that improve the lot of the LDCs and their populations. In the years 1990–96 most new bank lending from the DCs went to Asia, whereas the other regions had to rely more on foreign direct investment (as in Latin America) or on aid and multilateral loans from international organizations (sub-Saharan Africa) (*The Economist*, 15 February; *The Economist*, 1 March 1997).

Debt and instability

If the 1970s are remembered for two major oil-price shocks (in 1973 and 1979) and a persisting downturn in the world economy, the 1980s were marked by three global phenomena. One was the international debt problem which effectively undermined growth in many LDCs for most of the decade. Though a variety of international agreements renegotiating interest payments reduced the overall severity of the international debt crisis in the 1980s, debt burdens remain at historic highs and effectively undermine the possibility of future credit-led economic development (Corbridge, 1993, offers a thorough overview of this phenomenon). The absolute debt load of LDCs peaked in 1994 at around 39 per cent of their combined GDP after declining in a cyclical fashion from a previous high of around 37 per cent in 1986. By 1996 the figure stood at 35.6 per cent; hardly a dramatic improvement. So debt is a problem that still haunts many peripheral economies. Its impact varies widely, however, and is not a direct function of the absolute size of the debt. For example, as of 1996 Mexico had a huge debt of over US$160 billion but its debt–service ratio (interest and principal payments as a share of all exports) was 'only' 24.2 per cent, whereas, at the other extreme, Hungary with a debt of around US$40 billion had a debt–service ratio of 39.1 per cent (*The Economist*, 26 April, 1997, p. 110; World Bank, 1996, pp. 220–1). A second was the volatility of exchange rates, illustrated most vividly by the steep appreciation of the US $ until

global currencies
Currencies used in international transactions. The US dollar is the most important one today, though its use is challenged somewhat by the German mark and the Japanese yen.

1985 followed by its subsequent dramatic descent. These shifts had important effects on the imports and exports of countries whose currencies were 'pegged' in value to **global currencies** such as the US $. Commodities would become more or less expensive depending on the shifts in the value of the global currency. The net impact was negative for all LDCs except the Asian NICs. Finally, the US federal government and national trade deficits, emerging spectacularly after 1983, produced growth in the USA at the expense of stagnation elsewhere. Investment that could have gone to the LDCs under other circumstances was diverted to the USA to finance the deficits. This also raised interest rates on outstanding loans such as those held by heavily indebted LDCs. Consequently, high interest rates and low commodity prices rather than the debt loads incurred in the 1970s were probably the major barriers to growth in the LDCs in the 1980s (Rogoff, 1992). On balance the 1980s was not a good decade for the LDCs as a whole. The 1990s have been somewhat better but hardly stellar, given the declining terms of trade for primary commodities, the debt loads incurred but not yet paid off and increasing populations.

Core–periphery polarization?

From one point of view, the net effect of the changes in the global context for interdependence has been an increased division between the LDCs, on the one hand, and the DCs, on the other. National income and purchasing power statistics support this interpretation (Emmanuel, 1972; McMichael *et al.*, 1974). But, from another point of view, the periphery has in fact developed rapidly. This interpretation is supported by data on output, health and education (Warren, 1980; Schiffer, 1981). One way of reconciling these discordant interpretations is to argue that the post-colonial world economy has come to rest on increasingly diffused global production but has lacked a similar attainment of a global spread of consumption. The relatively low incomes available in LDC factories and plantations have put a cap on local purchasing power even as local labour forces were made more efficient (through improved health and education) and increased their output.

The problem with this reconciliation and the interpretations upon which it is based is that they are geographically over-aggregated. The experience of different groups of LDCs has been different, as suggested previously by the 'grouping' of countries. On the one hand, some of the Latin American countries, for example, are relatively large and have relatively high levels of per capita national income (e.g. Brazil, Argentina). Some of them did achieve considerable income growth in the 1960s and 1970s on the basis of industrialization to satisfy local markets (import substitution). Most of them, however, were heavy importers of oil (Mexico and Venezuela were exceptional) and their industrial sectors were generally uncompetitive in world markets. They were hit in the 1970s by the combination of oil price rises and their failure to switch to export-oriented manufacturing in the boom years of the late 1960s. They had to borrow to ease the oil-shock adjustment instead of paying for it with export earnings. They are now caught in a 'debt trap' of accumulated loans and compounded by the high interest rates of the early 1980s and the growth of trade barriers to the manufactures they export to the United States and Western Europe.

The sub-Saharan African countries, on the other hand, are much poorer on average than the Latin American countries and their low level of output in all sectors is undermined by their even faster rates of population increase. They are economies with small industrial sectors and a heavy dependence on primary commodities. As commodity prices have dropped in the 1980s they have been forced to borrow to maintain minimal levels of consumption. Their debt burden is similar to that of the most indebted Latin American countries. The consequence, as in Latin America, is a general reduction in the standard of living but in contexts where it is already desperately low.

Finally, Asian countries have, on average, managed the best over the past 20 years. They have been more successful in maintaining economic growth as a secular trend and adjusting to short-run cyclical downturns such as the world recessions of 1974–75, 1979–82 and 1989–93. The East Asian NICs, the most dynamic economies in the region, have been able to expand their export of manufactures. They now account for about three-quarters of LDC total exports of manufactures. Although, like the Latin American countries, they borrowed heavily in 1974–75 to adjust to the oil price increases, their export performance has allowed them to keep relatively good borrowing terms and adjust more easily to the massive interest rate increases of 1980–81. Yet, they are not without their own difficulties. Growth rates have slowed throughout East Asia in the mid-1990s indicating that there may be limits to an export-based strategy of economic development when established markets stagnate or decline and growth is more reliant on cheap labour and high savings than on technological and organizational innovation (Krugman, 1994).

ALTERNATIVE MODELS OF DEVELOPMENT

In the face of the failure of many LDCs to maintain, let alone increase, their production output and the consumption levels of their populations, the models of development upon which national development efforts have been based have been called into question. This coincides with the growing questioning of the American model (especially the lack of effective national trade and industrial policies) in the USA and the collapse of the Soviet model in the (former) Soviet Union and Eastern Europe. Neither these models nor the syncretic local versions that have appeared over the past 30 years appear to offer a way out of the 'development impasse'. The spread of more liberal and open trading policies in the 1980s did produce benefits for some countries, particularly the NICs. But these have been strictly limited geographically.

It is in this context that new models have appeared (and disappeared). Perhaps the three most important ones are based on (1) Japanese experience, (2) Chinese experiments in the 1960s and, to a much more limited extent, (3) Islamic economic practices. Each of these alternative models is noted briefly. There is no space here for extended critical evaluation.

Japanese experience is seen as relevant because Japan is the sole case of a country with a non-European population rising from periphery to core within the world economy (see Chapters 3 and 5). Even other DCs are now exhorted to imitate the

Japanese (Vogel, 1980). The Japanese model is seen as involving, among other things, national mobilization around well-defined economic objectives; an integration of business and government operations through finance and product targeting; export-orientation; capitalization of agriculture to increase agricultural self-sufficiency; and increased urbanization to profit from agglomeration economies of scale. In practice, of course, combining these elements is most difficult since the historical and geographical settings in which Japan developed cannot be reproduced at will. The ability of S. Korea and Taiwan to follow the Japanese example is probably as much related to their cultural and political character (ethnic homogeneity, strong family attachments, high investment in social and physical infrastructure) as it is to their export orientation.

The Chinese experiments of the 1960s with emphasizing agricultural development and national self-sufficiency have had a greater appeal in many LDCs (Harris, 1978). But, of course, the Chinese have themselves now departed from this road in pursuit of industrialization and world trade (Nolan, 1983). The model remains, however, and is an important part of the philosophy of oppositional and guerrilla movements in many parts of Africa, Asia and Latin America (e.g. Peru's 'Shining Path').

Finally, practices and beliefs drawn from the Islamic religion have become important in the Middle East, North Africa and parts of Asia (e.g. Indonesia). The prohibition of usury or 'excessive' interest charged on monetary loans is one of the more concrete and obviously appealing features of Islamic economics. But as yet no system of political economy based upon Islamic principles has been established in any country (including Iran). The conclusion of Katouzian (1983, p. 164), one of the leading authorities on Islamic economics, seems appropriate:

> While one may empathize with the desire to construct an indigenous ideology that can be identified with the Islamic beliefs and practices of its advocates particularly in view of the havoc caused by selective application of Western ideas under the late Shah [of Iran], it is no more to be expected that Islam can provide a comprehensive economic system than that the latter could be based on Christianity, Judaism, or any other traditional religio-political system.

Summary

In this chapter we have surveyed the dynamics of interdependence between the core and the periphery of the world economy from the colonial period to the present day. We have identified the following points as being of critical importance:

1. Existing economies were transformed into colonial ones through regional specialization in primary commodity production.
2. In the late nineteenth and early twentieth centuries a 'crossover' multilateral system of trade with Britain as its linchpin integrated the world economy.
3. The 'crossover' system was progressively displaced by direct investment from transnational firms (TNCs). American firms were especially important.
4. Colonialism created the conditions for wage labour and gave priority to improving communications, transportation and medical care. The European-style territorial state became accepted as the basic political unit for regulating economic activity.

5. Western values had paradoxical effects. On the one hand, values of work and private property were disseminated. On the other hand, new syncretic traditions were invented.

6. With decolonization, new states came into existence which attempted to encourage industrialization.

7. For many years much manufacturing in the LDCs was import substitution. Since the 1960s, however, TNCs have engaged in 'global sourcing': dispersing some production functions to appropriate sites in LDCs and exporting components/products back to the USA, Western Europe or Japan. Some NICs have developed their own export-oriented industries.

8. Many LDCs are still heavily dependent on the export of primary commodities, the prices of which are highly volatile.

9. Cartels and production agreements have largely failed to stabilize the production or prices of most primary commodities. The success of OPEC in the 1970s is the one exception.

10. The Cold War between the United States and the Soviet Union and the increased pace and internationalization of the world economy have placed serious constraints on development efforts. The global 'debt crisis' of the early 1980s has been another especially important constraint.

11. The integration of production within the world economy has not been matched by an integration of consumption. However, different regions of the periphery have had different experiences in this regard: the Asian countries (especially the East Asian NICs) have been most successful, the countries of sub-Saharan Africa least so.

12. New models of development, from Japan, China and Islam, have arisen to challenge the dominant Western/Soviet ones (or mixes thereof) because of the failure of the dominant ones to maintain or generate sustainable economic development.

The next two chapters take off from this general perspective on the transformation of the periphery to examine contemporary patterns of agriculture and industry in the periphery.

Key Sources and Suggested Reading

Achebe, C. 1975. *Morning Yet on Creation Day*. London: Faber.

Agnew, J. A. 1987. *The United States in the World Economy: A Regional Geography*. Cambridge: Cambridge University Press.

Auty, R. M. 1991. Third world response to global processes: the mineral economies. *The Professional Geographer*, **43**, 68–76.

Bairoch, P. 1982. International industrialization levels from 1750 to 1980. *Journal of European Economic History*, **11**, 269–333.

Christopher, A. J. 1984. *Colonial Africa*. Beckenham: Croom Helm.

Cook, P. and Kirkpatrick, C. 1997. Globalization, regionalization and third world development. *Regional Studies*, **31**, 55–66.

Cooper, F. 1980. *From Slaves to Squatters: Plantation Labour and Agriculture in Zanzibar and Coastal Kenya, 1890–1925*. New Haven: Yale University Press.

Corbridge, S. 1993. *Debt and Development*. Oxford: Blackwell.

Crossley, J. C. 1983. The River Plate countries. In H. Blakemore and C. T. Smith (eds), *Latin America: Geographical Perspectives*. 2nd edn. London: Methuen.

De Vroey, M. 1984. A regulation approach to interpretation of the present crisis. *Capital and Class*, **23**, 45–66.

Diakosavvas, D. and Scandizzo, P. 1991. Trends in the terms of trade of primary commodities, 1900–1982: the controversy and its origins. *Economic Development and Cultural Change*, **39**, 231–64.

Duignan, P. and Gann, L. H. 1985. *The United States and Africa: A History*. Cambridge: Cambridge University Press.

Dunning, J. H. 1983. Changes in the level and structure of international production: the last one hundred years. In M. Casson (ed.), *The Growth of International Business*. London: Allen and Unwin.

Economist 1992. Net financial flows to developing countries. 14 September, 120.

Economist 1997a. International credit. 15 February, 100.

Economist 1997b. Investment flows. 1 March, 108.

Economist 1997c. Foreign debt. 26 April, 110.

Edwards, C. 1985. *The Fragmented World: Competing Perspectives on Trade, Money and Crisis*. London: Methuen.

Eisold, E. 1984. *Young Women Workers in Export Industries: The Case of the Semiconductor Industry in South East Asia*. Geneva: Working Paper, ILO World Employment Programme.

Emmanuel, A. 1972. *Unequal Exchange: A Study of the Imperialism of Trade*. London: New Left Books.

Fiala, R. and Kamens, D. 1986. Urban growth and the world polity in the nineteenth and twentieth centuries: a research agenda. *Studies in Comparative International Development*, **21**, 23–35.

Gann, L. H. and Duignan, P. 1978. *The Rulers of British Africa. 1870–1914*. Stanford: Stanford University Press.

Gifford, P. and Louis, W. R. (eds) 1971. *France and Britain in Africa: Imperial Rivalry and Colonial Rule*. New Haven: Yale University Press.

Grant, R. J. and Agnew, J. A. 1996. Representing Africa: the geography of Africa in world trade, 1960–1992. *Annals of the Association of American Geographers*, **86**, 729–44.

Gugler, J. and Flanagan, W. G. 1978. *Urbanization and Social Change in West Africa*. Cambridge: Cambridge University Press.

Harris, N. 1978. *The Mandate of Heaven: Marx and Mao in Modern China*. London: Quartet.

Hobsbawm. E. J. 1968. *Industry and Empire*. New York: Pantheon.

Hopkins, A. G. 1973. *An Economic History of West Africa*. London: Longman.

Jackson, R. H. 1986. Conclusion. In P. Duignan and R. H. Jackson (eds), *Politics and Government in African States, 1960–1985*. Beckenham: Croom Helm.

Joekes, S. P. 1987. *Women in the World Economy*. New York: Oxford University Press.

Katouzian, H. 1983. Shi'ism and Islamic economics: Sadr and Bani Sadr. In N. R. Keddie (ed.), *Religion and Politics in Iran: Shi'ism from Quietism to Revolution*. New Haven: Yale University Press.

Krugman, P. 1994. The myth of Asia's miracle. *Foreign Affairs*, **73**, 62–78.

Langley, L. D. 1983. *The Banana Wars: An Inner History of American Empire, 1900–1934*. Lexington: University Press of Kentucky.

Latham, A. J. H. 1978. *The International Economy and the Undeveloped World, 1865–1914*. Beckenham: Croom Helm.

Lessard, D. R. and Williamson, J. 1987. *Capital Flight and Third World Debt*. Washington, DC: Institute for International Economics.

Lonsdale, J. 1985. The European Scramble and conquest in African history. *The Cambridge History of Africa*, Vol. 6, Ch. 12, Cambridge: Cambridge University Press.

Marks, S. and Rathbone, R. (ed.) 1982. *Industrialization and Social Change in South Africa*. London: Longman.

McMichael, P. *et al.* 1974. Imperialism and the contradictions of development. *New Left Review*, **85**, 83–104.

Mintz, S. 1985. *Sweetness and Power: The Place of Sugar in Modern History*. New York: Viking.

Nolan, P. 1983. De-collectivization of agriculture in China, 1979–82: a long-term perspective. *Cambridge Journal of Economics*, **7**, 381–403.

O'Loughlin, J. 1986. World power competition and local conflicts in the third world. In R. J. Johnston and P. J. Taylor (eds), *A World in Crisis? Geographical Perspectives*. Oxford: Blackwell.

Powell, A. 1991. Commodity and developing country terms of trade: what does the long run show? *The Economic Journal*, **101**, 1485–96.

Ranger, T. 1976. From humanism to the science of man: colonialism in Africa and the understanding of alien societies. *Transactions of the Royal Historical Society*, **26**, 115–41.

Rogoff, K. 1992. Dealing with developing country debt in the 1990s. *The World Economy*, **15**, 475–92.

Rondinelli, D. A. 1983. Dynamics of growth of secondary cities in developing countries. *Geographical Review*, **73**, 42–57.

Schiffer, J. 1981. The changing post-war pattern of development: the accumulated wisdom of Samir Amin. *World Development*, **9**, 515–37.

Sender, J. and Smith, S. 1986. *The Development of Capitalism in Africa*. London: Methuen.

Stein, A. A. 1984. The hegemon's dilemma: Great Britain, the United States, and the international economic order. *International Organization*, **38**, 355–86.

Stokes, E. 1959. *The English Utilitarians and India*. London: Oxford University Press.

Svedberg, P. 1991. The export performance of sub-Saharan Africa. *Economic Development and Cultural Change*, **39**, 549–66.

Taylor, P. J. 1985. *Political Geography: World-Economy, Nation-State and Locality*. London: Longman.

UNCTAD 1987. *Commodity Yearbook*. Geneva: UNCTAD.

UNCTAD 1994. *Commodity Yearbook*. Geneva and New York: UNCTAD.

UNIDO. 1981. *A Statistical Review of the World Industrial Situation, 1980*. Vienna: United Nations Industrial Development Organization.

Vogel, E. 1980. *Japan as Number 1: Lessons for America*. New York: Harper and Row.

Wachtel, H. M. 1987. Currency without a country: the global funny money game. *The Nation*, **245** (26 December), 784–90.

Warren, B. 1980. *Imperialism, Pioneer of Capitalism*. London: Verso.

Wolf, E. R. 1982. *Europe and the People without History.* Berkeley: University of California Press.

World Bank. 1991. *World Development Report, 1991.* New York: Oxford University Press.

World Bank. 1996. *From Plan to Market: World Development Report, 1996.* New York: Oxford University Press.

Yates, R. L. 1959. *Forty Years of Foreign Trade.* London: Allen and Unwin.

Picture credit: World Bank

AGRICULTURE: THE PRIMARY CONCERN?

'Development' is often equated with the structural transformation of an economy whereby agriculture's share of the national product and of the labour force declines in relative importance. Agriculture has often been viewed as a 'black box from which people, and food to feed them, and perhaps capital could be released' (Little, 1982, p. 105). This perspective, long dominant among planners and politicians, and common to both American and Soviet models of development, reflected the low-income elasticity of food (demand increases very little with higher incomes), the secular global trend towards higher labour productivity in agriculture (same output can be produced by fewer workers because of technology, fertilizers, etc.), the limited multiplier or stimulative effect of agriculture, and the secular tendency for the barter terms of trade to turn against countries that export primary products and import manufactured goods.

However, it is almost certain that the world's population will rise to at least 6 billion by the early years of the next century. It is equally certain that about 70 per cent of the growth in population between now and then will take place in the LDCs. Consequently, these countries in particular will need to increase their food production to supply the additional people and to increase their standard of living. At the same time they face two major constraints: much land is unsuitable for agricultural purposes (Fig. 2.5) and their involvement with the world economy often reduces their food self-reliance without sufficient compensation in other sectors.

The purpose of this chapter is to describe the contemporary state of agriculture in the 'periphery' of the world economy. To this end, the chapter is organized as follows: a first section establishes the importance of agriculture as an economic sector and stresses the dual trends of increased agricultural production for the world market and decreased food self-reliance; a second section discusses the general relationships between land, labour and capital in the periphery with special attention to efforts at rural land reform; third, the capitalization of agriculture in the periphery by multinational enterprises is described; fourth, and finally, the role of science and technology in agriculture in the periphery, especially in the form of the so-called Green Revolution, is assessed.

AGRICULTURE IN THE PERIPHERY

The countries of the periphery have all been significantly involved with modern commercial farming since the beginning of Western colonization in the sixteenth century. But subsistence and production for local markets have remained of great, if decreasing, importance. Malassis (1975) identifies four types of agricultural systems in the periphery: (1) the 'customary' farm involving common ownership of land for both cultivation and grazing; (2) the 'feudal or semi-feudal' estate, hacienda and latifundia; (3) 'peasant agriculture', including minifundia (small, subsistence farms), commercial farms and share-cropping; and (4) capitalist plantation or mechanized agriculture based on wage labour. These four types of farm organization produce three types of commodity: (a) commercial foods, primarily cereals for the domestic market; (b) peasant foods, primarily for personal use; and (c) export crops, where the major market is overseas. The historical trend in agriculture in most countries of the periphery has been from (1) and (2) to (3) and, especially, (4) in farm organization and from (a) and (b) to (c) in types of agricultural commodity.

However, the three continents of the periphery – Africa, Latin America and Asia – differ in terms of agricultural organization and performance. Above all, sub-Saharan Africa is, or has been until recently, abundant in land and sparse in population; Asia is largely short of land relative to population; and Latin America contains both areas with large populations and areas with few inhabitants. Agriculture is also of much greater relative importance in sub-Saharan Africa and Asia than in Latin America, both in terms of employment and contribution to national product (Hopkins, 1983).

Women's work

It is also important to recognize that agriculture is overwhelmingly more important as a source of employment to women than to men. Indeed, the 'gender dimension' is not a secondary consequence of variations in agricultural organization but 'a fundamental organizing principle of labour use' (Joekes, 1987, p. 63). Regional differences are apparent, however, indicating the contingencies of resource endowment and carrying capacity. There are far more women involved in agriculture in Africa, relatively speaking, than elsewhere. In 1980, 87 per cent of all women in the labour force in sub-Saharan Africa were involved in agriculture, compared to 70 per cent in India, 74 per cent in China, 66 per cent in other low-income Asian countries and 55 per cent in middle-income Asian countries. In Latin America the comparable figure is a very low 14 per cent. This reflects the greater degree of mechanization (and export crop orientation) in Latin American agriculture and higher levels of female rural to urban migration compared to other regions (Joekes, 1987). Official figures probably miss many of the subsistence and peasant agriculture activities carried out predominantly by women (Beneria, 1981). Labour force participation data usually involve very narrow definitions of agricultural activity focused on land cultivation and large-scale livestock keeping (Hill, 1986).

Forms of agricultural organization

Forms of agricultural employment and organization also tend to differ by world region. 'Mechanized agriculture' and export crops have become of greatest importance in Latin America. 'Green Revolution' agriculture has become most widespread in producing wage and peasant foods in lowland Asia with pockets in Latin America and North Africa. 'Resource-poor' agriculture producing a range of crops predominates in sub-Saharan Africa and areas of poor soils and drainage elsewhere. Production differences reflect these organizational and endowment differences.

While per capita food production in the periphery has not matched that of the core, and in many cases has not kept up with population increases, spectacular growth in the production of specific crops for export to the core was characteristic of the 1970s and 1980s. For example, in Latin America, the production of sugar increased by over 200 per cent in El Salvador, Guatemala and Honduras between 1965 and 1977. Beef production in the Dominican Republic grew at 7.6 per cent per annum between 1970 and 1979. Sorghum, unimportant in Brazil before 1970, averaged 253 000 metric tons per annum in that country by 1979. 'By the late 1970s, it was estimated that commercial agriculture, largely centred in the large-farm sector, accounted for half of all agricultural production, utilized nearly a third of the cultivated area, and employed a fifth of the work force in Latin America as a whole' (Grindle, 1986, p. 81).

The expansion of export production and regional specialization has been most characteristic of agriculture in Latin America. In sub-Saharan Africa, however, export crops have failed to maintain global market shares even as total agricultural production increased (see Table 8.4; Sender and Smith, 1986). This reflects both declining productivity in the export sector and government attempts to direct investment into industrialization rather than agricultural commodities. Food production has been dismal, particularly in the context of rapid population increase. In Asia, both productivity and production have increased enormously because of fertilizers and the application of new technologies, but most growth has been in cereals (especially rice and wheat) production rather than 'special' export crops such as those of growing importance in Latin America (e.g. fruits, beef) (Lele, 1984). The problems for the Asian countries are their high land/population ratios and the competition they face from agriculture in the United States and Western Europe in the crops (such as wheat and rice) in which their growth has been concentrated. US, EC and Japanese subsidies to and market protection of agricultural production deprive Asian (and other LDC producers such as Argentina) of both higher prices and international markets. Lower production of cereal crops in the core of the world economy would produce higher world prices (through a decrease in the amount produced) and greater access of LDC producers to DC markets.

Problems

Each of the three major regions of the periphery, therefore, faces distinctive problems with respect to its agriculture. For Latin America it is the expansion of

Table 9.1: Food production per capita for selected countries (1987 = 100)

	1975	1980	1985	1990
China	70.7	77.7	94.5	108.2
Malaysia	63.3	77.7	94.4	115.8
India	94.8	93.0	104.0	112.2
Nigeria	106.4	98.8	98.1	103.4
Bangladesh	114.9	109.4	104.5	105.1
Indonesia	90.1	84.0	93.1	100.5
Egypt	90.1	84.0	93.1	100.5
Mexico	94.6	102.0	102.6	103.9
Ghana	135.0	101.8	102.3	87.4
Philippines	110.3	117.8	101.4	100.4
Sri Lanka	103.1	119.4	118.4	106.7
Haiti	100.9	99.4	100.7	90.9
Ethiopia	116.7	117.0	102.7	99.7
US	100.9	102.8	111.0	103.7

Source: World Bank (1992).

export crops at the expense of local food crops. As a consequence, food imports are often necessary (Grindle, 1986). For sub-Saharan Africa it is the total deterioration of agriculture in the face of population pressure on marginal land, low productivity, government bias against investment in agriculture, and fluctuations in export earnings. Food imports are now an absolute necessity. For Asia, production of cereals has increased greatly but prices have been low because of global 'gluts'. Hence, increased agricultural production has not generated the capital necessary for investment in other sectors such as industry. When prices increase, local populations must pay the increase or substitute other cereals that are imported, more often than not, from Western Europe or the United States. Between 1975 and 1990 food production per capita increased substantially and consistently only in China and Malaysia among all LDCs (Table 9.1).

Although the world as a whole produces sufficient food for everybody, approximately 780 million people in the LDCs, one in five of the population, is chronically undernourished. Perhaps as many as 2 billion people fill themselves daily with adequate food calories but lack a diet balanced in needed nutrients. Hunger and inadequate diets are especially serious in Africa, where 33 per cent of the population (1989) is chronically undernourished. Comparable figures are 13 per cent in Latin America and 19 per cent in Asia. In these world regions conditions have improved since 1970 when 19 per cent and 40 per cent, respectively, were chronically underfed. In Africa there has been little or no improvement (35 per cent in 1970). The remarkable improvement in Asia owes much to improved rural health care, which protects people from falling sick and losing income or work and subsequently disrupting family food-supply, and increased crop yields (FAO, 1992).

Another way of putting the food problem would be to compare food production per capita in the three regions. In this perspective, Asia has seen an impressive 70.8 per cent increase from 1961 to 1995 and Latin America has experienced a 31.4 per cent increase. In Africa food production per capita has *dropped* by 11.6 per cent over the same period (Sachs, 1996).

In large parts of the periphery today agriculture is a vulnerable sector: either oriented externally or subject to the vagaries of world market prices without the protection and subsidies enjoyed by agriculture in the core. Yet it is absolutely vital. Vast numbers of people are still employed in or are immediately dependent on agriculture. And, whatever the model of economic development adopted, any hope of improving living standards in general depends upon increasing agricultural production.

LAND, LABOUR AND CAPITAL

Agriculture in the contemporary periphery rests upon a foundation of agrarian history, and recent changes can only be understood in this context. Central to agrarian history the world over has been the impact of market forces on land-holding patterns and the structure of rural social relationships. Though rural areas are often characterized as static and traditional, the historical record shows frequent changes in agricultural practices and labour relationships in response to global and domestic political-economic conditions. But some features of land-holding systems and rural life have persisted from the period of incorporation into the world economy. In this section the mix of 'old' and 'new' in the agricultural organization of different parts of the contemporary periphery (Latin America, sub-Saharan Africa, Asia) will be examined.

Latin America

In Latin America, conquest and colonial domination created patterns of subsistence and commercial agriculture based on large landholdings. After independence, this characteristic, and its corollary, an exploited and powerless peasantry, became firmly entrenched as the region was firmly tied into the world market as a producer of primary commodities. Between the 1850s and 1930s the various countries of Latin America came to depend on the export of one or two primary export commodities to the industrial countries – first Britain and later the United States. The older hacienda system, though complex and varied in its particulars from place to place, went into decline and was replaced by a plantation system that already had a considerable history in the sugar plantations of NE Brazil and the Caribbean (Table 9.2).

The growth of export-oriented agrarian capitalism was associated with the emergence of a politically powerful landed élite linked to foreign investors and commercial agents dealing in primary commodities. Agriculture for domestic consumption was largely ignored and through control over governments the agricultural élite was able to increase its hold over land, labour and capital.

Table 9.2: Land, labour, capital and markets: haciendas and plantations

	Haciendas	Plantation
Markets	Relatively small and unreliable, regional, with inelasticity of demand; attempt to limit production to keep prices high.	Relatively large and reliable. European, with elasticity of demand; attempt to increase production to maximize profits.
Profits	Relatively low; highly concentrated in small group.	Relatively high; highly concentrated in small group.
Capital and technology	Little access to capital, especially foreign. Operating capital often from Church. Technology simple, often same as that of peasant cultivators.	Availability of foreign capital for equipment and labour. Direct foreign investment late in nineteenth century. Relatively advanced technology, with expensive machinery for processing.
Land	Size determined by passive acceptance of indigenous groups. Attempt to monopolize land to limit alternative sources of income to labour force; much unused land. Relatively cheap. Unclear boundaries.	Size determined by availability of labour. Relatively valuable with carefully fixed boundaries.
Labour	Large labour force required seasonally; generally indigenous; informally bound by debt, provision of subsistence plot, social ties, payment in provisions.	Large labour force required seasonally; generally imported; slavery common; also wage labur.
Organization	Limited need for supervision; generally hired administrators/managers, absentee landlord.	Need for continual supervision and managerial skill. Generally resident owner/manager.

Source: Grindle (1986), Table 3.1, p. 30.

The concentration of land-holding and the marginalization of peasant agriculture did not occur without resistance. Agrarian uprisings and social banditry were widespread. In Mexico the 1910 uprising was a major impetus to the Revolution; strikes were extremely common in the corporate plantations of coastal Peru in the period 1912–28; in Colombia rural violence by agrarian tenant syndicates directed against commercial coffee producers lasted well into the 1930s. The 1930s also was a period of rural unrest in the Brazilian northeast and in El Salvador among dispossessed peasants and unemployed plantation workers (Duncan and Rutledge, 1977; Wolf, 1968; Landsberger, 1969).

When the world economy collapsed in the 1930s so too did export-oriented agriculture. This spurred the emergence of active nationalists, often in the military, who wanted to increase industrialization and diminish reliance on the export of primary commodities. Between 1930 and 1934 there were 12 forcible takeovers of power – from Argentina to Peru to El Salvador. Mexico, Brazil, Argentina, Colombia, Chile and Uruguay all instituted import-substitution industrial strategies. These led to a massive movement of people off the land. For the region as a whole, in 1920, only 14 per cent of the population lived in urban areas, but by 1940 the proportion had risen to 20 per cent. In Argentina, Chile and Uruguay urban percentages reached 35–45 per cent of the population. One major consequence of this was a decline in the hold of the land-holding élite over national politics in some countries as urban professional and working classes grew in size and influence (Grindle, 1986).

This change, however, can be exaggerated, Many countries continued to rely on the export of one or few primary commodities – the Central American and Caribbean countries, but also Argentina, Colombia, and Chile – and rural land remained concentrated in the hands of the landed élite. What was different was the emergence of nationalist and populist movements committed to industrialization rather than export agriculture.

Pursuing policies of import substitution had important effects on agriculture. For one, manufacturing surpassed agriculture in its contribution to gross domestic product in a number of countries (Argentina, Brazil, Chile, Mexico, Uruguay and Venezuela) in the 1940s. Much of the new capacity was concentrated in or near the capital cities of the states which were its major sponsors (Buenos Aires, Rio de Janeiro, Santiago, etc.).

Industrialization required a 'draining' of agriculture for resources (cheap food, raw materials) and capital (foreign exchange, taxation). As a consequence, a premium was placed upon efficiency in agricultural production. This was thought to require large holdings, the spread of technological innovation, and capitalization (heavy capital investment). Between 1940 and 1960 there was a massive migration of people from the countryside to the cities as a consequence of mechanization and the expansion of large landholdings at the expense of small tenants and proprietors.

In the 1960s import substitution became increasingly expensive as the 'easy phase' emphasizing light consumer goods was played out and the prodigious expense of moving into heavier capital goods became apparent. In a process that accelerated during the 1970s, import substitution was slowly displaced by a new development model based on export promotion. According to this model, agriculture had been neglected and, although no substitute for industrialization, more efficient production of domestic food crops and increased agricultural exports were important in both maintaining political stability and obtaining foreign exchange. After 1965 public investment in rural areas and agriculture increased in a large number of Latin American countries.

Government policies have discriminated heavily in favour of the larger landowners. The geographical distribution of official credit, research and extension, infrastructure mechanization and Green Revolution inputs reflects the geography of landholding. In Peru, for example, about half the credit supplied by the Agricultural Development Bank between 1940 and 1965 went to cotton growers, who were among the wealthiest coastal agricultural exporters. Food crop producers – largely peasants – were mainly ignored by the bank (Frankman, 1974; Durham, 1977). In Mexico in 1970, mechanization was used on 25.7 per cent of the crop area of farms of more than 5 hectares but was used on only 4.3 per cent of the crop area of farms under 5 hectares in size (Grindle, 1986). In Brazil all government policies have tended to reinforce the emphasis upon commercial agriculture in the south and east regions at the expense of the northeast and small-scale producers everywhere (Gomes and Perez, 1979).

This is not to say that large-scale capitalist agriculture has completely displaced peasant production. Far from it. A large section of the agricultural labour force is still 'part-peasant' in that it supplements its wage-earnings with the produce of its often less-than-subsistence plots. This serves to sustain capitalist agriculture through reducing the costs of reproducing a labour force (Taussig, 1978). In many parts of Latin America, therefore, large-scale capitalist agriculture and small-scale peasant production still uneasily coexist. The past is still present.

Sub-Saharan Africa

In sub-Saharan Africa, unlike Latin America (or Europe), access to labour not land was always the basis of economic and political power. From 1830 to 1930 agriculture in sub-Saharan Africa underwent an incredible expansion in the form of small-scale commercial farming. Some commercial farming had existed prior to this period, for example in the Hausa-Fulani and Mandinka states of northwest Africa, but the introduction of new crops and the expansion of existing ones into previously uncultivated areas increased the scale and geographical distribution of commercial agriculture. Of special importance were such crops as cocoa, cotton, coffee, groundnuts and oil palm, which were grown mainly for export markets. They spread along with European traders, the introduction of foreign capital, the shifting objectives of native farmers and traders, and finally colonial rule. This was the 'cash crop revolution' (Tosh, 1980) that brought Africa into the world economy and capitalism into Africa.

Colonial rule involved massive intervention in existing agriculture through forced labour and taxation. Taxation in particular provided a fresh stimulus to cash-cropping. In some parts of Africa, especially the east and south, taxation also encouraged labour migration to mines, plantations and industries established by European settlers (Berg, 1965). In West Africa, however, labour migration pre-dated colonial rule. It was of a seasonal nature and involved the integration of farming in the interior with migration to more fertile but labour-deficient coastal areas. In West Africa cash-cropping by small-scale farmers and long-distance labour migration at harvest time were indigenous phenomena that increased in intensity after the onset of colonial rule. Elsewhere, cash-cropping and labour migration were relatively novel and related much more to either European settlement (as in South Africa, Zimbabwe or Kenya) or European initiatives in mining and plantation agriculture (as in Zambia and Zaire) (Hart, 1982; Swindell, 1985).

Another distinctive feature of West Africa as compared, for example, to Kenya was that the production of food and cash crops was complementary rather than competitive (Bates, 1983). Even today food crops such as plantains, cocoyams and peppers are grown to provide shade for young cocoa trees. Moreover, the period of peak labour demand for cocoa harvesting (November–February) complements the peak labour demand periods for the cereal-growing areas to the north (May–July and February–March). Cocoa farms, therefore, have rarely faced a maximum price for labour and the commercial cocoa industry can coexist with the market for labour in foodcrop production.

In Kenya, however, the European settlers specialized in the production of food crops and their production cycle matched that of subsistence producers. They consequently had to compete for labour with the subsistence sector. In addition, the establishment of estates or plantations in Kenya involved the confiscation of land from subsistence producers and the subsidy of commercial production at the expense of the subsistence sector (Bates, 1983).

The rate of agricultural production slowed markedly during the 1930s and the Second World War. It was only in the 1950s, when world prices for many export crops increased as the industrial countries entered into their long boom of the 1950s and early 1960s, that there was a rapid expansion in export crop production. But the increase in demand for Africa's export crops was short-lived, peaking as early as 1956. Since then cash-cropping and commercialized livestock farming have been concentrated in the districts where they were dominant 30 years ago. With the exception of sugar, most new planting (of cocoa, coffee or tea) has taken place within the areas which were already the major producers in the early 1950s (O'Connor, 1978).

In those districts in which agricultural production has intensified or expanded, it has involved different types of farming. For example, in the Ivory Coast plantations have been the major agent of growth, whereas in Ghana, Kenya and Sudan it has been small-scale peasant cash-crop production that has been responsible for most growth. Indeed, in Kenya the small-scale farming sector has largely replaced the plantation sector as the most dynamic in terms of commercial production.

Total agricultural production (cash crops and food staples) increased substantially in sub-Saharan Africa over the period 1979–90 (Table 9.3). However, the rate of population increase over the region as a whole has meant that there has been very little or no increase in per capita terms. Most African governments have adopted policies which seek to depress food prices to feed their burgeoning populations. This often leads them to set higher prices for large-scale producers because of presumed efficiencies (and political influence?). Penalizing the food production sector is meant to both stimulate export-crop production and feed increasingly large urban populations. In fact it has discouraged farmers, especially the mass of small-scale farmers, from increasing their production through investment in increased productivity.

The trade policies of DCs and the advice their experts offer have also contributed to the problems of African agriculture. North America, Western Europe and Japan may practise fairly free trade in the manufactured goods and services in which they may have comparative advantages but they are relentlessly protectionist about foodstuffs; precisely the sector in which African countries can offer competitive products. For example, US government subsidies to its sugar, tobacco and groundnut farmers lead to lower prices for US-produced crops than would be the case without the subsidies. This deprives African producers of potential markets. With respect to advice, Africa has been on the receiving end of some of the worst advice ever offered by people from one part of the world to another. The litany of disasters resulting from advice offered by foreign experts is much too long to provide here. Two examples must suffice. In Burkina Faso and elsewhere in the dry Sahel region of North-Central Africa the UN Food and Agricultural Organization (FAO)

Table 9.3: Index of total and (per capita) agricultural production: selected African countries (1979–81 = 100)

	1979		1990	
Ethiopia	102.97	(104.85)	107.23	(84.30)
Ghana	98.85	(101.46)	140.64	(100.69)
Ivory Coast	96.26	(100.12)	129.31	(88.39)
Kenya	99.67	(103.54)	150.49	(104.11)
Malawi	99.27	(102.67)	118.35	(83.59)
Mozambique	97.45	(100.15)	108.01	(83.42)
Nigeria	95.77	(99.00)	155.51	(112.40)
Senegal	89.31	(92.18)	131.75	(99.93)
Tanzania	99.64	(103.28)	120.53	(83.25)
Zambia	97.34	(101.07)	127.32	(86.49)
Zimbabwe	90.80	(93.72)	132.25	(97.25)
Africa	97.15	(100.06)	124.79	(92.78)

Source: FAO (1990).

encouraged local farmers into growing potatoes. A bumper crop resulted, which then rotted unsold in local markets where potatoes were seen as an exotic crop without any history in local diets. By Lake Turkana in East Africa, Norwegian experts persuaded Turkana cattle herdsmen to give up their cattle and take up fishing only to find out that the cost of chilling the fish exceeded what they could bring in city markets. Not only was the fishing equipment a wasted investment but the Turkana were now without their cattle. They ended up on food aid provided by the surpluses bought up by the USA and other governments as a result of overproduction brought about by their subsidy programmes to cereal producers and dairy producers! (Thomas, 1996).

But countries differ in the relative extent to which farmers must bear the brunt of tax and price-setting policies. It all depends upon the political base of governing élites and the origins of marketing organizations (Bates, 1983). In Ghana and Zambia, for example, urban-based politicians have put the burden on small-scale farmers to a much greater extent than the rural-based politicians of Kenya. In Ghana the Cocoa Marketing Board is a patronage organization whereas in Kenya marketing organizations are controlled by producers. Interestingly, the increase in total agricultural production has been higher in Kenya than in Ghana and Zambia (see Table 9.3). In the former this has benefited both food production for domestic consumption and increases in sales of export crops.

Three trends have nevertheless been fairly general over the past 30 years. One has been the increased importance of wage labour, especially with respect to export crops. This has further 'monetized' the rural economy and reduced the degree of reliance on domestic groups (families) as sources of farm labour. This in turn has reinforced the role of long-distance migration in agricultural labour and given some districts the specialized role of 'migrant labour reserve' for other districts in which export agriculture is important. For example, even with restrictions on international migration, Burkina Faso in West Africa has been a major source of temporary and permanent migrants to Ivory Coast and Ghana (Fig. 9.1).

A second trend has been the changing role of women in African agriculture. Women have become central to the production of food crops on small-scale farms such as that dominate throughout Sub-Saharan Africa. As Swindell (1985, p. 179) puts it:

> As men have become more involved in commercial cropping and non-farm occupations, so women have become increasingly responsible for the cultivation of food staples. This is especially true in those areas where the out-migration of men is persistent, and it could be argued that the expansion of commercial cropping and the industrial labour force has been built on the backs of women farmers.

The third trend has been growth in agricultural production through extending areas under cropping or grazing rather than through raising yields. Green Revolution technologies (high-yield varieties, fertilizers, etc.), mainly addressed to cereal production, have been either inappropriate or not widely adopted in sub-Saharan Africa. Whatever the cause, however, commercial agriculture has become extensive rather than intensive. This has led to farming on poor soils in areas with unreliable rainfall and the displacement of subsistence agriculture onto ever more marginal terrain. Sen (1981) implicates this trend as a major factor in the famines that have afflicted many parts of Africa over the past 20 years. Civil wars, poor food-

Figure 9.1

External migration streams in West Africa, early 1980s

Source: Swindell (1985), Fig 4.3

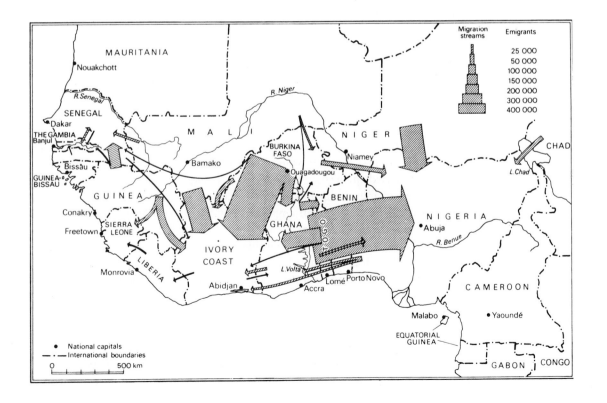

distribution networks and the degradation of soils through lack of crop rotation have also played some part.

Though much of African agriculture has become increasingly commercialized, it remains largely small-scale and still involves domestic groups or families. The level of agricultural production, however, has not kept up with the world's highest rates of population increase. In many countries there are now major national food deficits. At the same time government policies in many countries have had the effect of discouraging agricultural production both for food staples and export crops. But in most countries farming must remain the dominant activity for the foreseeable future if only because an increase in agricultural productivity is a prerequisite for industrial development. At present the growth of industry through import substitution is limited by the small size of most domestic markets and these can only grow if the incomes of farmers rise.

Asia

Asian agriculture presents a more complex picture than agriculture in Latin America or sub-Saharan Africa. On the one hand are the world's highest rural population densities but on the other hand the populations are organized in agricultural systems with quite different and distinctive features. The major contrast, at least until recently, was between China, where there is no export agriculture to speak of, and the rural economy has been organized around 'collective' ownership (from 1954 to 1979), and those countries such as Malaysia and the Philippines, where export agriculture (rubber and sugar, respectively) is important and share-cropping tenancy (renting with payment in kind to landlords) predominates outside the plantations. But in general there is a high incidence of tenancy in Asian countries and share-cropping is its major form, especially in those areas where rural population densities are very high such as Bangladesh, Java, Central Luzon (in the Philippines), the West Zone of Sri Lanka, and eastern and southern India (Hossain, 1982).

Along with the preponderance of tenants goes an extreme concentration of landholding, although less on average than in Latin America (Table 9.4). Half the farms in India cover less than one-tenth of the total agricultural area; in Pakistan one-third of the farms account for 3.5 per cent of the total area. Some of the figures in Table 9.4 suggest that the proportion of small farms (less than one hectare) has recently increased in several Asian countries. Other evidence from India suggests two types of change in historical patterns of rural social structure: the growth in some areas of the class of self-employed cultivators or rich peasants, favoured by 1950s' land reform (e.g. Gujarat) and the transformation of large landowners into capitalist farmers employing migrant labourers (e.g. Punjab) (Rudolph and Rudolph, 1987). Both of these changes are signs of increasing commercialization of agriculture even as share-cropping tenancy persists in 'marginal' areas to provide labour reserves for seasonal and cyclical purposes at little or no cost to the commercial sector.

In the colonial period, governments concerned themselves either with plantation agriculture or with raising taxes from other forms of agriculture. In India the British created a class of landed aristocrats called Zamindars as revenue collectors for the government. The Zamindars, however, did not have any real interest in improving

Table 9.4: Land distribution in Asia and Latin America

A. Asia

Country	Year	% of farms under 1 hectare	% of total area covered by farms under 1 hectare
Bangladesh	1960	51.6	15.2
	1974	66.0	24.0
India	1961	39.8	6.8
	1970/71	50.6	9.0
Indonesia	1963	70.1	28.7
Korea	1963	73.3	45.0
	1974	67.0	58.3
Philippines	1960	11.5	1.6
	1970	13.6	1.9
Pakistan	1960	32.9	3.5
Malaysia (West)	1960	45.4	15.2
Thailand	1963	18.5	2.5

B. Latin America

Country	Year	Subfamily farms[a] % of total farms	Subfamily farms[a] % of total area	Small holdings[b] Av. size (ha.) 1960	Small holdings[b] Av. size (ha.) 1970	Small holdings[b] % change 1960–70
Argentina	1960	43.2	3.4	–	–	–
Brazil	1950	22.5	0.5	2.46	2.16	–12.2
Chile	1960	36.9	0.2	1.40	1.67	+19.3
Colombia	1955	64.0	4.9	1.64	1.64	0
Ecuador	1954	89.9	16.6	1.72	1.50	–12.8
Guatemala	1950	88.4	14.3	–	–	–
Peru	–	88.0	7.4	1.70	1.44	–15.3
Uruguay	–	–	–	2.64	2.71	+2.6
Venezuela	–	–	–	2.17	2.24	+3.2
El Salvador	–	–	–	1.67	1.56	–6.6
Nicaragua	–	–	–	3.10	2.36	–23.9
Jamaica	–	–	–	1.54	1.25	–18.8

Notes:
[a] 'Subfamily farms' were defined as 'farms large enough to provide employment for less than two people with the typical incomes, markets and levels of technology now prevailing in each region'.
[b] 'Small holdings' were defined as enterprises of less than 5 hectares except in El Salvador (less than 10 hectares) and Jamaica (less than 25 acres).
Source: Loup (1983), Table 3.4, p. 115.

agriculture. Over time they and other intermediaries became an immense burden upon actual cultivators whose rents included not only revenue for the government but also income for the various intermediaries (Hossain, 1982, p. 149). After independence India, Pakistan and other countries in South and Southeast Asia where this system prevailed abolished intermediary tenures. However, many of the old intermediaries continued to cultivate their holdings through tenants and share-croppers on the same exploitative terms as before. Only in China, South Korea and Taiwan did land redistribution lead to an effective abolition of the power of large landlords (Loup, 1983; Perkins and Yusuf, 1984).

Since independence, however, total agricultural production has increased at rates at least commensurate with population growth in most Asian countries. Unfortunately, much of the growth has been concentrated in export crops or cereals (wheat, rice) rather than across the board. Moreover, the unequal social structure of most rural areas has ensured an upward drift of the benefits of increased production. Rural poverty has increased as agricultural production has increased (Jones *et al.*, 1982; Loup, 1983).

A major source of increased production of cereals (especially wheat) since the 1960s has been the Green Revolution. This had its most significant impact in the two Punjabs (in India and Pakistan) and the Indian state of Haryana where irrigation facilities could be utilized. Benefits have accrued disproportionately to large farmers and the technologies involved (new seed varieties, heavy applications of chemical fertilizers) cannot be applied in areas without irrigation facilities: 80 per cent of the cultivated area in India, 90 per cent in Bangladesh (Loup, 1983).

In general, over the past 40 years most Asian governments have not favoured agriculture. Many have pursued pricing and credit policies similar to those noted earlier for sub-Saharan Africa. This seems also to be true at least for considerable periods in the case of China (Hsu, 1982; Lardy, 1984). Indian development plans until the late 1970s were systematically biased against the agricultural sector. Yet there is a direct relationship between agricultural yields and a price structure that favours the agricultural sector (Timmer and Falcon, 1975). The countries with the highest ratios of product prices (e.g. rice) to input costs (e.g. fertilizer cost) are also where yields are highest. The three countries with the highest rice yields in Asia – Japan, S. Korea and Taiwan, also have perhaps some of the poorest soils in Asia. Government policies (especially subsidies for inputs such as fertilizers) and egalitarian rural social structures (all farmers are rewarded) are the most plausible causes of the differences in crop yields. One negative effect of this, however, is a high level of water pollution produced by the heavy use of subsidized fertilizers.

According to the World Bank, three-quarters of the world's 'absolute poor' (those unable to maintain a minimum nutritional standard) live in Asia and more than four-fifths of them live in rural areas. During the 1970s the number of rural unemployed increased and the real wages of agricultural workers stagnated or decreased in most countries of the region. The number of landless workers also increased: from 22 per cent (1961) to 38 per cent (1973) in Bangladesh and from 25 per cent (1961) to 38 per cent (1971) in India (Asian Development Bank, 1977). Even in the face of improved agricultural production in the 1980s and a dramatic diminution in the number of chronically undernourished, particularly in India and China, it is hard to avoid the general conclusion of Loup (1983, p. 31) that 'The verdict is disastrous. During recent decades the situation of the rural masses of non-Communist Asia has at best stagnated and at worst has deteriorated. Whatever assumptions we retain, there is a striking contrast between the present picture and the euphoria created 15 years ago by the beginnings of the Green Revolution!'

RURAL LAND REFORM

In Latin America and Asia the landholding and tenurial systems have been periodically 'reformed' as a result of pressure from peasant movements, government attempts to make agriculture more efficient and productive, and external pressures from TNCs and international development agencies. Certain models have sometimes been followed depending on whether efficiency or equity has been the overriding goal. In the former case the Taiwanese and South Korean experiences are emphasized, in the latter the Chinese experience is often the model. However, in practice, agricultural reform, especially land reform, is overwhelmingly a socio-political process rather than a technical one of choosing a model and then following it.

At one time or another, but especially between 1960 and the early 1970s, virtually every country in Latin America and Asia passed land reform laws (Thiesenhusen, 1989). A wide range of arguments have been proposed to justify a role for land reform in agricultural development. There are perhaps four justifications that have been most common and they have appealed differentially to different social groups. The first of these is a 'conservative' argument: land reform is a minimal concession for political stabilization; second, the 'liberal' argument: land reform is needed to create a class of capitalist farmers and expand the domestic market; third, the 'populist' argument: small farms are more efficient (and equitable) than large ones; and fourth, the 'radical' argument: peasants are rapidly being dispossessed of their status as independent producers and are prisoners of cheap food policies and agro-export policies, consequently land reform towards collective production (collective farms, state farms) is necessary, if insufficient, for economic development (de Janvry, 1984).

Most actual land reform policies have been of the 'liberal' type, concerned with creating a reform sector. Thirty-five land reforms are classified in Table 9.5, including those in the same country when a land reform programme was later redefined (e.g. Chile). All the diagonal reforms are redistributive ones in the sense that they either increase the size of the reform sector without changing the non-reform sector (1, 7, 13 in Table 9.5) or involve expansion of the reform sector (25). Reforms 2, 3, and 4 are oriented towards eliminating 'feudal' (or other pre-capitalist) remnants from agriculture rather than redistributing land. In each case the transition to capitalism is dominated by (2) a landed élite, (3) farmers or (4) peasants.

The only possible reforms, as opposed to drastic changes, once a capitalist agriculture has been established are either shifts in the type of agrarian structure (8, 9, 14) or distributive reforms within a given type (7, 13, 19). All reforms can give way to counter-reforms: Chile essentially switched to (12) from (3) after the 1973

Table 9.5: A typology of land reforms

	Post-land reform				
Mode of production in whole society	Capitalist	Capitalist	Capitalist	Capitalist	Socialist
Mode of production in agriculture	Semifeudal	Capitalist	Capitalist	Capitalist	Socialist
Land tenure	Semifeudal estates and reform sector	Capitalist estates and reform sector	Capitalist farms and reform sector	Peasant farms	Socialist farms
Semifeudal estates	(1) Mexico, 1917–34; Taiwan, 1949–51; Colombia, 1961–67; Chile, 1962–67	(2) Bolivia, 1952–; Venezuela, 1959–; Philippines, 1963–72; Equador, 1964–; Peru, 1964–69; Colombia, 1968	(3) Mexico, 1934–40; India, 1950–; Guatemala, 1952–54; Egypt, 1952–66; Iran, 1962–67; Chile, 1967–73	(4) South Korea, 1950–; Taiwan, 1951–63; Iraq, 1958–	(5) China 1949–52
Capital estates	(6)	(7) Costa Rica, 1962–76	(8) Peru, 1969–75; Philippines, 1972–79	(9)	(10) Cuba, 1959–63; Algeria, 1961–71
Capitalist farms	(11) Guatemala, 1954–	(12) Chile, 1973–	(13) Mexico, 1940–; Dominican Republic, 1963–; Egypt, 1961–	(14)	(15) China, 1979–
Peasant farms	(16)	(17)	(18)	(19)	(20)
Socialist farms	(21)	(22)	(23)	(24)	(25) Cuba, 1963–; China, 1952–78; Algeria, 1971–77

Pre-land reform: Semifeudal — Capitalist; Capitalist — Capitalist; Socialist — Socialist

military coup, Guatemala returned to (11) from (3) after the military coup of 1954. The Chinese, Cuban and Algerian cases are ones of land reform involving collectivization that were part of more 'radical' social-political change. But since the early 1980s China has shifted from (25) to (15).

The most widespread and successful (in the sense of lasting) land reforms have been those facilitating the creation of a capitalist agriculture (1–5 in Table 9.5). In Latin America the combination of anti-feudal land reforms with more spontaneous development of capitalism has both removed most feudal remnants and put an end to reform efforts (Grindle, 1986). A similar conclusion can be drawn for Asia (Jones et al., 1982). Reform efforts generally ended in the early 1970s. By and large they cannot be said to have lived up to promise for the needs of the bulk of the rural population irrespective of the nature of the reform undertaken. However, in some cases, such as Taiwan and South Korea after the Second World War and China's land privatization since 1978, rural land reform appears to have served as a prerequisite for later industrialization by increasing crop yields and through increased rural earnings providing capital for industrial investment.

THE CAPITALIZATION OF AGRICULTURE

Spontaneous change, therefore, has now become much more important than reform in agricultural development. Over the past 20 years there has been a substantial increase in direct and indirect investment by transnational corporations (TNCs) in the agriculture of a number of peripheral countries. In many countries, TNCs, attracted by cheap land and labour, appropriate physical conditions, improved infrastructure, and a decline in the relative profitability of other sorts of investment, have increased their involvement in export-oriented agriculture and the production and distribution of seeds, pesticides and fertilizers. Thailand, for example, which exported no pineapples in the early 1970s had by 1979 become the major world exporter after Hawaii because the US company Castle and Cooke had moved a major part of its pineapple operations out of Hawaii. Similarly, the Philippines, which exported no bananas in the 1960s, had become one of the world's major exporters by the mid-1970s. This was again due almost entirely to new multinational investment (Turton, 1982; Jones et al., 1982; McMichael 1993; Watts 1994). So, just as TNCs that specialize in manufacturing use global sourcing, agricultural TNCs have turned to multiple sites of production to lower labour costs, gain year-round supplies for seasonal crops (for example, strawberries in January in Western Europe from Chile) and avoid labour and environmental regulations. Over the past 30 years the global food industry has been one of the world's fastest growing industries.

Of great importance, however, was the prior emergence in Western Europe, Japan and North America of a highly capital-intensive agriculture serving a food system in which consumers increasingly demanded high-value products (such as lean beef, chicken products and fresh fruit and vegetables) at the same time as marketing and distribution were concentrating in the hands of large-scale wholesalers and supermarket chains. Economies of scale could be realized within large

capitalization
The process whereby capital-intensive inputs such as technology are deployed by large firms and replace labour-intensive methods associated with smaller-scale production.

vertically integrated firms that supplied the new wholesalers and direct retailers. Global sourcing is an extension into the periphery of a shift towards industrialized agriculture that was well under way by the 1950s in the United States and Western Europe with beef cattle lodged in 'lots' for fattening and chickens stacked on top of one another in battery houses. The recent demand for 'organic produce' and very fresh fruit and vegetables and worries about contamination of the beef food chain (prompted by the outbreak of 'mad cow' disease in Britain), however, may signal the limits of the globalization of food production when consumer tastes and demands in urban and export markets resist the imposition of mass-produced items. Different food products now have different food systems associated with them. Only some are amenable to global sourcing (Fine, 1994).

It is in Latin America that the **capitalization** of agriculture by TNCs has been both most extensive and intensive. Of the six countries usually identified as the 'new agricultural countries' in which agricultural investment has been concentrated, four are in Latin America – Brazil, Mexico, Argentina and Chile. The other two are Hungary and Thailand. These countries are analogous to the NICs, in that their governments have promoted agricultural investment for urban and export markets, focusing on such high-value food products as meat, fruit and vegetables (Friedmann 1991; McMichael 1996). Sometimes control is exercised directly by purchase of land and involvement in production. For example, between 1964 and 1970 US-based TNCs purchased 35 million hectares of agricultural land in Brazil alone (Feder, 1978). Increasingly, however, TNCs and international development agencies (World Bank, the US Agency for International Development (AID), etc.) are encouraging traditional rural élites to become commercial élites practising mechanized farming of export crops that are processed and marketed by the TNCs or by contracting out to peasant producers. These strategies reflect both fear of the revolutionary potential of peasant movements in traditional agrarian social structures and the need for TNCs to keep a low profile lest they become the targets of nationalization drives.

Agribusiness

The impact of 'agribusiness' investment in the agriculture of the periphery, therefore, is not restricted to the development of export enclaves or plantation enclaves as was characteristic of an earlier phase in the development of the world economy. Rather, its most important effect is probably the way in which it channels capital to a class of rural capitalists and thus consolidates TNC control over entire national agricultural systems. The penetration of peripheral agriculture by international agribusiness is, in effect, just another aspect of the New International Division of Labour (Burbach and Flynn, 1980).

Between 1966 and 1978 US investment in Latin American agriculture expanded from US$365 million to US$1.04 billion, growing from 15 per cent to 21 per cent of total US direct foreign investment in Latin America. This investment was heavily concentrated in Argentina, Brazil, Mexico and Venezuela, where the growing urban middle and upper classes provided a domestic supplement to US demand for so called 'luxury foodstuffs' (meat, fruits and vegetables). As demand grew for the fertilizers, pesticides, herbicides, improved seeds and agricultural machinery needed by the 'new' agriculture, TNCs such as du Pont, W. R. Grace, Monsanto, Exxon and Allied Chemical were increasingly involved in local production.

TNCs and foreign portfolio investment capital were involved in a variety of ways. In the state of Sinaloa in Northern Mexico, for example, 20–40 per cent of the credit for agricultural production in the 1970s came from north of the border. In Argentina the amount of foreign capital in beef production decreased while it increased in the packing and processing industries. In Mexico and Central America contract production now links national producers with TNCs. Foreign banks have become major agricultural lenders. For example, the San Francisco-based Bank of America became heavily involved in Guatemala in the 1970s, lending for major development projects, such as converting forest to pasture for beef production, and providing speculative export loans (Nairn, 1981; Grindle, 1986).

The local impacts of agribusiness

The consequences are manifold. At a global level there has been a marked reorientation of Latin American export agriculture from Europe to the United States. Before the Second World War exports were strongly oriented to Europe. At a national level there has been an extraordinary expansion of some crops at the expense of other crops, especially traditional food staples. Some crops that were not widely produced in the 1960s grew at enormous rates in the 1970s: sorghum in Brazil (125.5 per cent per annum), Venezuela (68.9 per cent) and Colombia (20.4 per cent); soybeans in Paraguay (83.3 per cent), Argentina (63.9 per cent), and Brazil (59.0 per cent); and palm oil in Ecuador (29.6 per cent) (Grindle, 1986).

The food staples have been replaced by more profitable products destined for affluent urban and foreign markets. In Chile fruits and livestock have replaced wheat and sugar beet; sorghum has replaced corn in Mexico and Brazil; livestock have replaced the basic crops throughout the region as indicated by statistics showing the vast expansion of permanent pasture lands at the same time croplands have either decreased in area (as in Mexico and Venezuela) or increased only moderately (as in Costa Rica, Colombia, Panama and Honduras). In some places increased livestock production has also stimulated the expansion of feed-grain production, often on land that formerly produced the food staples of middle- and low-income groups. Livestock production has also produced widespread deforestation and erosion of land that could be productive under other uses (Williams, 1986; Grindle, 1986; Brockett, 1988).

Shortfalls in food staple production have necessitated the increased import of basic food items. Until 1973 agricultural exports grew steadily even if they did not keep pace with imports (Grindle, 1986, p. 92). Since then, however, economic stagnation in the United States and Western Europe has reduced demand for Latin America's agricultural exports (such as beef and 'January' strawberries) at the same time that the cost of imported food (and other products such as fertilizers and machinery) increased appreciably.

The penetration of foreign agribusiness has also had important effects on rural populations. One effect has been the increased concentration of land holdings in the hands of capitalist farmers and TNCs such that

313

> Throughout the region, tenants and sharecroppers were replaced by agricultural workers, and permanent workers were displaced by part-time labourers. Given these changes, landowners could minimize the costs of maintaining a labour force through periods when it was not needed and expand cropping or live stocking areas by taking over lands that had been assigned to resident labourers, tenants, and sharecroppers. Labour costs were thus reduced for the entrepreneur, and the available pool of labourers, forced to provide for their own maintenance during inactive periods, was enlarged.

(Grindle, 1986, p. 98)

Another effect has been to increase the need to borrow and hence the indebtedness of surviving peasant farmers and part-time labourers. Debt is nothing new for peasant farmers. As the meaning of subsistence changed in a monetized economy to include 'urban goods' and processed foods, so did the importance of money. In the past, money was obtained through the sale of labour for cash wages or sale of market crops. Debt arose because of the need to store and transport crops and pay for inputs before cash was available. Often yields and cash wages were so low that more debt was incurred merely to survive. Today debt is also incurred by the necessity of competing against the capitalist-export sector for land, inputs and water resources (Pearse, 1975; Warman, 1980).

The cycle of indebtedness

In order to manage the higher debt load, peasants must farm their land more intensively. This only exacerbates the problem. Traditional farming methods such as crop rotation and fallow periods are replaced by monoculture to grow the most remunerative crop. This process leaches and depletes the soil, leading to poor harvests and soil erosion. As a consequence, more fertilizers and new seeds are required, thus deepening the cycle of indebtedness. Warman (1980, p. 238) describes the cycle of indebtedness that has followed the increased 'capitalization' of agriculture in central Mexico:

> The peasant has to combine several sources of credit, on occasion all of them, in order to bring off the miracle of continuing to produce without dying of starvation. He does it through a set of elaborate and sometimes convoluted strategies. Some people plant peanuts only in order to finance the fertilizers for the corn crop. Others use official credit to finance planting a cornfield or for buying corn for consumption in the months of scarcity, while they resort to the local bourgeoisie or the big monopolists in order to finance a field of tomatoes or onions. Many turn to usurers [money lenders] to cover the costs of an illness or a fiesta. . . . Given what they produce in a year, what is left after paying the debts does not go far enough even for food during the dry season, much less for starting a crop on their own. For them, obtaining a new loan is a precondition for continuing cultivation, one that must be combined with the sale of labour if they are to hold out to the next harvest. Each year the effort necessary to maintain the precarious equilibrium increases, and it seems to be a spiral that constantly demands more work, as well as the daring and inventiveness to find it. Creating employment, inventing ways of working harder, is part of peasant leisure.

Peasants, then, are survivors as much as victims. Increasingly, wage labour has come to provide a major portion of family income even for peasants who own land. Often this has involved temporary long-distance migration. In Guatemala, for example, the coffee, cotton and sugar harvests involve the seasonal migration of more than 300 000 highland Indians (Miro and Rodrigues, 1982). Temporary wage labour on nearby plantations and capitalist farms, however, is perhaps the major form of adaptation.

Drug crops

In some areas peasants have also supplemented their incomes by switching to the cultivation of drug crops. The market for these crops in the United States and Europe has grown exponentially since 1970 and the crops can be grown in remote areas on low-grade soils. Given the illegality of drug crops in world trade, remoteness becomes a virtue rather than the liability it is in more legitimate trade. Inaccessible corners of Turkey, Lebanon and Myanmar (Burma) are all important sources of cocaine and heroin destined for American and European markets. In three Latin American countries, Peru, Colombia and Bolivia, cocaine, heroin and marijuana exports are estimated to bring in US$600 million per annum. In Bolivia, cocaine exports exceed the total value of all legitimate exports. In Peru, cocaine is the country's largest export earner (Stone, 1988). Of course, much of the proceeds goes to 'drug barons', public officials and intermediaries. But for many peasants the drug traffic is one of the only ways they have of paying their debts and, thus, responding to the disruptions consequent upon the capitalization of agriculture by TNCs and foreign investment. Profits from drugs also fuel the insurgencies of ethnic and political opponents of existing governments. The main routes of surreptitious export change frequently in response to both new alliances between producers and intermediaries and successful efforts by police forces at intercepting the drugs before they hit the streets of American and European cities. The 'laundering' of profits from the international drug business is seen by some commentators as a major activity in some off-shore banking centres. The drugs business is not new. It has ancient roots. In the nineteenth century opening up China to the export of opium from India was one of the main causes of the war between Britain and China that was as a result called the Opium War. Illicit though it may now be, the global trade in drugs fits into the long history of the trade in stimulants as an important part of the growth of the modern world economy (see Chapter 8).

The case of the beef boom in Central America

An interesting case study in the capitalization of Latin American agriculture is the so-called 'beef boom' in Central America in the 1970s and early 1980s (Williams, 1986). This led to the emergence of Central America as a major supplier of beef to the United States when it had been previously relatively insignificant. It resulted from the tremendous increase in demand for beef in the United States because of the emergence of the fast-food franchises such as McDonald's and Burger King catering to a population increasingly given to 'eating out'. The new franchises were not particularly demanding of high-quality beef. What they wanted was quantity that could be formed into patties of equal size and weight by sufficient grinding and tenderizing. But the quantity needed was so huge that the fast-food chains (and 'TV dinner' makers) needed to look beyond the USA for sources of supply. Sources such as Australia, New Zealand and Canada were subject to severe quota limitations that were part of intensive 'tit-for-tat' trade negotiations on the part of the US government in the GATT. South America was 'out' because of the prevalence of hoof (foot)-and-mouth disease there. Central America was favoured by US government policy to help 'friendly' governments diversify their exports in the face of the perceived 'geostrategic threat' from Cuba and the Soviet Union in the region. By 1979 Central America had acquired 93 per cent of the share of the US beef quota available to LDCs.

A number of TNCs and individuals found it profitable to respond to the demand for beef from Central America. Some very large US companies became involved through subsidiaries and joint ventures. For example, R. J. Reynolds owns huge grazing ranches in Guatemala and Costa Rica through its Del Monte subsidiary and it directly processes and markets its beef through a variety of outlets: 'Ortega' beef tacos, 'Chun King' beef chow mein and 'Delmonte' Mexican foods. It also sells beef through 'Zantigo' Mexican Restaurants (Kentucky Fried Chicken). One of the largest firms in the Central American beef business is Agrodinamica Holding Company, a company formed in 1971 with 60 per cent of the stock owned by wealthy Latin Americans and 40 per cent of the stock owned by the American ADELA Investment Company. This operation controls thousands of acres of pastures in Central America, owns numerous packing plants, and runs a Miami (Florida) beef-import house and wholesale distributor (Williams, 1986).

Other TNCs have become involved in supplying the beef business with inputs (grass seed, barbed wire, fertilizers, feed grains and veterinary supplies). Pulp and paper companies such as Crown Zellerbach and Weyerhauser invested in cardboard-box factories to supply packing houses with containers for shipping the beef. Finally, fruit companies with access to large blocks of land turned them into money-making properties.

TNCs, however, were not the only beneficiaries. Wealthy families with access to large amounts of 'marginal' and forest land have turned them into profitable pastures. Some very powerful families, for example the Somozas in Nicaragua before 1979, have tapped profits from every stage of the beef-export business. Some urban-based professionals (lawyers, bankers, etc.) have also become involved as 'weekend ranchers' of peripheral areas previously untouched by commercial agriculture (Williams, 1986; Brockett, 1988).

The massive displacement of peasants by ranchers and cattle, however, has met with tremendous resistance. As R. G. Williams (1986, p. 151) puts it: 'The receding edge of the tropical forest became the setting of a conflict between two incompatible systems of land use, one driven by the logic of the world market, the other driven by the logic of survival.' The violence and civil war throughout much of Central America in the 1970s and 1980s bore no small relationship to the expansion of the beef-export business.

Another important consequence is deforestation. Much of the loss of forest in Central America, the Amazon Basin of Brazil and in Southeast Asia is due to the extension of ranching as well as timber extraction and the burning of timber as fuel wood. A related stimulus to the incredible pace at which tropical forests have been disappearing since the 1970s has been the need to pay off the debts incurred in expensive industrialization campaigns. The opening of forest land to capital-intensive agriculture has been one strategy for swapping natural resources for income to repay debts. Five of the world's 'mega-debtors', Brazil, India, Indonesia, Mexico and Nigeria, all rank among the top 10 deforesters (George, 1992, pp. 9–14). The capitalization of agriculture in the periphery, therefore, has had correlates other than increased productivity, the establishment of 'comparative advantage' in export crops and the increased import of food crops.

SCIENCE AND TECHNOLOGY IN AGRICULTURE

The beef-export boom in Central America would not have been possible without the importation of techniques of 'scientific agriculture'. In this context this involved creating 'new' breeds of cattle by combining 'beefier' attributes with high resistance to pests and tropical heat, transforming pasture management by sowing higher-yield grasses and fertilizers, enhancing water supplies by digging new wells and ponds, and providing better veterinary care to cattle herds.

The past 40 years have witnessed an intensive drive on the part of international development agencies (such as the Food and Agriculture Organization (FAO) of the UN and the World Bank), some national governments, and agribusiness to introduce scientific farming into agriculture in LDCs. The results have been controversial. From one point of view, yields have been increased and, especially in parts of Asia but to a degree also elsewhere, agricultural productivity and production have been significantly increased. Of particular importance have been the new wheat, maize and rice varieties associated with the so-called 'Green Revolution'. It is generally acknowledged that the gains from these new varieties (and the fertilizers and irrigation they require) have been concentrated in certain districts of India, Pakistan and Sri Lanka, the Central Philippines, Java in Indonesia, peninsular Malaysia, northern Turkey and northern Colombia. In addition to increased yields the new techniques can involve an increase in demand for labour in land preparation, fertilizer application, and harvesting and increases in the wages of agricultural labourers (as in the Indian Punjab). Doubts are sometimes expressed, however, about the sustainability of these trends in yields and labour use (Wortman and Cummings, 1978; Hayami, 1984).

From another point of view, scientific agriculture is largely an instrument of commercialization and capitalization rather than a mechanism for improving agricultural productivity and production *per se*. This is not to say that new seed varieties, fertilizers, etc., are always inappropriate; rather, that it all depends on the socio-political context in which they are applied. In particular, research efforts in scientific agriculture have been heavily biased towards certain commodities that are either most important in the industrialized countries or significant in world trade. The very small amount of research on important food staples such as cassava, coconuts, sweet potatoes, groundnuts and chickpeas is especially noteworthy. The 'research system' gives high priority to export crops such as cattle, cotton and sorghum and to those such as rice and wheat which have 'wide adaptability': ability to transfer a new variety from one region to others (Evenson, 1984; Eicher, 1984). Wide adaptability can be criticized, however, for its potential in reducing genetic variety and making crops more vulnerable to disease.

A more frequent criticism of scientific agriculture, particularly in its manifestation as the Green Revolution, is that it primarily benefits larger, more prosperous farmers who have readier access to the necessary inputs and credit sources. At the same time it encourages the 'debt cycle' among poorer peasants and part-time labourers discussed earlier (Pearce, 1979; Yapa, 1979; Griffin, 1974). Moreover, the new varieties require increased dependence on the acquisition of energy-intensive inputs (such as fertilizers and agricultural machinery), largely controlled by TNCs.

The substitution of feed crops, crops for feeding animals rather than direct human use, and export-food crops for local food crops, has been one important recent impact of scientific agriculture. Some observers refer to this as the 'second green revolution', meaning that it has produced a new wave of crops whereas the earlier trend produced greater yields of staple crops. This is not only biased in favour of farmers with capital, it also can lead to the neglect of food crops fundamental to local diets. As a result, while exporting increasing quantities of meat and fruits, some countries find themselves having to import beans, wheat and maize to feed their rural populations (McMichael, 1996, p. 104).

Evidence from such diverse settings as Mexico, India and Bangladesh suggests that where capital-intensive agriculture is introduced into areas with an uneven distribution of resources it exacerbates the condition of the rural poor by marginalizing subsistence systems, such as share-cropping, and encouraging the polarization of land control between a class of capitalist farmers, on the one hand, and the mass of the rural population, on the other (Schejtman, 1982; Jones, 1982). The impact of scientific agriculture, therefore, cannot be separated from issues of social structure.

Science and rice

For one-half of the world's population, overwhelmingly in Asia, the cycle of life revolves around rice. In Vietnam, a child's first food, after mother's milk, is rice gruel. In Taiwan death is symbolized by chopsticks stuck in a mound of cooked rice. Getting a good job in Singapore is an 'iron rice bowl', and unemployment is a 'broken rice bowl'. The characters for Toyota and Honda, the great car companies, mean in Japanese, 'bountiful rice field' and 'main rice field', respectively. For people in places where rice has long been the main staple of everyday diet, rice means just about everything that is important: birth, death, power, wealth, virility, fertility, vitality, and so on. The oldest recorded cultivation of rice occurred in what is today Thailand in 4000 BC, although the crop is thought to have originated in Africa. Its cultivation spread widely but rice became the staple crop in Southeast and East Asia. Elsewhere, wheat and other cereal grains tended to be more important. The great advantage of rice lies in its yields which, on average, are twice as large as wheat. Today, rice feeds more people than any other crop. Although more wheat is harvested annually than rice, over 20 per cent of that harvest goes to feed animals. Virtually all of the annual rice harvest (527.4 million metric tons in 1995) goes to feed people, mostly in Asia, where more than 60 per cent of the world's population lives. In East Asia, between 30 and 70 per cent of daily calories come from rice. Rice has what botanists call 'developmental plasticity': it can grow in a wide variety of circumstances. But it flourishes best in the humid tropics. The three largest producers, China, India and Indonesia, produce and consume 60 per cent of the world's rice. With only 4 per cent of the world's rice in world trade, a stable local supply is crucial to the food supply of most Asian countries. All of the world's exports, about 12 million tons, would not meet demand from India for more than 2 months. From the 1930s to the 1950s rice yields in Asia stagnated, while improved health care led to a doubling of the population. The application of chemical fertilizers did little to improve the situation. The established types of rice grew, but they grew too tall, fell under their own weight, and rotted in the flooded fields in which they were cultivated. A new strategy came in the early 1960s as a result of research on new hybrid varieties of rice carried out at the International Rice Research Institute (IRRI) in Los Banos, Philippines. IR8, one of the first new varieties, was spectacularly successful in raising yields. It grew faster – maturing in 130 rather than the usual 180 days – and allowed farmers to harvest two or even three crops per year from the same land. It also produced twice as much rice as either of the parent varieties. This variety and subsequent ones were so successful in doubling the world's rice crop that they were called 'miracle varieties'. They and new wheat varieties led to the declaration of a 'green revolution' in which the war on hunger and famine was said to have been won. This was premature. By the 1980s the IRRI had engineered 250 new varieties of rice that are planted in 106 countries. But at the same time world rice production has flattened out and the population has kept on growing. A simple answer might be to just plant more land in rice. In Asia, however, little or no land is left for expansion. So the pressure is on to increase yields even further through more varieties better fitted to specific ecological conditions and resistant to pests. Nearly 25 per cent of rice crops are destroyed by insects and diseases. The question of the moment is whether or not yields can be increased indefinitely. The levelling out of production might suggest that the limits to scientific agriculture in rice production have now been reached.

Summary

Since 1960 GNP growth rates have been faster in the less developed countries than in the developed countries (5.5 per cent per annum compared to 4.2 per cent). In addition, despite large rates of population growth, the per capita incomes of the periphery taken as a whole have grown at about 2.5 per cent per annum. Agricultural production has increased at similar rates, in contrast to the stagnation of the colonial period in many Asian and some African countries. Food production per capita in Latin America and Asia grew by 5 and 10 per cent between 1960 and 1980, although these rates slipped somewhat in the 1980s. Only in countries with birth rates of 3 per cent or more, as in parts of sub-Saharan Africa, or where there were major social upheavals, such as Central America, Bangladesh,

Cambodia and Vietnam, is this picture particularly misleading. Throughout the periphery the incidence of chronic hunger and malnourishment did decline between 1970 and 1990, in the absence of wars and natural disasters.

At the same time, however, the incidence of rural indebtedness and poverty and the loss of land for food production to meet local demand have increased enormously. This is because increased agricultural production in the context of the modern world economy is no guarantee that the people involved in achieving it will see its fruits. This chapter has attempted to show how this can be the case by detailing the effects of progressive commercialization and capitalization. When subsistence uses and food-staple production are displaced by export crops, increased agricultural production does not necessarily benefit rural populations. Far from it. They often find themselves ensnared in webs of poverty and indebtedness that are the direct product of modern scientific agriculture in contexts where there are few alternatives to agricultural employment. In reaching this conclusion the argument of this chapter has involved making the following major points:

1. Agriculture is often given a subsidiary role in models of development followed by governments even when it is a vital source of sustenance and employment.
2. The three continents of the periphery – Latin America, Africa and Asia – differ significantly in terms of agricultural organization and performance.
3. Agriculture is overwhelmingly more important as a source of employment to women than to men, especially in sub-Saharan Africa.
4. There is a long history of commercial agriculture in the periphery. Until recently, however, it was a plantation or export-enclave sector surrounded by a largely subsistence sector.
5. Governments have not tended to favour agriculture. For a variety of reasons, their pricing and credit policies have tended to drain agriculture in favour of the industrial–urban sector.
6. Rural land reform has tended to encourage the development of capitalist agriculture rather than benefit the interests of peasant farmers.
7. Rural land reform and the recent activities of governments and multinational enterprises have produced a much more widespread commercialization and capitalization (increasingly capital-intensive type) of agriculture. This has been most marked in Latin America but can also be seen elsewhere.
8. 'Scientific' agriculture has tended to reflect and reinforce the capitalization of agriculture even as it has increased yields for a limited number of agricultural products, mainly a few staples such as rice and wheat and those in export trade.

Key Sources and Suggested Reading

Asian Development Bank. 1977. *Rural Asia: Challenge and Opportunity.* New York: Praeger.

Bates, R. H. 1983. *Essays on the Political Economy of Rural Africa.* Cambridge: Cambridge University Press.

Beneria, L. 1981. Conceptualizing the labour force: the underestimation of women's activities. *Journal of Development Studies,* 17, 10–27.

Berg, E. J. 1965. The development of the labor force in sub-Saharan Africa. *Economic Development and Cultural Change,* 13, 394–412.

Brockett, C. D. 1988. *Land, Power, and Poverty: Agrarian Transformation and Political Conflict in Central America.* Boston: Unwin Hyman.

Burbach, R. and Flynn, P. 1980. *Agribusiness in the Americas*. New York: Monthly Review Press.

De Janvry, A. 1984. The role of land reform in economic development: policies and politics. In C. K. Eicher and J. M. Staatz (eds), *Agricultural Development in the Third World*. Baltimore: Johns Hopkins University Press.

Duncan, K. and Rutledge, I. 1977. Introduction: patterns of agrarian capitalism in Latin America. In K. Duncan and I. Rutledge (eds), *Land and Labour in Latin America*. Cambridge: Cambridge University Press.

Durham, K. F. 1977. Expansion of agricultural settlement in the Peruvian rainforest: the role of the market and the role of the state. Paper presented at the Latin American Studies Association, Houston, TX, 2–5 November.

Eicher, C. K. 1984. Facing up to Africa's food crisis. In C. K. Eicher and J. M. Staatz (eds), *Agricultural Development in the Third World*. Baltimore: Johns Hopkins University Press.

Evenson, R. E. 1984. Benefits and obstacles in developing appropriate agricultural technology. In C. K. Eicher and J. M. Staatz (eds), *Agricultural Development in the Third World*. Baltimore: Johns Hopkins University Press.

FAO. 1990. *FAO Production Yearbook*. Rome: UN Food and Agricultural Organization.

FAO. 1992. *World Food Conference: Report*. Rome: Food and Agricultural Organization, December.

Feder, E. 1978. *Strawberry Imperialism*. Mexico City: Editorial Campesina.

Fine, B. 1994. Towards a political economy of food. *Review of International Political Economy*, **3**, 519–45.

Frankman, M. T. 1974. Sectoral policy preferences of the Peruvian government, 1946–1968. *Journal of Latin American Studies*, **6**, 289–300.

Friedmann, H. 1991. Changes in the international division of labor: agro-food complexes and export agriculture. In W. Friedland *et al.* (eds), *Towards a New Political Economy of Agriculture*, Boulder, CO: Westview Press.

George, S. 1992. *The Debt Boomerang: How Third World Debt Harms Us All*. London: Pluto Press.

Gomes, C. and Perez, A. 1979. The process of modernization in Latin American agriculture. *CEPAL Review*, **8**, 55–74.

Griffin, K. 1974. *The Political Economy of Agrarian Change*. Cambridge, MA: Harvard University Press.

Grindle, M. S. 1986. *State and Countryside: Development Policy and Agrarian Politics in Latin America*. Baltimore: Johns Hopkins University Press.

Hart, K. 1982. *The Political Economy of Agriculture in West Africa*. Cambridge: Cambridge University Press.

Hayami, Y. 1984. Assessment of the Green Revolution. In C. K. Eicher and J. M. Staatz (eds), *Agricultural Development in the Third World*. Baltimore: Johns Hopkins University Press.

Hill, P. 1986. *Development Economics on Trial: The Anthropological Case for the Prosecution*. Cambridge: Cambridge University Press.

Hopkins, M. 1983. Employment trends in developing countries, 1960–80 and beyond. *International Labour Review*, **122**, 461–78.

Hossain, M. 1982. Agrarian reform in Asia – a review of recent experience in selected countries. In S. Jones *et al.* (eds), *Rural Poverty and Agrarian Reform*. New Delhi: Allied.

Hsu, R. C. 1982. Agricultural financial policies in China, 1949–80. *Asian Survey*, **22**, 638–58.

Joekes, S. P. 1987. *Women in the World Economy*. New York: Oxford University Press.

Jones, S. 1982. Introduction. In S. Jones *et al.* (eds), *Rural Poverty and Agrarian Reform*. New Delhi: Allied.

Jones, S. *et al.* (eds), 1982. *Rural Poverty and Agrarian Reform*. New Delhi: Allied.

Landsberger, H. A. 1969. *Latin American Peasant Movements*. Ithaca, NY: Cornell University Press.

Lardy, N. R. 1984. Prices, markets, and the Chinese peasant. In C. K. Eicher and J. M. Staatz (eds), *Agricultural Development in the Third World*. Baltimore: Johns Hopkins University Press.

Lele, U. 1984. Rural Africa: modernization, equity, and long term development. In C. K. Eicher and J. M. Staatz (eds), *Agricultural Development in the Third World*. Baltimore: Johns Hopkins University Press.

Little, I. 1982. *Economic Development: Theory, Policy, and International Relations*. New York: Basic Books.

Loup, J. 1983. *Can the Third World Survive?* Baltimore: Johns Hopkins University Press.

Malassis, L. 1975. *Agriculture and the Development Process*. Paris: UNESCO.

McMichael, P. 1993. Agro-food restructuring in the Pacific Rim: a comparative-international perspective on Japan, South Korea, the United States, Australia, and Thailand. In P. McMichael (ed.), *The Global Restructuring of Agro-Food Systems*. Ithaca, NY: Cornell University Press.

McMichael, P. 1996. *Development and Social Change: A Global Perspective*. Thousand Oaks, CA: Pine Forge Press.

Miro, C. A. and Rodrigues, D. 1982. Capitalism and population in Latin American agriculture. *CEPAL Review*, **16**, 51–71.

Nairn, A. 1981. Guatemala. *Multinational Monitor*, **2**, 12–14.

O'Connor, A. M. 1978. *The Geography of Tropical African Development: A Study of Spatial Patterns of Economic Change since Independence*. 2nd Edition. Oxford: Pergamon.

Pearce, A. 1979. *Seeds of Plenty, Seeds of Want*. London: Oxford University Press.

Pearse, A. 1975. *The Latin American Peasant*. London: Cass.

Perkins, D. and Yusuf, S. 1984. *Rural Development in China*. Baltimore: Johns Hopkins University Press.

Rudolph, L. I. and Rudolph, S. H. 1987. *In Pursuit of Lakshmi: The Political Economy of the Indian State*. Chicago: University of Chicago Press.

Sachs, J. 1996. Growth in Africa. *The Economist*, 29 June, 19–21.

Schejtman, A. 1982. Land reform and entrepreneurial structure in rural Mexico. In S. Jones *et al.* (eds), *Rural Poverty and Agrarian Reform*. New Delhi: Allied.

Sen, A. 1981. *Poverty and Famines: An Essay on Entitlement and Deprivation*. Oxford: Clarendon Press.

Sender, J. and Smith, S. 1986. *The Development of Capitalism in Africa*. London: Methuen.

Stone, C. 1988. Drugs from the Third World. *World Press Review*, January, 64.

Swindell, K. 1985. *Farm Labour*. Cambridge: Cambridge University Press.

Taussig, M. 1978. Peasant economies and the development of capitalist agriculture in the Cauca Valley, Colombia. *Latin American Perspectives*, **18**, 62–91.

Thiesenhusen, A. 1989. *Searching for Agrarian Reform in Latin America*. Boston: Unwin Hyman.

Thomas, T. 1996. Africa for the Africans. *The Economist*, 7 September, Survey.

Timmer, C. P. and Falcon, W. P. 1975. The political economy of rice production and trade in Asia. In L. G. Reynolds (ed.), *Agriculture in Development Theory*. New Haven: Yale University Press.

Tosh, J. 1980. The cash crop revolution in Africa: an agricultural reappraisal. *African Affairs*, **79**, 79–94.

Turton, A. 1982. Poverty, reform and class struggle in rural Thailand. In S. Jones *et al.* (eds), *Rural Poverty and Agrarian Reform*. New Delhi: Allied.

Warman, A. 1980. *'We Come to Object': The Peasants of Morelos and the National State*. Baltimore: Johns Hopkins University Press.

Watts, M. 1994. Life under contract: contract farming, agrarian restructuring and flexible accumulation. In P. D. Little and M. J. Watts (eds), *Living Under Contract: Contract Farming and Agrarian Transformation in Sub-Saharan Africa*. Madison, WI: University of Wisconsin Press.

Williams, R. G. 1986. *Export Agriculture and the Crisis in Central America*. Chapel Hill: University of North Carolina Press.

Wolf, E. R. 1968. *Peasant Wars of the Twentieth Century*. New York: Harper and Row.

World Bank 1992. *World Tables 1992*. Baltimore: John Hopkins University Press.

Wortman, S. and Cummings, R. W. 1978. *To Feed This World*. Baltimore: Johns Hopkins University Press.

Yapa, L. 1979. Ecopolitical economy of the Green Revolution. *Professional Geographer*, **31**, 371–6.

CHAPTER 10

Picture credit: World Bank

INDUSTRIALIZATION: THE PATH TO PROGRESS?

In the 1950s and 1960s the development strategies of many LDCs placed considerable emphasis upon manufacturing industry, which was considered to be the 'noble' or leading sector of economic development. More recently, as the industrialized countries have 'lost' some branches of manufacturing to locations in the global periphery and some peripheral countries have embarked on aggressive export-oriented development strategies, it seems that efforts at industrialization can pay off. Indeed, an earlier complacency about the immobility of capital at a global level has now given way in some quarters to predictions about 'the end of the Third World' (e.g. Harris, 1986). But what exactly has been the result of several decades of industrialization in the periphery?

In the periphery over the past 30 years value added in manufacturing (MVA) has risen at a rapid pace. The increase is relative, however. The LDCs supplied 8.2 per cent of world MVA in 1960 and still only approximately 21 per cent in 1994 (UNIDO, 1981; World Bank, 1996). Moreover, industrialization is highly concentrated. From 1966 to 1975, four countries, representing 11 per cent of the population of the LDCs, accounted for over half the increase of the LDCs' MVA. Eight countries (Brazil, Mexico, Argentina, South Korea, India, Turkey, Iran, Indonesia) with 17 per cent of the total population produced about two-thirds of the increase. From 1975 to 1994, however, Argentina dropped out and China, South Africa, Thailand and Taiwan were added to the list. The presence of China and India on this list raises the proportion of the population of the LDCs in the high-growth category to 60 per cent.

At the same time, growth rates of MVA have been lowest in the poorest countries. In sub-Saharan Africa the growth of manufacturing has been particularly slow. In 1975 manufactured production represented 5 per cent of the GDP of the African LDCs as opposed to 16 per cent for the Asian ones and 25 per cent of those in Latin America and the Caribbean (Loup, 1983). By 1994 the comparable figures were approximately 13 per cent for the African LDCs, 27 per cent for Asia (30 per cent in East Asia, 15 per cent in South Asia), and 21 per cent for Latin America and the Caribbean (World Bank, 1996).

In all LDCs, irrespective of their growth rates, industrial production has been characterized by a particular expansion of heavy industries: iron and steel, machinery and chemicals. Over the entire period 1950–94 this expansion was more rapid than the growth of food processing or textile, clothing and shoe industries. This point needs emphasizing because of the tendency to assume, because of increasing exports of goods such as clothing and shoes to Europe and the United States, that light and consumer goods industries have grown the most. However, since the 1960s the industrial mix of many underdeveloped countries has undergone significant change. After the Second World War, industrialization, even if involving foreign investment by TNCs, was largely concerned with import substitution. Since the early 1960s the possibilities of substitution have dwindled in the face of mounting costs of establishing heavy industries and as new sub-contracting and global sourcing strategies of TNCs have replaced the older strategy of direct establishment of subsidiaries. In this new global context a fundamental reorientation has taken place in the most industrialized LDCs in East Asia and Latin America. In these countries an increasing proportion of industry is oriented towards exporting manufactured goods, mostly to the developed countries of the core (Jenkins, 1988).

The shift towards an export orientation was facilitated by rapid economic growth (and increasing consumer incomes) in the developed countries (especially in Western Europe and Japan) and the liberalization of world trade beginning in the 1960s. Above all, however, it reflects a change in national industrialization strategy. The role of the state remains central. Like import substitution, export promotion involves a strong managerial role for the state in adjusting to new global pressures. The states that have been most successful in doing so, such as South Korea and Taiwan, are now the leading industrializers (Tyler, 1981). But the ideological and institutional context has changed fundamentally. The focus has moved from self-sufficiency in a world of national economies to gaining competitive advantage in a global economy organized on principles of market-access (Biersteker, 1995; Agnew and Grant, 1997).

In this chapter the progress of industrialization in the LDCs is examined in four complementary ways. First, the national and global stimuli to industrialization are described. Particular attention is directed towards the role of industrialization in national ideologies of modernization, the practical basis to the demand for industrialization, and the global context for the shift from import substitution to export orientation. Second, the problems facing industrialization in the periphery are reviewed, focusing on the limits to industrialization posed by certain national-level and global constraints. Third, the geographical pattern of industrialization is surveyed at global, regional and urban scales. Fourth, and finally, Brazil, Cuba and China are profiled for their particular experiences of industrialization over the recent past.

NATIONAL AND GLOBAL STIMULI TO INDUSTRIALIZATION

The central attention given by many peripheral countries to industry is partly a result of the prestige of this sector, which is widely considered the hallmark of development. Although the notion of 'industrialization-in-general' can be criticized on grounds of vagueness and lack of attention to the specific mix of industries and their relation to the needs of the mass of the population (e.g. Corbridge, 1982), industrialization figures prominently in most national ideologies of modernization. Perhaps China for part of the 1960s was something of an exception, at least in theory (see Murphey, 1980). But even there the lure of industrialization as a development strategy proved stronger than ideological commitment to rural-agricultural development.

Interestingly, in the core of the world economy, particularly in Western Europe, industrialization has always given rise to various 'discontents'. A certain contempt for the production of worldly goods shows up even in the writings of the classical economist Adam Smith. Later concerns have been more with the nature of industrialization, in particular the 'balance' between heavy and consumer goods industries. These ideas have their most vocal expression today among those East Europeans who decry the 'overemphasis' of the now-defunct Communist governments on heavy industry and Latin American complaints about the 'lack' of heavy industry (Hirschman, 1992). However, the association between industrialization and modernity is now strongly established. Only environmental activists, a rare breed in most LDCs, question the unrelenting priority to industrial development.

Three myths

Three ideas of mythic proportions are at the centre of the claim that national industrialization is the path to modernity, even though they are of questionable empirical validity. Interestingly, they all involve negative views of agriculture as much as positive endorsements of industry and they all imply a simple sectoral logic of development as movement from agriculture to industry (Loup, 1983, pp. 155–8).

First, agriculture is viewed as having more limited stimulus effects on other economic activities than industry. In other words, industry is seen as providing **multiplier effects** that agriculture cannot provide. The best refutation of this particular idea is the key role that agriculture played in the early industrialization of Europe and the continued importance of agriculture in the economies of the developed world. Of course, in each case investment was required to develop forward linkages to consumer industry (food processing, etc.) and backward linkages to input providers (fertilizers, etc.). In each case farmers were also important as consumers of industrial products, when not penalized by low prices for their products, and significant financers of industrial investment, through savings and taxation. Adelman (1984) has proposed that precisely these stimulative features of agriculture can be used to substitute 'agriculture-demand-led-industrialization' (ADLI) for import substitution and export-led models.

Second, farmers have a reputation for conservatism whereas industrialists (and workers) are viewed as agents of modernization. Imprisoned in ancient and traditional cultures, farmers, especially peasant farmers, are without dynamism and rationality. Yet, again, this idea is easily refuted by evidence from all over the world. For example, as shown in Chapter 9 there is a strong link between producer prices

multiplier effects
The extra industries, firms, incomes and employment generated by a new activity can be said to result from that activity's multiplier effects. Such effects can be localized and give rise to a growth pole or occur in a more diffuse manner.

and yields of rice in different Asian countries. Corn (maize) production in Thailand, bean production in the Sudan, and wheat production in India and Pakistan have all increased as prices have increased and decreased when prices have declined. These are hardly indications of conservatism and lack of responsiveness to commercial incentives.

Third, for many governments industry is seen as the only productive sector. Only in industry, the argument goes, are there increasing marginal returns through economies of scale in production. Moreover, the average productivity of workers in industry is higher than that of those in agriculture. However, the productivity of other factors, capital in particular, is probably higher in agriculture. In most of the countries for which data exist, the gross marginal capital/output ratio is lower and hence productivity is higher in agriculture than in other sectors (Szczepanik, 1969). In the United States, the only country with a sufficiently long statistical series, the total productivity of all factors has increased faster in agriculture than in other sectors (Samuelson, 1964).

Practical rationales for industrialization

But whatever the empirical merit of the three ideas, they have become firmly entrenched and associated with modernization through industrialization. Manufacturing industry is widely viewed as the path to progress and it figures prominently in most national development ideologies and plans. These ideologies, whatever precise roles they reserve for 'private' business and state direction, have been reinforced by certain practical problems facing most governments. There are perhaps three that appear most important. One of them concerns the terms of trade in exchanging primary commodities for manufactured goods. As suggested in Chapter 8 there are good grounds for pessimism about the growth potential in general of primary production (raw materials and agricultural exports) because of the deteriorating barter terms of trade with manufactured goods. However, for specific primary commodities and specific countries, investment in primary production can be preferable. Nevertheless, by and large, governments have not been persuaded of this. They can even point to the case of the OPEC cartel, the most successful attempt in the history of the world economy to bolster the price of a primary commodity, to illustrate the limitations of primary production. From its dominant position in 1973–74, OPEC has become less and less able to govern the world price of oil. This reflects both adjustment strategies in consumer countries (energy conservation, shifts to non-OPEC suppliers) and the emergence of political conflicts and different production strategies among member countries (the Iran–Iraq war of the 1980s; conservation of oil reserves vs. rapid production, e.g. Saudi Arabia vs. Nigeria). It is easy to infer from the experience of OPEC the long-run limitations of a development strategy based upon primary production for world markets.

Second, many LDCs have massive unemployed and underemployed populations concentrated increasingly in urban areas. Deteriorating living conditions in the countryside (see Chapter 9) and the availability of better public services in urban areas have encouraged large-scale rural–urban migration, often in the absence of industrialization. To survive in the cities, people engage in a wide range of 'informal' economic activities as street vendors, shoeshine boys, stall keepers, public letter writers, auto mechanics, taxi drivers, sub-contractors, tailors, drug dealers and prostitutes (Mattera, 1985; Sanders, 1987). Sometimes these activities can be linked to industrialization of a formal variety through sub-contracting, but often they cannot. International migration, both temporary and permanent, is sometimes an alternative for those with better education and greatest initiative. But, among other things, this produces a 'brain drain' that poor countries can ill afford. It is in this context that expansion of employment in manufacturing industry can often become an important national imperative.

A third incentive for industrialization comes from the state-building activities of national élites. National industrialization can be a 'prestige' goal around which national populations can be mobilized. All governments are also under pressure to industrialize in order to compete with other countries. Pressure comes from both domestic élites, especially the military, and from foreign allies and patrons. Some of the emphasis on heavy industries undoubtedly derives from this pressure. The significant growth of military industries in the LDCs is directly related to it (Neuman, 1984). Industry is also an important instrument of political favouritism and patronage. Governments can reward 'loyal' social and ethnic groups and punish 'disloyal' ones by directing industrial activities towards some places and away from others. Industrialization, therefore, often involves political stimulation, of both 'noble' and 'ignoble' varieties.

The national industrialization drives that took place in the aftermath of decolonization had limited effects until the late 1960s, except in those countries with large domestic markets and long-sustained import substitution policies (e.g. Brazil, Mexico). The spread and intensification of industrialization since the late 1960s have coincided to a certain extent with the declining rate of profit in the industrialized countries (see Chapter 3, pp. 98–9) and the consequent shift in strategy by TNCs from high wage/high consumption forms of production in the industrialized countries (Fordism) towards spatially decentralized forms of production in which low-wage labour forces are important in certain phases (see Table 10.1). This suggests that the changing global context has been fundamentally important in stimulating the recent expansion of manufacturing industry in some LDCs. In other words, the New International Division of Labour in manufacturing and its associated spatial decentralization of many production activities (largely related to the international product-cycle model) are closely related to the 'crisis' of capital accumulation in the industrialized countries. A new geography of manufacturing employment has been the result (Table 10.2).

The extent to which cost advantages for TNCs (particularly low-cost labour) have entirely spurred the growth of the NICs can be exaggerated. In the first place, TNCs do not dominate production in the NICs. Even in Brazil, a country that for many years had had a high proportion of TNC investment relative to local sources, foreign direct investment has declined since the 1970s (Stallings, 1990). Second, governments in the NICs have controlled foreign investment, preferring investment by banks and other financial interests to foreign direct investment. 'Productive capital has largely remained under the control of local corporations; most foreign capital has entered countries either as bilateral or multilateral aid or as financial

Table 10.1: Labour conditions in global manufacturing

Country	Average hourly earnings (1987) in manufacturing (US$)	Industrial disputes 1974–83*
United States	13.46	312
United Kingdom	9.07	470 (1974–82)
West Germany	16.83	26
Japan	11.34	59
Brazil	1.49	n.a.
Mexico (maquiladoras)	0.84	n.a.
Hong Kong	2.11	10
South Korea	1.69	2
Singapore	2.37	2
Taiwan	2.33	2 (1974–82)
China	0.37	n.a.
Sri Lanka	0.25	383

*Average annual working days lost from disputes per 1000 non-agricultural workers
Source: Peet (1987), p. 780; US Department of Labor, Bureau of Labor Statistics (1987).

Table 10.2: The changing geography of manufacturing employment: paid employment in manufacturing (millions of people)

Region/country	1974	1981–84	Change
North America	22.0	21.4	−0.6
Japan	12.0	12.1	+0.1
Western Europe*	35.2	28.2	−7.0
'Centre'	35.2	28.2	−7.0
South Asia[+]	5.6	6.9	+1.3
Southeast and East Asia[†]	6.3	9.6	+3.3
Latin America[§]	7.0	8.7	+1.7
'Periphery'	18.9	25.2	+6.3

*Austria, Belgium, France, Federal Republic of Germany, Italy, Netherlands, Spain, Sweden, Portugal, United Kingdom.
[+]Bangladesh, India, Sri Lanka.
[†]Hong Kong, Republic of Korea, Malaysia, Philippines, Singapore, Thailand, Taiwan.
[§]Brazil, Venezuela, Mexico.
Source: Peet (1987), p. 781.

capital' (Webber and Rigby, 1996, p. 453). Nevertheless, export-based industriali-zation, as opposed to import substitution, does allow firms to pay lower wages than they would have to if the local economy had to absorb all of the supply. This has broken the bond between production and consumption upon which organized capitalism was based. The expansion of industry in the periphery, therefore, has some basis in cost advantages, even if these are not ones that accrue largely to TNCs, and it has had the consequence of increasing the competitiveness of world markets across a wide range of manufacturing industries (Gereffi, 1995).

But perhaps the best, and most simple, evidence that costs alone cannot explain the spread of industrialization to the periphery is that export-oriented industrialization did not concentrate where wage rates were lowest. It has been concentrated only in certain countries with other characteristics. The most successful ones, the East Asian NICs of South Korea, Taiwan, Singapore and Hong Kong, were ones in which the state and local industrial capital were closely interlocked and receptive to massive foreign investment (Evans and Alizadeh, 1984). Moreover, they are the ones in which much of the foreign investment was in the form of private bank credit rather than direct investment by TNCs (Goldsbrough, 1985). They thus maintained more local control over investment decisions. In addition, the privileged geopolitical arrangements of South Korea and Taiwan with the United States were important in opening American markets and making available American aid and investment in return for the East Asian countries' forward role in 'containment' of mainland China and the Soviet Union (Hamilton, 1983). Since the late 1980s there has been an increased level of foreign direct investment in East Asia, much of it flowing between East Asian countries rather than emanating from outside of the region (Griffith-Jones and Stallings, 1995). Indeed, between 1991 and 1997 private financial flows to LDCs grew rapidly after stagnating in the 1980s but 'official' (governments, World Bank, etc.) investment tailed off having been more important than private flows throughout the 1980s (*The Economist*, 1 March 1997). This has led some commentators to see a foundation being laid for a future boom in peripheral industrialization. However, though the global context may well trigger the *possibility* for industrialization, especially of the export-orientated type, it never guarantees its realization. Various contingencies – political, economic, social – intervene to determine where it takes place.

THE LIMITS TO INDUSTRIALIZATION IN THE PERIPHERY

There are a number of constraints that will probably limit the spread and intensification of the export-oriented industrialization that has lain behind the impressive growth of the NICs since the 1970s.

Profit cycles

In the first place, the declining rates of profit in the industrialized countries – between the 1960s and the 1980s – may well have been cyclical rather than secular. There is some evidence that corporate profitability in the USA, for example, began to rebound in the late 1980s after a 15-year downturn (Uchitelle, 1987). This may mean that there will be less incentive to locate production facilities or engage in subcontracting in the LDCs. However, the new figures (4.9 per cent average return in manufacturing in the USA in 1986 compared to 3.5 per cent in the late 1970s and 8 per cent in the early 1960s) may reflect the falling dollar after 1985 helping

American-based manufacturers in foreign markets and the increased return on foreign investments rather than improving domestic profitability (e.g. the European operations of General Motors lifted what would otherwise have been an even more dismal rate of profitability in the late 1980s). Webber and Rigby (1996, p. 346) argue that stagnant wage rates and lack of investment in new technologies have fuelled the return to profitability. But the picture is a complex one, with not only countries but also industrial sectors (electronics, automobiles, etc.) having different trajectories in profit rates over time.

In addition, some of the giant American TNCs that pioneered the 'global shift' in production through the creation of foreign assembly operations in the LDCs, such as IBM and GE, have become lacklustre performers, challenged in the most profitable production lines by more innovative smaller firms with currently more localized patterns of production in industrial districts because of their reliance on immediate response from independent sub-contractors. If the big firms follow in this direction, then there may well be a retreat from the trend to global sourcing that occurred in the 1970s and 1980s.

Protectionism

Second, there are inherent limits to the generalization to other LDCs of the export successes of the NICs. To begin with, the period 1950–80 was one of unusually high growth in world trade. Whereas global trade expanded only at 1 per cent per annum during 1910–40, the period 1953–73 saw an increase of total trade of 8 per cent per annum, and of 11 per cent per annum for manufactures (Tuong and Yeats, 1981). The consequence was an increased interdependence in the world economy as trade barriers were lowered. Since the late 1970s, however, protectionism has been a major response in the industrialized countries to their declining rates of economic growth and increased unemployment, domestic inflation and balance of payments deficits (Yoffie, 1983). It is the LDCs that have become the major advocates of global trade liberalization. Cline (1982) estimated that if all the LDCs had the same export intensity as South Korea, Taiwan, Singapore and Hong Kong, there would be a shift from 16.4 to 60.4 per cent of aggregate DC manufactured imports originating in the LDCs. Given the reliance of the NICs on DC markets, the likelihood of this expansion without protectionist responses seems extremely unlikely. As Cline (1982, p. 89) concludes: 'It is seriously misleading to hold up the East Asian G4 [Gang of Four, i.e. Hong Kong, Taiwan, South Korea, Singapore] as a model of development because that model almost certainly cannot be generalized without provoking protectionist response ruling out its implementation.'

Of course, Cline's generalization from the current NICs to all LDCs represents an extreme scenario. There is still considerable scope for building up export markets in both DCs and other LDCs (Loup, 1983). In particular, there is evidence that the established NICs are now transferring some of their more labour-intensive industries to countries that have a short-run comparative advantage. For example, Taiwanese firms have subsidiaries in Malaysia and China, and Hong Kong firms have subsidiaries in the China and the Philippines. Among the LDCs, consequently, adjacent countries are not necessarily condemned to see their initially advantaged neighbours permanently monopolize the positions they have won in world markets, even though they must of necessity start out as subservient to them.

Access to technology

Third, the TNCs that have been involved in setting up subsidiaries or engaging sub-contractors in the NICs must now respond to protectionist pressures in the DCs where their final markets are concentrated. This will involve substituting radical new automation technologies for low-wage labour. Even if the technology diffuses evenly at a global scale, production costs will decline more steeply in the DCs than in the NICs (Rada, 1984). In addition, new scale economies associated with batch production for customized as opposed to mass markets and new external economies associated with a close geographical integration of component suppliers and final assembly in 'last minute', 'just-in-time' and 'zero inventory' production systems will reduce the attractiveness of global sourcing (Ayres and Miller, 1983). There is already evidence that the assembly of electronics circuits is being brought back to DCs from LDCs and the introduction of new technologies and new production systems may further reduce the need for cheap, unskilled labour (Kaplinsky, 1984).

It is important to note again, however, that the growth of the NICs is not uniquely due to the activities of TNCs. Domestic firms have played important roles, particularly in such export sectors as textiles, clothing and shoes. This is especially the case in the East Asian countries where the share of manufactured exports attributable to TNCs is in the range of only 5 to 15 per cent (Loup, 1983). Only in Latin America are the expansion of TNCs and the growth of exports closely related (Evans, 1987). But the control of TNCs over most of the new automation technologies will limit their transfer even when the TNCs are not locally dominant (Kaplinsky, 1984).

Debt and creditworthiness

Fourth, the success of export-oriented industrialization has been tied to the growth of enormous debt loads underwritten by foreign direct investment, official aid agencies and multinational banks. Among the most problematic features of the world economy in the late 1990s are the absence of any new actors willing to finance industrialization in the LDCs and the appearance of a global 'credit shortage' because of the emergence of Eastern Europe (especially the former East Germany) and the former Soviet Union as competitors in world financial markets. The institutions which did play a central role in the past are largely unwilling to do so now. Perhaps the only institution that has substantially increased its lending and encouraged others to do likewise is the International Monetary Fund (IMF). But the IMF still lends relatively little compared to the size of LDCs' current account deficits and attaches conditions to loans that many countries with strong state direction find undesirable and damaging to long-run development (Griffith-Jones and Rodriguez, 1984; Lewis, 1988).

Even in the 1970s, lending, especially private bank lending, was concentrated in the LDCs with relatively high per capita incomes and those with already impressive growth records. The four largest borrowers (Mexico, Brazil, South Korea, and the Philippines) accounted for over 60 per cent of total accumulated non-OPEC LDCs' debt to international banks in December 1982. Much of the money lent came from recycled 'petrodollars' in the context of declines in credit demand from traditional clients in the DCs in the wake of the 1974–75 recession. This situation is not likely to repeat itself even though in the 1990s private lending from banks and to stock markets to East Asia and Latin America did increase after declining precipitously in the 1980s (Griffith-Jones, 1980; Seiber, 1982; Griffith-Jones and Stallings 1995).

R & D capacity

Fifth, the new leading sectors of industrial growth in the world economy are the so-called 'information technologies'. The USA, Japan and three European countries (France, West Germany and the United Kingdom) account for nearly 85 per cent of the world demand for these technologies and their products. Consequently, success in the global information-processing and electronics industries requires a research and development, production, marketing and political lobbying base inside all three of the regions of demand. Given the importance of politics – especially in the form of lobbying against protectionist threats – established TNCs and countries with large public sector investments in information technologies are at a distinct advantage. Those companies and countries struggling to gain entry beyond the labour-intensive, assembly level will be faced with formidable barriers, not least the absence of local protection in the face of US-led attempts in such international institutions as the GATT over the past 10 years at 'deregulation' of LDC domestic markets (Tunstall, 1986).

Polarization and instability

Sixth, the non-city state NICs (the experience of Singapore and Hong Kong is not immediately relevant to most LDCs) with the highest growth rates – South Korea and Taiwan – have enjoyed a unique set of circumstances that are not generalizable to other peripheral settings (Corbridge, 1986). In particular, they inherited the transport and education infrastructure imposed by Japanese colonialism. Later, massive US aid in the 1950s – for geopolitical purposes – stabilized their economies and tight controls on imports plus government allocation of foreign exchange and capital promoted export-led growth (Cumings, 1984). In addition to land reform (see Chapter 9) in both South Korea and Taiwan, state enterprises were a key part of national growth strategies in the 1970s, accounting for 25 per cent to 35 per cent of total fixed investment (Bradford, 1987). In South Korea – following the practice of Japan – interest subsidies and other incentives that rewarded performance were employed to induce private companies to develop major industries and focus on exports.

Crucially, relatively equal income distributions in Taiwan and South Korea – where the richest 20 per cent receive, respectively, 4.2 and 7.9 times the incomes of

Table 10.3: International variations in the concentration of incomes (higher figures indicate greater inequality).

	Year	Ratio of income of lowest fifth/income of highest fifth
Low-income countries		
Kenya	1992	18.3
China	1992	7.1
India	1992	5.0
Sri Lanka	1990	4.4
Middle-income countries		
Brazil	1989	32.1
Guatemala	1989	30.0
South Africa	1993	19.2
Chile	1994	17.4
Colombia	1991	15.5
Russia	1993	14.5
Costa Rica	1989	12.6
Malaysia	1989	11.7
Thailand	1992	9.4
South Korea	1989	7.9
Tunisia	1990	7.8
Philippines	1988	7.4
Poland	1992	6.5
Indonesia	1993	4.7
Taiwan	1989	4.2
Slovenia	1993	3.9
High-income countries		
Britain	1988	9.6
USA	1985	8.9
France	1989	7.5
Sweden	1981	4.6
Japan	1979	4.3

Source: World Bank (1996).

the poorest 20 per cent – have allowed governments to adopt policies pursuing efficiency and growth with limited social and political unrest. By contrast, Latin American countries such as Brazil, where the comparable income differential exceeds a factor of 32, run larger government budget deficits through financing redistributive programmes to the poor that help prevent political unrest while yielding to the demands of the rich to limit taxation (Fishlow, 1972; Menzel and Senghaas, 1987). In such circumstances the extreme inequality of incomes, stimulated according to some accounts by the greater degree of penetration by TNCs and their stimulus to service rather than manufacturing jobs (Evans and Timberlake, 1980; Rau and Roncek, 1987), limits government fiscal flexibility (Table 10.3).

The weight of population growth

Seventh, and finally, it seems that export-oriented industries can make and have already made an important contribution to the creation of new employment in certain small countries. However, it would probably be mistaken to believe that these industries can make anything other than a marginal contribution to employment in the periphery as a whole. As of 1980, the World Bank estimated the total number of direct jobs created by export industries in the LDCs as between 2 and 3 million, about 10 per cent of total industrial employment in these countries, or, in other terms, less than 0.5 per cent of the total labour force of some 850 million people (World Bank, 1979). Even considering multiplier effects, the total number of direct and indirect jobs created came to between 5 and 10 million, around 1 per cent of the total labour force. As this labour force is growing today at an annual rate of 2.2 per cent, the number of workers added every year to the total labour force is more than twice as large as the entire labour force employed in jobs created by export-oriented manufacturing. Across all LDCs, therefore, export industries cannot possibly absorb the additional labour force arriving each year or even reabsorb those unemployed due to cyclical shifts in demand for export products. Even restricting attention to eight NICs and potential NICs (Brazil, Mexico, Egypt, India, the Philippines, South Korea, Taiwan, and the former Yugoslavia), total jobs created during the 1960s by the export of manufactured goods represented only 3 per cent of total employment (Tyler, 1976). However, for small countries with high levels of industrial exports the picture is somewhat different. Thus in Taiwan in 1969 one job in six was created by manufactured exports and in South Korea in 1970 one job in 10 was created by exports of all kinds. These numbers doubled in the 1980s. Moreover, for the city states such as Hong Kong and Singapore, with their dynamic producer-service (finance, organization, research) sectors as well as large concentrations of export-orientated manufacturing industries, the figures are even higher. By 1990 in Taiwan, perhaps the most consistently successful NIC over the past 20 years, the 20 million population had a GNP per capita twice that of Greece or Portugal (US$8800), universal schooling and health care, and the world's largest reserves of foreign exchange. Taiwanese business was a major investor throughout Asia and contributing to the industrialization of coastal China even though 'officially' still on a war footing against the mainland (see the later section on China).

But it is not only the absence of large dependent peasant and urban populations that gives the small states of East Asia distinct advantages in relative job creation from export-orientated manufacturing. These countries have concentrated until recently on labour-intensive export industries, such as textiles, clothing and electrical and mechanical assemblage. These industries use a great deal of labour per unit of production and this labour is unskilled. They also employ disproportionately large numbers of women whose social position can be exploited to suppress wages and limit union organization. As Joekes (1987, p. 90) puts it: 'Women's lesser education and their expectation (born of past experience) of receiving little training make them apparently suited to unskilled occupations and, most importantly, prepared to stay at such unskilled jobs, however monotonous they may be.'

Moreover, these countries initially managed to limit capital-intensive production in order to maximize the return on their higher labour/capital ratios in export-oriented manufacturing (Krueger, 1978). This sets them apart from other NICs that have much lower capital/labour ratios in export industries and higher levels of capital-intensive production for domestic markets (e.g. Brazil). It also leads to an improvement in income distribution as well as employment creation, since the heavy reliance on unskilled labour implies that relatively more of the incomes generated by export industries will go to the poorest segments of the population and less to those classes rich in capital or technical skills. The peculiar historical development of the East Asian NICs in job creation by export industries, therefore, is not readily duplicated by larger countries (or some smaller ones e.g. Puerto Rico (Santiago, 1987)) in which incomes and wealth are divided unequally and in which powerful socio-political classes have a vested interest in maintaining the status quo.

THE GEOGRAPHY OF INDUSTRIALIZATION IN THE PERIPHERY

As stated earlier, during the late 1960s and 1970s a number of LDCs underwent a rapid process of industrialization, financed in part by the export of capital from the developed world. These countries, the NICs, were the ones where growth was fastest. But growth was not limited to them. What is most important is to grasp the dynamics of industrialization in the late 1960s and 1970s. The 1980s did not see the same dynamism, with the exception of coastal China (see pp. 356–9).

Trajectories of industrialization

Figure 10.1 illustrates one way of doing this. This shows the groupings of countries resulting from applying Sutcliffe's (1971) three 'tests' of industrialization: Test One, at least 25 per cent of GDP in industry; Test Two, at least 60 per cent of industrial output in manufacturing; Test Three, at least 10 per cent of the total population employed in industry. This last test measures the impact of the industrial sector on the population as a whole. Seven groupings and two paths to industrialization result from applying the three tests. The groupings are:

A   Fully industrialized countries that pass all three tests.
B   Countries which pass the first two tests but with limited 'penetration' into the population or the economy as a whole. These are semi-industrialized countries.
A/B Borderline cases (e.g. Greece).
C   Countries which pass the first and third tests. A large industrial sector (in mining or oil) affects the population widely, but manufacturing is weak.

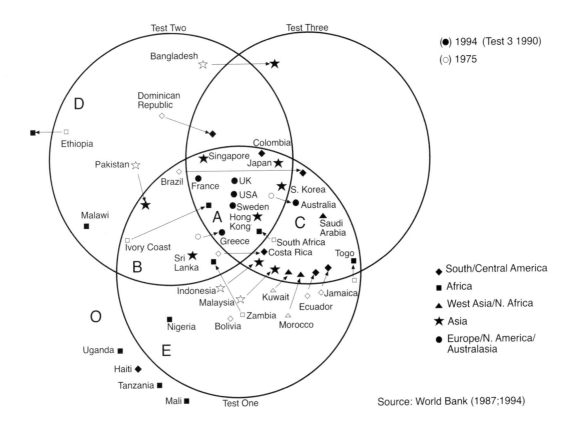

Figure 10.1
Three 'tests' of industrialization, 1975 and 1994

D   Countries which pass the second test only. A small industrial sector dominated by manufacturing.
E   Countries which pass the first test only. A substantial industrial, but non-manufacturing, sector has limited impacts on the population.
O   Other countries i.e. non-industrialized countries failing all three tests.

There are two possible paths to industrialization given this categorization. They are E → C → A, where a mining enclave expands to involve the total population and manufacturing develops later; and D → B → A where manufacturing leads to increases in industrial output and later towards a more industrial labour force. A third path, perhaps from labour-intensive rural industry towards manufacturing and an increasing industrial labour force, is possible but without any real-world examples.

If data for 1994 are used as well as data for 1975 a number of shifts are discernible. A number of D → B → A moves are clearly visible; most of the NICs (see below) fit this model (Fig. 10.1). The E → C → A path, characteristic in the past of the USA, Australia and South Africa, and the expected route, perhaps, of the members of OPEC and other mineral rich economies, is also of importance, but as

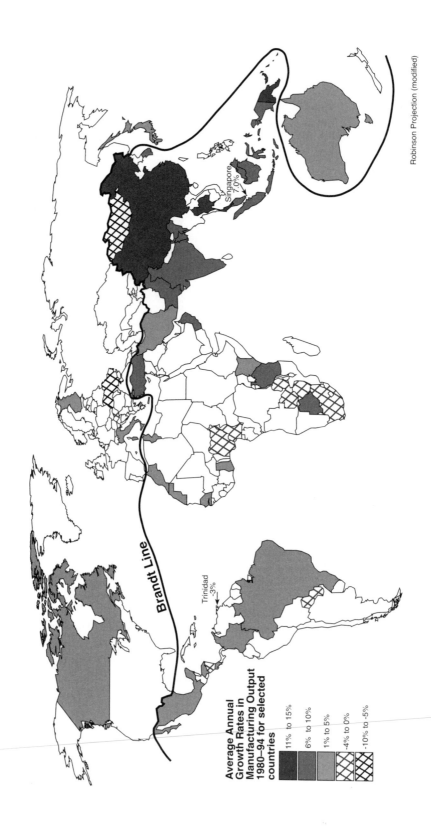

Average Annual
Growth Rates in
Manufacturing Output
1980–94 for selected
countries

11% to 15%
6% to 10%
1% to 5%
-4% to 0%
-10% to -5%

Brandt Line

Singapore
7.0%

Trinidad
-3%

Robinson Projection (modified)

Figure 10.2
The geography of growth in manufacturing output, 1980–1994

yet at an early stage. There are also a number of D → B → C moves indicating a weakening rather than strengthening of manufacturing; a new phenomenon since 1975 (Crow and Thomas, 1985). It is clear that, although there are some important examples of industrialization in the periphery, most LDCs are not industrializing, are industrializing quite slowly or are deindustrializing (losing manufacturing industry).

One problem with this type of analysis is that it provides minimum thresholds for defining industrialization but misses the point that after achieving these levels economies typically begin to lose jobs in manufacturing and experience an expansion of their producer and financial service industries. Manufacturing in itself is then no longer the driving force behind economic growth. This is what has happened in all of the so-called industrialized countries or DCs, where services are now a much more important component of their economies than is manufacturing. For example, in the USA in 1993 services accounted for 72 per cent of GDP, manufacturing had fallen to 23 per cent, and only 18 per cent of jobs were in manufacturing (*The Economist*, 20 February 1993).

The conclusion about the general lack of industrialization at the periphery is reinforced by focusing more directly on rates of growth in manufacturing output. Figure 10.2 shows that the high rates of growth in the period 1980–94 were relatively concentrated geographically. There is no 'exclusive' list of NICs, as different indicators lead to the inclusion of different countries (see Menzel and Senghaas, 1987). Also, there is some volatility over time. The 1980s was not a good decade for the NICs as a whole. In 1979 the OECD recognized 10 countries as NICs (Brazil, Mexico, South Korea, Taiwan, Hong Kong, Singapore, Spain, Portugal, Greece and the then Yugoslavia) on the grounds of 'fast growth of the level and share of industrial employment, an enlargement of export market shares in manufactures, and a rapid relative reduction in the per capita income gap separating them from the advanced industrial countries' (OECD, 1979). In Figure 10.2, only two of these NICs have growth rates exceeding 6 per cent per year (Singapore and South Korea), one is between 2 and 3 per cent (Mexico), one is between 1 and 2 per cent (Brazil), one is −1.1 per cent (Greece) and five are unknown. Of the other countries with high growth rates (5 per cent or more), some are major oil exporters (Saudi Arabia, Indonesia) or starting from tiny industrial sectors (Oman, Botswana). Some are 'new' NICs (e.g. China, Thailand, India); at least for the time being and if their low incomes per capita are ignored.

Export-processing zones

The NICs such as Taiwan, South Korea, Mexico and, now, China, have been especially active in export-oriented industrialization (Fig. 10.3). As we have noted previously, some of this industry has been attracted to and is concentrated geographically in 'export-processing zones' (EPZs) or 'free trade zones' (also see Chapter 2). These are limited areas in which special advantages accrue to investors. These include duty-free entry of goods for assembly, limited restrictions on profit repatriation, lower taxation, reduced pollution controls, and constraints on labour organization (strikes banned, etc.). The intention of EPZs is to attract local and

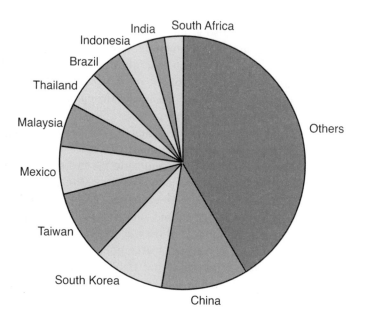

Figure 10.3

The share of certain NICs in total LDC exports, 1994, by value

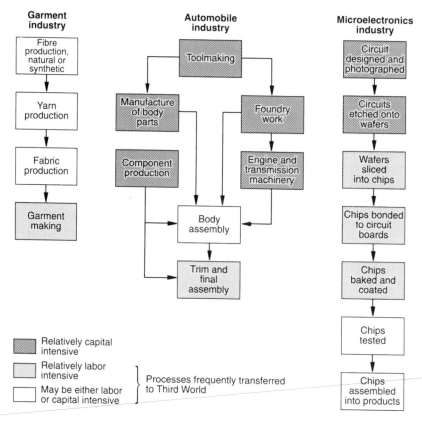

Figure 10.4

Labour processes in three manufacturing industries

Source: Crow and Thomas (1985), p. 49

international capital to set up factories oriented to export production. By 1984 there were 68 zones in 40 LDCs (Thrift, 1986). As of 1987 most of the EPZs were still clustered around major markets as they had been in 1979: North America,

Western Europe and Japan. South Korea, Taiwan, Mexico, Malaysia, Haiti and Brazil have the largest employment in EPZs. EPZ production is concentrated in textiles and clothing, microelectronic assembly, and the assembly of cars and bicycles. More specifically, it is the labour-intensive stages of production that are characteristic of the EPZs (Fig. 10.4).

Much of the industry in EPZs is owned by TNCs and most of the employment is in microelectronics and textiles and clothing. In Mexico in 1978, 60 per cent of the **maquiladoras**, the assembly plants located in EPZs along the US border since the programme of tax-breaks for American business began in 1965, were engaged in electrical and electronic assembly and 30 per cent were involved in textiles and clothing (Hansen, 1981). In the Asian EPZs, over 60 per cent of the employment is in electronics with the clothing and footwear industries second. The build-up in electronics has been especially marked since the early 1970s (Scott, 1987). The total amount of employment in EPZs, however, is relatively small. In Mexico *maquiladora* employment in 1980 was only around 110 000, although, by 1992 it had grown to 500 000. There are perhaps 700 000 workers directly employed in EPZs in Asia. Most of the workers are unmarried women between the ages of 17 and 23. About 85 per cent of employment in Mexican EPZs is of such young women earning around 40 per cent of the hourly American wage rate.

341

maquiladoras
Literally, 'mills' in Spanish. These are export-processing zones in Mexico, mainly near the US border

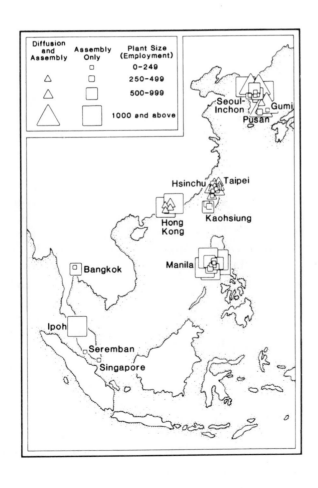

Figure 10.5

Locally-owned semiconductor plants in Southeast Asia, 1985

Source: Scott (1987), Fig. 3, p. 50.

Within Asia young women account for 88 per cent of zone employment in Sri Lanka, 85 per cent in Taiwan and Malaysia, 75 per cent in South Korea, and 74 per cent in the Philippines (Morello, 1983). Young women are preferred as workers because their wages tend to be lower than men's and because they are considered 'nimble fingered' and 'more able to cope with repetitive work' (Armstrong and McGee, 1986). Certainly, very little training is required and wages are very low. Often an exploitative trainee system is used in which 'trainees' are paid only 60 per cent of the local minimum wage and are repeatedly fired and rehired so as to obtain a permanent 40 per cent reduction in the wage bill (Thrift, 1986).

Many of the EPZs also do not appear to have had major stimulative effects outside their boundaries. Links to local economies are generally limited especially in Latin America (Thrift, 1986; South, 1990). In the Mexican case, the recent North American Free Trade Agreement (NAFTA) with the United States and Canada may well reduce the need for the special border EPZs as the whole of Mexico becomes a giant EPZ! However, Scott (1987) shows that in some Asian cases the activities of TNCs have led to the growth of both 'diffusion facilities' owned by local firms engaged in higher-level (not solely assembly) operations and locally owned sub-contract assembly houses. The former are concentrated in South Korea, Taiwan and Hong Kong, the latter are found in Thailand, Malaysia, Singapore, the Philippines, Hong Kong, Taiwan and South Korea (Fig. 10.5).

Barbie: American icon and global product

The famous (and impossible) physique of the Barbie doll says 'Made in America' but the box it comes in says 'Made in China'. But tracing the doll's production path raises questions about how its place of origin can be identified and how the globalization of production ties together disparate locations in core and periphery. Barbie is made from plastic injected into moulds at two factories in south China adjacent to Hong Kong, one in Indonesia and one in Malaysia. Barbie has never been made in the USA. The first doll was produced in Japan in 1959. As costs rose in Japan, production was moved to other sites in Asia. At one time the producer, Mattel, had Barbie factories in Taiwan, Hong Kong and the Philippines. In 1988, after a strike, Mattel closed its two Philippine factories with a loss of 4000 jobs. The plastic is made from ethylene, refined from Saudi-Arabian oil, that is turned into pellets by a firm in Taiwan. Barbie's nylon hair comes from Japan. Her cardboard packaging is made in the United States. The manufacturing and packaging are managed from Hong Kong. The production story begins, however, in the 'commodity management centre' of the Mattel Corporation, one of the world's largest toy companies, in El Segundo, California, near the Los Angeles International Airport. That is where information about commodity prices and wage rates is used to decide upon the best locations to buy the plastic resins, the cloth, the paper and other materials and bring them together at a final point of assembly.

At one time, Japan and Taiwan were the main toy makers to the world economy. As their economies diversified into more capital-intensive production, they became the suppliers of the plastics that hitherto had come from the

USA and Western Europe. Production shifted to lower-wage sites, such as those in China. Making Barbie is extremely labour-intensive. Workers must operate plastic moulds, sew clothing, and paint the details on the dolls. A typical Barbie requires 15 separate paint stations. These tasks cannot be performed by machines. Thus, the two Barbie plants in China employ about 11000 workers, mainly unmarried women between 18 and 23 from peasant families in poor regions of interior China brought to work at the factories for 2–5 years.

So, Barbie is made in China. In the trade ledgers – where country trade deficits and surpluses are defined – Barbie is one of its exports. In fact, a whole number of firms in different countries contributed to its production and reaped their profits from the final product. The Chinese firms and workers obtain only about 35 cents out of the US$2 export value placed on each Barbie when she leaves Hong Kong. Barbie retails in the USA for around US$9.99. In 1995, Barbie accounted for US$1.4 billion in sales for Mattel. She is sold in 140 countries at the rate of two dolls per second. Over 40 per cent of the dolls are sold in Europe and Japan (Tempest, 1996).

Transnational investment patterns

Direct investment by American TNCs has shifted dramatically over time. In 1969, most investment in electronic assembly was in Hong Kong, South Korea, Taiwan and Singapore. By 1983 the Philippines and Malaysia had become relatively more important. Direct foreign investment, therefore, is extremely footloose and sensitive to marginal shifts in wage rates, local fiscal conditions, and political 'instability'. To spread their risks, many of the 30 US TNCs engaged in electronic assembly operations in Asia have plants in a number of different locations in different countries (Table 10.4). Motorola, for example, has seven different manufacturing facilities in six Asian countries (including Japan).

In textiles and clothing there has also been a tendency for the most labour-intensive activities to move away from Hong Kong, Taiwan and South Korea to other parts of Asia. China, Bangladesh, Sri Lanka, and Indonesia have become especially important. 'The clothing industry uses little capital and is very mobile. All you need is a shed, some sewing machines, and lots of cheap nimble fingers' (Economist, 1987, p. 67). Partly this has been a product-cycle effect, as wage rates for unskilled labour have increased in the old NICs. Perhaps of greatest importance in this geographical shift, however, has been the imposition of quotas by the USA and European countries on imports from established producers. This has set businessmen from Hong Kong and South Korea, the major figures in this industry, searching for countries with higher quotas and low production. For example, Hong Kong is allowed only a 0.6 per cent annual increase in its shirt exports to the EC, while Sri Lanka is allowed 7 per cent. China has been allowed a large expansion simply because the rich countries are keen to expand their exports of capital goods (e.g. steel, weapons) to such a large market and do not want to invite retaliatory action on their exports in the face of a clothing quota decrease.

Table 10.4: Corporate affiliations of US-owned assembly plants in Southeast Asia, 1985

	Hong Kong	Indonesia	Korea	Malaysia	Philippines	Singapore	Taiwan	Thailand	Sub-total
Advanced Micro Devices				1	1	1		1[1]	3
Ampex							1		1
Analog Devices					1		1		1
ATT						1		1	2
Commodore	1								1
Data General	1							1	1
Fairchild	1	1	1		1	1			5
General Electric						1			1
General Instrument					1		1		2
Gould[2]			2		1				3
GTE							1		1
Harris Semiconductor				1					1
Hewlett Packard				1		1			2
Intel				1	1				2
Intersil						1			1
Microsemiconductor	1								1
Monolithic Memories				1					1
Mostek				1					1
Motorola	1		1	2	1		1		6
National Semiconductor	1	1		2	1	2		1	8
Raytheon					1				1
RCA					1		1		2
Signetics			1					1	2
Siliconix	1						1		2
Silicon Systems						1			1
Sprague	1				1		1		3
Teledyne	1								1
Texas Instruments					1	2		1	5
Western Digital				1					1
Zilog						1			1
Sub-total	8	2	5	14	11	11	8	4	63

Notes:
[1] Plant under construction
[2] American Microsystems and Korean Microsystems.
Source: Scott (1987), Table 6, p. 151.

During the 1980s Japanese sources displaced American and European ones as the major investors in the countries of Southeast Asia. Even as Japanese foreign direct investment in the USA and Europe followed an incredible growth rate during 1985–90, Japan's stock of investment in Asia, already higher than that of the USA or Europe in 1985, grew faster than either of the other two. This has led to talk of an East Asian economic bloc forming under Japanese dominance. In fact the evidence suggests that the region has much less of an intra-regional bias in trade and direct investment flows than does Western Europe. The Japanese yen has become a more important currency in trade and central bank holdings in Southeast Asia than previously. But as of 1988 more of Japan's trade with the EC was denominated in yen than was its trade with Southeast Asia (Frankel, 1991; Kawai, 1990). As yet East Asia in general and Southeast Asia in particular, including such countries as Malaysia, Indonesia, Thailand and the Philippines, do not form an incipient trading bloc diverting trade to other destinations through some deliberate strategy of economic regionalism.

However, there is undoubtedly a degree to which regional proximity to a historically dominant trading and investment partner underpins the success of the old NICs and the emerging NICs close by. Most of the successful new export-oriented economies are adjacent to other ones and at least one important already-industrialized economy. But though necessary, this is not a sufficient condition. Other factors such as underlying productivity conditions, political regime strength and stability, adaptability (rather than factor endowments or resources), and a mix of import-substitution and export-orientated industrialization are among the most vital ingredients (Gibson and Ward, 1992). The economies of East Asia, including the emerging NICs of Southeast Asia, have major advantages with respect to all of these factors. It will be interesting to see if the formerly centrally planned economies of Eastern Europe can gain competitive advantage in like manner from their proximity to Western Europe and reasonably well-developed infrastructures and well-trained labour forces.

Regional linkages and industrial evolution

It should be emphasized, however, that although the EPZs and the geographical division of labour within East Asia are symptomatic of the importance of 'off-shore production' for final markets in the developed countries, it would be mistaken to see peripheral industrialization solely in these terms. For one thing, as Scott (1987), for example, shows, local markets for electronic components and semiconductor devices have grown rapidly in East Asia over the past decade. Even US-owned branch plants now ship about 18 per cent of their production to consumers in East Asia. This has encouraged the establishment of marketing, sales and after-sales service facilities in the region, especially in Hong Kong and Singapore. These two centres now function as nodes in a global system of producer services located in major world cities (Daniels, 1985; Cohen, 1981). They have also become important global banking and financial centres as a result of their coordinating roles in East Asia manufacturing industry (Forbes and Thrift, 1987). One factor facilitating this has been the international networks between 'ethnic Chinese' groups (often family or kinship based) in East Asia and North America (Goldberg, 1987).

Also of great importance, however, local firms have developed a wide range of industries oriented towards producing final products for exports. Since the early 1960s, for example, South Korea has pursued an aggressive export-oriented industrial policy. In the early 1960s, familiarity with manufacturing acquired during the earlier import-substitution period was used to develop a number of 'infant' export industries by obtaining manufacturing licences, loan capital and imported business 'know how' (Table 10.5). These were mainly labour-intensive activities. But even with low wages it was not until the late 1960s that productivity and quality were high enough to make these industries, textiles, clothing and footwear, internationally competitive. The South Korean government subsidized this process by using currency depreciations to boost exports, making tax concessions and providing cheap loans.

During a second stage, 1966–71, other 'infant' industries were encouraged as the initial group achieved international competitiveness. These were more technologically advanced industries such as electronic assembly and shipbuilding. In shipbuilding, production went from 25 000 gross tons in 1970 to 996 000 gross

Table 10.5: Stages in South Korea's export-oriented industrial development

	1961–66	1966–71	1971–76	1976–81	1981
Infant industries	Textiles Clothing Footwear	Electronic assemblies Shipbuilding Fertilizers Steel	Motor vehicle assembly Consumer electronics Special steels Precision goods (watches, cameras) Turn-key plant building Metal products	Automative components Machine tools Machinery assembly Simple instruments Assembly of heavy electrical machinery Semiconductors	
Industries becoming competitive		Textiles Clothing Footwear	Metal products Electronic assembles Shipbuilding Fertilizers Steel	Motor vehicle assembly Consumer electronics Special steels Precision goods Turn-key plant building Metal products	Automotive components Machine tools Machinery assembly Simple instruments Assembly of heavy electric machinery Semiconductors
Self-sustaining industries			Textiles Clothing Footwear	Electronic assemblies Shipbuilding Fertilizers Steel	Motor vehicle assembly Consumer electronics Special steels Precision goods Turn-key plant building Metal products

Source: Linge and Hamilton (1981), Table 1.9, p. 33.

tons in 1975; the ships built were also increasingly large and simple (such as supertankers and bulk carriers) with greater potential for automated production. By 1976 the Korean shipbuilding industry was globally competitive (Linge and Hamilton, 1981).

In the early 1970s a third 'wave' of industries was in the process of creation. For instance, the automobile industry, which began in South Korea in 1967, produced 83 000 units in 1977 and 3.5 million units in 1985. By the mid-1980s, therefore, South Korea had acquired a wide range of internationally competitive and self-sustaining industries. And, despite the increasing technological sophistication of each wave of innovation, considerable emphasis is still placed on the original labour-intensive industries; although some of these are increasingly decentralized to other Asian locations. The South Korean policy of widening its manufacturing base is the most successful model of the kind of development policy being pursued by all the NICs. Peripheral industrialization, therefore, is not just the off-shore processing or assembly work for TNCs that a single-minded focus on EPZs would imply (Rohwer, 1993).

Agglomeration and new industrial complexes

Peripheral industrialization is organized at the intra-national and urban levels as well. Coastal and metropolitan areas have been favoured locations, because of infrastructural advantages and ease of external access. Even in countries with little export-orientated industry this pattern is evident. In Nigeria, for example, nearly 65 per cent of the country's industrial employment is concentrated in Lagos and five other coastal states, leaving the remaining 35 per cent to the 13 hinterland states (Ayeni, 1981; Abumere, 1982). In East Asia and Latin America much of the new manufacturing industry of the past 30 years is found in the major metropolitan areas. Both foreign and indigenous investment tends to be attracted by the amenities, basic infrastructure and political access characteristic of the larger urban areas (see 'Production Networks and Regional Motors' p. 106 for why this is the case). When urban growth is already concentrated in a primate city, such as in Bangkok in Thailand, recent growth has tended to reinforce primacy. This seems to be especially true of foreign direct investment (Forbes, 1986).

The case of Japanese direct investment in East Asia is illustrative. It is highly concentrated in the national capitals and their immediate vicinity (Figure 10.6). This is the case whether investment is measured by number of firms, employment or capital, although capital concentration is most pronounced (Table 10.6). Metropolitan concentration ranges from nearly 100 per cent in Thailand to 62 per cent in Malaysia with respect to capital; and from 99 per cent in Thailand to around 30 per cent in Malaysia with respect to firms and employment. In Taiwan and South Korea there is a relatively high proportion of Japanese investment in regional centres. What distinguishes Malaysia is that even its rural periphery has some share of Japanese direct investment (Fuchs and Pernia, 1987). So there is some variation even if metropolitan concentration is the norm.

In a number of urban areas there are also incipient industrial complexes redolent of Silicon Valley in California and other high-technology complexes in developed countries; what Amin and Thrift (1992) call 'neo-Marshallian [industrial district] nodes in global networks' (also see Chapters 2, 3 and 7). They are made up of both foreign-owned (US and Japanese) and locally owned assembly plants and a surrounding constellation of linked activities. Hong Kong, Manila, Seoul, Singapore, Penang, and Taipei all have such complexes. Scott (1987) uses the Manila case to illustrate how a complex can arise in a context of low general economic

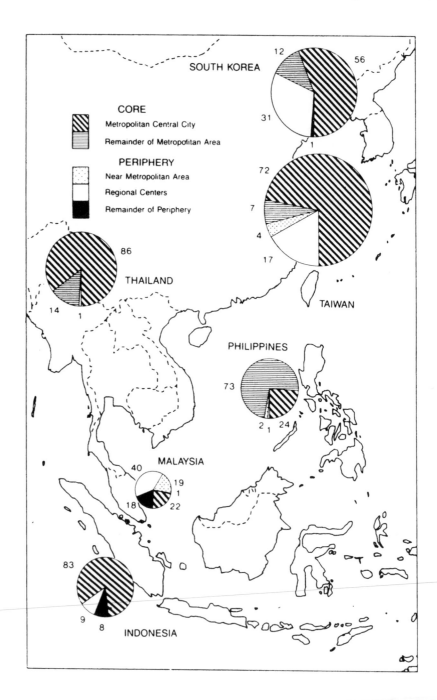

Figure 10.6

Location of employees in Japanese firms in selected Asian countries, percentages in different locations

Source: Fuchs and Pernia (1987), p. 99

development. The Manila complex consists of a core of nine major US-owned semiconductor branch plants. These are served by 14 locally owned sub-contract assembly houses and three specialized capital-intensive 'test and burn' facilities. Lastly, the Manila complex has a number of specialized tool and die and metal shops serving the semiconductor industry. Some of these are 'captive' to or totally dependent on the US-owned assembly plants but most are independent, local operations. All of these units cluster together to minimize transactional costs and gain joint access to the Manila International Airport. They are also at the centre of a metropolitan labour market that provides production workers who have considerable experience with the norms and rhythms of assembly work and a pool of technicians and engineers with the necessary skills (Scott, 1987).

Attempts have been made in many LDCs to decentralize manufacturing activities to regional growth centres (see Chapter 7 for a discussion of the models of regional change upon which these ideas are based). For example, beginning in 1967 the government of Indonesia established a set of tax incentives for 'priority sectors' that

Table 10.6: Spatial distribution of Japanese direct investment by size: Pacific Asia, 1987

Size of investment and location	Taiwan	South Korea	Malaysia	Philippines	Thailand	Indonesia
Number of firms (%)						
Metro centre	53	51	37	44	81	83
Rest of metro	9	14	1	50	18	0
Metro periphery	7	1	25	3	0	1
Regional centre	30	33	21	3	1	8
Rural periphery	1	1	16	0	0	8
All locations	100	100	100	100	100	100
Number of employees (%)						
Metro centre	72	56	22	24	86	83
Rest of metro	7	12	1	73	14	0
Metro periphery	4	0	10	2	0	0
Regional centres	17	31	40	1	1	9
Rural periphery	0	1	18	0	0	8
All locations	100	100	100	100	100	100
Estimate capital (%)						
Metro centre	81	90	61	24	99	98
Rest of metro	4	3	1	76	1	0
Metro periphery	4	0	12	0	0	0
Regional centres	11	7	23	0	0	1
Rural periphery	0	0	4	0	0	1
All locations	100	100	100	100	100	100
Firms (N)	175	154	126	97	184	158
Employees (N)	114	64	25	45	51	44
Estimated capital (US$ million)	592.3	37.8	7.2	3.1	1.4	129.3

Source: Fuchs and Pernia (1987), Table 5.3, p. 98.

were to be located in 11 industrial zones. The idea was to decentralize suitable industries away from the island of Java in general and the political capital of Jakarta in particular. The policy has failed. At a time when Indonesia's MVA growth rate was the highest in the world (1973–81) – 14.6 per cent per annum – more than half of both domestic and foreign investment was in Java (56.8 per cent of foreign and 64.6 per cent of domestic investment). Only resource-based industries have pulled some investment away from the centre even in the presence of significant tax incentives (Suhartono, 1987). Similar failures are reported from Latin America (Gore, 1984), Africa (Moudoud, 1986), and other parts of Asia (Lavrov and Sdasnyk, 1982).

Paralleling the concentration of manufacturing in coastal regions and metropolitan areas has been the growth of a proportionate 'service' sector. Within this sector there are essentially two groups: a minority employed in commercial and bureaucratic activities and a huge 'army' of unemployed and underemployed street vendors, day labourers, domestic workers, and other low-wage workers. While one quarter or more of the labour force in the NICs and 'middle income' LDCs is involved in industrial activity, over one-third is engaged in 'services'. In poorer countries the proportions are almost even. This suggests that rather than resolving the problem of urban unemployment and underemployment, recent industrialization has been unable to employ rural–urban migrants at a rate commensurate with their rate of movement from the countryside (Petras, 1984).

The only countries in which the labour force categorized in services has declined relative to that employed in manufacturing are the so-called socialist or collectivist ones. Cases of collectivist (or state-socialist) industrialization in the periphery (Vietnam, North Korea, etc.) indicate a distinctive pattern relative to that of industrialization based on an export-orientation of the advanced capitalist countries. Above all, the pattern probably reflects restrictions placed on rural–urban migration and government import-substitution policies. It is important to note, however, that governments in some of the collectivist countries, especially China, have recently begun to encourage the growth of the service sector (Shunzan, 1987) and to engage in export-orientated industrialization through the creation of EPZs and joint ventures with TNCs (Leung and Chin, 1983; Wong, 1987).

The possibility of industrialization throughout the periphery under present global conditions is limited. The process of industrial growth is not a linear diffusion process spreading throughout the globe. Indeed, there are a number of cases not only of a lack of any growth at all (largely, but not entirely, in sub-Saharan Africa) but also of stagnation and even (as in Argentina) of deindustrialization. The causes of stagnation are numerous: for example the migration of labour from Yemen to the oil-rich states of the Persian Gulf; the civil war during the late 1970s in Nicaragua; the pursuit of extreme 'free-market' monetarist policies by governments in Argentina. Consistent industrialization in the face of global downturns and increasingly competitive world markets seems to require at a minimum, in Petras's (1984, p. 199) words, 'a cohesive industrializing class linked to a coherent policy, promoting an internal market and selective insertion in the international market'.

PROFILES OF PERIPHERAL INDUSTRIALIZATION

Brazil

Brazil was incorporated into the world economy as an exporter of primary commodities, agricultural exports such as coffee and industrial raw materials such as iron ore. Today the country is characterized by an immense land area; 180 years of political independence; a large population (150 million in 1990); and a large and diversified industrial economy that has grown mainly since the Second World War.

Several stages in Brazil's industrialization can be distinguished (Cavalcanti *et al.* 1981; Baer *et al.*, 1987):

1. Until the Second World War there was an emphasis on import substitution of non-durable goods for the domestic market and industrial raw materials processing for export.
2. During the Second World War Brazil accumulated large foreign exchange reserves by exporting industrial raw materials without a corresponding increase in imports.
3. After the war (1950–62) these reserves combined with pent-up demand led to a substantial increase in the import of consumer durables. The rapid rise in imports depleted exchange reserves, leading the government to place restrictions on imports. The import restrictions coincided with the global expansion of TNCs which, attracted by the potential of Brazil's domestic market, began to establish branch plants in order to avoid limitations on their sales. A major example would be the large automobile manufacturers (General Motors, Volkswagen, etc.) which all established subsidiaries in Brazil.
4. Although the markets for Brazilian-produced consumer durables increased beyond the national boundary, by the late 1960s the Brazilian government was keen to expand them further because of its ambitious industrialization plans and, later, because of the need to pay for vastly more expensive oil imports. This coincided with the massive restructuring of TNC operations (especially in the automobile industry response to Japanese competition). In Brazil, government underwriting of new plant and technology encouraged the TNCs to become more export-oriented. For example, Volkswagen embarked on the manufacture and export of specifically 'developing country' models. In other cases, the strategy of international production has led to world-wide sourcing of components such that Brazil now exports parts for assembly elsewhere: Brazilian four-cylinder engines are shipped by Ford to Canada for installation in vehicles that are then sold in the United States (Mericle, 1984); Fiat uses Brazilian engines in its 127 model, and Volkswagen exports Brazilian engines and transmissions to its German factories (Jenkins, 1987). Despite these efforts, exports as of 1996 remained at the low level of 7 per cent of GDP, lower even than India (which has had until recently a highly protectionist trade policy) and only 25 per cent of Mexico's level.
5. During the 1980s rapid industrialization based upon the mineral and hydroelectric resources of the Eastern Amazon basin provided a further

element to the evolution of the Brazilian economy. One large-scale development scheme sponsored by the Brazilian government, the Greater Carajas Programme, has been especially important. The cornerstone of this scheme is the so-called Iron Ore project involving a large iron mine, a 780-kilometre railway and a new deepwater port. The Carajas mine contains the world's largest reserves of high-grade iron ore. But the scheme is envisaged as a fully fledged development programme; it is not simply a mining enterprise. Other industries, using the vast mineral wealth and timber resources of the region have been attracted. The Brazilian government's fiscal incentives policy allows investment credits and tax-breaks to both domestic and foreign firms.

Within this changing political-economic context there have been two major spurts of economic growth: the first between 1948 and 1961, when average growth was around 7 per cent per annum; and the second between 1968 and 1981, when growth averaged 8.9 per cent per annum. In the late 1980s growth rates went up again after three successive years of decline only to fall again between 1989 and 1992 (Rohwer, 1991) with moderate expansion again in the mid-1990s (*The Economist*, 17 May 1997). This boom-and-bust picture represents in modern form the old European image of Brazil as a volatile and unstable place that originated in the sixteenth century. Sugar boomed first in the Northeast but it dwindled as an export earner when faced with competition from the Caribbean. Gold came next in the southern state of Minas Gerais in the eighteenth century. As the gold ran out, coffee took over. By the mid-nineteenth century coffee accounted for half of Brazil's export earnings.

The difference today is that Brazil is a major industrial power with an infra-structure and internal markets on a par with most DCs and an ability to export (but not an export performance) that compares to the East Asian NICs. Brazil is the fifth largest country in area in the world. It is sixth largest in population. With a GDP of US$749 billion in 1996, Brazil is the eighth largest market economy. Among LDCs, Brazil produces more than India. East Asia's NICs (e.g. Taiwan, South Korea) are export-oriented states with a total GDP that is a fraction of what Brazil produces for its internal market alone. Mexico is the only challenger to Brazil among LDCs in size of GDP – but then only with a GDP of around US$335 billion. More specifically, Brazil's steel production is the seventh largest in the world, about one-quarter that of the United States; Brazil is the world's ninth largest producer of cars; the second largest producer of iron ore and the eighth largest producer of aluminium. In other words, by virtue of its position in the international production league table, Brazil is an industrial colossus.

But Brazilian industrialization has a number of features associated with it that severely tarnish the image of miraculous growth. In the first place, the current productive structure of the Brazilian economy reflects an incredibly inegalitarian consumption structure. In a simulation exercise Locatelli (1985) has shown that a more egalitarian distribution of income (similar to that of Britain) would result in a 16 per cent growth in industrial employment. This would happen because the purchasing power of poorer groups would increase the demand for goods with greater labour-intensity in production (also see Baer, 1983; Baer *et al.*, 1987). However, labour policies adopted in the late 1960s to make industries more internationally competitive have reduced worker incomes relative to productivity. The share of wages in final prices has been in steady decline since the 1960s. This has led to large-scale labour unrest. In the motor vehicle industry, for example, workers 'face unstable employment, harsh supervision, limited promotion prospects and intensive work' (Humphrey, 1984 p. 109). Outside of manufacturing industry working and living conditions are even more precarious and marginal.

Second, the clientelistic nature of Brazilian politics, in the form of bloated government payrolls, pay-offs to gain influence and industries located on the basis of political influence, seems to have retarded industrialization in recent years. Between 1940 and 1980 Brazil's total output expanded every year but one (1942). Since 1980 the political direction of the economy, privileging on geopolitical grounds the expansion of industry in Amazonia and using vast government resources to do so, has retarded industrial growth in the country as a whole. Failure to reduce public spending has encouraged both inflation and public-sector borrowing. With the signing of a free-trade agreement with Argentina, Paraguay and Uruguay in 1991 (Chile joined as an associate member in 1996), the Brazilian government did throw in its lot with an organization (Mercosur) devoted to free trade and regional cooperation. As yet, however, Brazil still tends towards imposing unilateral trade barriers whenever domestic politics indicate that a Brazilian-based firm will suffer from less fettered trade (*The Economist*, 29 June 1996; Reid, 1996). The most positive change has been the dramatic reduction in inflation effected by the administration of President Cardoso since 1994. At 10 per cent in 1996, this is still high by the standards of North America or Western Europe, but is the lowest rate since the 1950s (*The Economist*, 17 May 1997). Price stability helps business planning by encouraging firms to extend their time horizons and the poor by reducing the resources need to hedge against inflationary spirals.

Third, the Brazilian economy is dominated by TNCs to a much greater extent than are the economies of the other NICs. Brazil has one of the largest stocks of foreign direct investment of any country in the world: US$32 billion in 1991. Multinational penetration has affected most industrial sectors but it is especially concentrated in the more dynamic ones. There is some evidence that the falling cost of labour in the period 1968–81 and the restructuring of TNCs were particular incentives for expanded TNC operations (Cunningham, 1986). This has led to considerable debate over the future 'survival' of Brazilian private enterprise. Yet there is evidence that Brazilian private concerns can be successful both alone and in 'triple alliance' with multinational and public sector interests (Evans, 1979; Baer *et al.*, 1987). What is clear is that the Brazilian state played a major role in encouraging TNC investment, especially in the 1970s (Cunningham, 1986). The 1990s have seen a revival of foreign direct investment, particularly in the form of mergers and takeovers of Brazilian-owned firms by TNCs (*The Economist*, 9 November 1996).

Fourth, Brazil is massively indebted to multinational banks. In the aftermath of the 1973–74 oil price shock, Brazil's international indebtedness increased markedly. Between 1975 and 1980 external debt grew at an annual rate of 18.4 per cent (Tyler, 1986). The borrowing this reflected, though it went primarily to finance balance of payments deficits, was also channelled into major industrial and infrastructure projects, especially in the Amazon Basin and elsewhere in the interior. Contracting such indebtedness was possible because of the high dollar liquidity of the large US and European banks during the 1975 recession releasing the massive petrodollar deposits from oil exporters for global lending. Much of the borrowing involved variable interest rates linked to the US prime interest rate, which was low in the mid-1970s but much higher by the end of the decade. Tyler (1986) was optimistic that the Brazilian debt load could be rescheduled (because of its sheer size) without lasting damage to Brazil's economy. Others (e.g. Cunningham, 1986), however, suggest that the 1970s was a decade that provided a unique 'one-time' opportunity for the industrial restructuring that did take place but that it cannot probably continue. 'The superimposition of multinational bank finance upon an existing nexus of foreign direct investment established by other multinational corporations (especially in manufacturing)' (Cunningham, 1986, p. 61) may not return, especially with so many other countries (especially in Eastern Europe and the former Soviet Union) vying for loans and investment.

Fifth, and finally, industrialization in Brazil is extremely concentrated geographically. In 1977 70.9 per cent of all factories were in the Southeast 'core' region with 58.4 per cent of these in the São Paulo and Rio de Janeiro metropolitan areas (Cavalcanti et al., 1981). The poorest region, the Northeast, with about 29 per cent of the population has only 9.6 per cent of the factories but 50 per cent of the country's poor (Thomas, 1987). Though projects for 'opening up' the Amazon Basin to industrialization have taken huge public investments, they have not as yet significantly disturbed the historic regional pattern of development even as they have led to massive disruption of the native ecosystem (Anderson, 1990). Industrialization has thus deepened historic spatial inequalities in living standards. This has been especially true since the late 1960s as industries directed to both export and the domestic market have conspired to encourage industrial development in the coastal–metropolitan regions of the South and Southeast.

Cuba

Cuba represents one of the few remaining centrally planned economies. Heavily dependent upon the United States from the early 1900s until 1959, after the Revolution of that year it moved increasingly close in political allegiance to the former Soviet Union. This involved exchanging a free-wheeling economy based on tourism, gambling and sugar for an economy modelled closely on that of the former Soviet Union and its European satellites. One thing is clear, at the same time in the 1980s that Latin America in general had been experiencing negative or low growth rates, public austerity programmes and declining industrial wage rates, Cuba increased spending on social welfare, widened the availability of consumer goods and maintained robust growth rates. In 1972 the future prospect had looked bleaker. The terms of trade for sugar exports (Cuba's major primary commodity) had deteriorated badly and the country's hard currency debt of US$3 billion was roughly three and a half times the value of exports that generated hard currencies. The constraints facing Cuba seemed remarkably similar to those of the rest of Latin America. Despite its socialist political system and incorporation into the Soviet trade bloc (CMEA) Cuba still seemed to be subject to the vagaries of the world economy and dependent upon capital from the 'capitalist world' (Turits, 1987).

The question raised by Cuba's surprising resilience – in the late 1980s it reduced its total debt as well as met its rescheduled debt repayments on time – was whether this was due to the nature of its political economy, increased dependence on Soviet aid and oil provided at subsidized prices, or some of each. From one point of view, Cuba remained in an almost classic mode of dependency, having exchanged one dominant partner (the United States) after the 1959 Revolution for another (the Soviet Union). Certainly, Cuba's choice of a development strategy based on sugar maintained a high level of structural dependence and a high degree of vulnerability. Yet there was only limited evidence to suggest that dependence on the Soviet Union shaped and limited the Cuban economy in any way analogous to Cuba's former dependence on the USA. For example, Cuba has not been subject to long-term capital outflows, foreign ownership of the means of production nor increasing economic inequality; all features of pre-revolutionary Cuba and much of Latin America today. But Cuba was fortunate that the Soviet Union took so much of its sugar and provided considerable aid. With the collapse of Communist party dominance in the Soviet Union and the disintegration of CMEA since 1991, Cuba has faced increasing economic difficulty. This would suggest that dependence on the Soviet Union was greater than had been thought.

Though sugar exports remained the engine for surplus generation and foreign exchange, the Cuban economy from 1970 onwards was increasingly driven by a growing industrial sector. In the 1970s import substitution, sugar-processing and agricultural machinery industries were built up by so-called 'debt-led' growth. By 1981 34.9 per cent of investment was in industry compared to 16.7 per cent in 1966 (Brundenius, 1984). A radical reallocation of labour from agriculture to industry and services followed. The percentage contribution of agriculture to GDP declined from 18.1 per cent in 1961 to 12.9 per cent in 1981 (Brundenius, 1984; Zimbalist, 1987).

Debt-led growth was possible for Cuba because, even though blockaded and ostracized by the United States, credit and capital were increasingly available in the 1970s and this coincided with a dramatic climb in world sugar prices, which improved Cuba's credit rating. Western imports increased fourfold in the years 1973–75. Trade with capitalist countries reached over 40 per cent of total trade in 1975, compared to 20 per cent in 1967, and less than 14 per cent as of 1983 (Turits, 1987). The four years 1975–78 culminated in a trade deficit with capitalist countries of 2.1 billion pesos, compensated for by increased foreign borrowing.

The burden of debt service impelled renewed attention to import reduction and export production. But the investment of the 1970s had paid off in terms of new and growing industries such as capital goods, consumer durables, chemicals, medicines (Cuba produces 83 per cent of its needs), electronics, computers and steel. For example, where engineering and capital goods industries accounted for 1.4 per cent of total industrial production in 1959, this sector accounted for 13.2 per cent of industrial output in 1983 (Zimbalist and Eckstein, 1987), due to an annual average growth rate of 16.6 per cent from 1970 onwards. Cuba also currently produces a large share of its new sugar-cane harvesters, buses, refrigerators and other durables (Edquist, 1985). From 1976 to 1980 Cuba also introduced 115 new export products that sell mainly in other LDCs.

In the 1980 debts were met by rescheduling payments, increased hard currency earnings and large-scale Soviet aid. Nevertheless, it is important to note, Soviet aid to Cuba was considerably less per capita, and of more immediate benefit in terms of the conditions of existence of the population, than, for example, American aid to Israel or Puerto Rico (Turits, 1987). Cuban planners have tended to reduce investment in industry and agriculture in order to maintain high levels of employment and provision of social services. The income distribution in Cuba is extremely egalitarian by world standards, particularly so when compared to other Latin American countries.

The decision in the 1960s to emphasize agricultural development in general and sugar in particular, led to a focus on industries that would have favourable effects in rural areas. As a result there was considerable investment in sugar mills. Much of this type of industry was located in small and medium-sized cities. In an effort to spread economic development throughout the country much of the new industry was given to Oriente province at the eastern end of the island. As a consequence, urban growth has been more rapid there than elsewhere (Gugler, 1980). The chief objective throughout the 1960s was to weaken the urban primacy of Havana and slow down its growth. After 1970, however, and with the switch towards an industrializing strategy Havana again experienced growth. This is because of its advantages in infrastructure, skilled labour, port facilities and market potential. Some 70 per cent of all industrial activities other than sugar-processsing are now located in the Havana metropolitan area (Reitsma and Kleinpenning, 1985).

The major problem now facing Cuba is that non-sugar exports will have to increase if the country is to have sustained growth and a more independent political economy. In particular, Cuba has lost its major guaranteed customer for sugar, the Soviet Union. It is also unlikely that debt-led growth will work again (as for Brazil). Some research suggests that an internal decentralization of planning controls and an external decentralization of export markets might jointly produce more sustained industrial development in the future (White, 1987; Zimbalist and Eckstein, 1987). But this might be at the expense of the truly impressive levels of social services enjoyed by its 11 million people if expenditures on production are increased without foreign borrowing. An export-oriented growth strategy is unlikely while the Cuban government is still led by Fidel Castro and the American government continues to enforce a blockade against Cuban participation in the world economy. But without it, or a new 'sugar daddy' like the former Soviet Union, the prospects for maintaining the current economic system appear bleak.

China

China is the economic development success story of the 1980s and 1990s. With the largest population of any country in the world, 1.19 billion in 1994, China had a growth in GNP per capita between 1985 and 1994 that averaged 7.8 per cent each year (against a 3.4 per cent average for all low-income LDCs). In the 1980s and early 1990s its rates of growth in GDP, agriculture, industry, manufacturing and services were among the highest in the world. The rate of growth in manufacturing was particularly impressive: 14.9 per cent between 1980 and 1994. Only Indonesia, Thailand and South Korea came close to matching this figure (World Bank, 1996). As a result, China in 1994 was a little more advanced than South Korea was in 1970 in terms of output per head of some basic industrial products such as electricity, steel, cement, cotton fabrics and cotton yarn. China with a fifth of the world's population is now only a generation behind the older NICs of East Asia in conventional measures of economic development. Already China has the world's fourth largest national economy, after the USA, Japan and Germany; approximately one quarter that of the USA in 1992 (Rohwer, 1992).

Much of the growth of manufacturing has been concentrated in coastal China, especially in the zones around Shanghai and Hong Kong that have been opened to foreign investment. Originally 'experimental', as the Communist leadership in Beijing worked out a new model for maintaining political control over China while trying to absorb foreign capital, technology and management practices, the Special Economic Zones, such as Shenzhen on the border with Hong Kong, have become the nodal points for reforming the Chinese economy as a whole. This new model is not without its enemies, especially among those who are nostalgic for the autarchic China of the 1960s in which agriculture and national economic self-sufficiency were the priorities. But under the leadership of Deng Xiaoping from 1978 to 1996 China embarked on a thoroughgoing reorientation of its economy: privatizing collective farms, closing or privatizing parts of state industry, dismantling central planning in favour of private entrepreneurship and market mechanisms and fully integrating China into the world economy. Economic growth has been elevated above the class struggle.

Some 75 per cent of Chinese, 800 million people in 1990, live on the land. The first economic reforms concentrated on privatizing peasant agriculture, taking a leaf out of the book of their East Asian neighbours in privileging rural land reform. This increased farm production and rural incomes tremendously. This provided capital for industrialization and improvements in infrastructure in some restricted areas. 'World standards' were brought to bear on the economy through the establishment of the Special Economic Zones (SEZs), that introduced foreign technology and management. The dramatic success of Guangdong province in southern China has transfixed managers and bureaucrats elsewhere. The three SEZs in Guangdong (and another in Fujian province across the strait from Taiwan) have attracted capital and expertise from outside, particularly from Hong Kong and Taiwan. Much of the investment is in joint ventures or intercorporate alliances (such as those referred to in Chapter 3) between foreign firms and local enterprises operated by local governments and cooperatives. But throughout China the industrial structure increasingly resembles that of its East Asian neighbours rather than its former 'ideological friends' in Eastern Europe and the former Soviet Union; Chinese manufacturing output is now dominated by a large number of small firms with mixed state and private (often foreign) participation rather than giant state companies as in the past.

Guangdong represents the most obvious example of industrial transformation. The province's industrial output, largely of goods such as clothes, shoes and toys, rose 15 per cent per year during the 1980s (for an example, see the earlier case study of Barbie). Exports from the province, funnelled mostly through Hong Kong, accounted for almost two-thirds of Guangdong's output and fully one-third of China's total exports in 1990. With 63 million people the province experienced a growth in GDP of around 12.5 per cent per year in the 1980s. For comparison, Thailand (population 55 million), a frequently cited example of an 'emerging' NIC, had a GDP growth of 7.5 per cent during the same period (*The Economist*, 5 October 1991).

From one point of view, this and similar development in coastal China represent the reintegration of Hong Kong and Taiwan with their continental hinterlands after a long period of separation. Low labour and land costs allied to high labour productivity have been major attractions. China's high export quotas to the EC and the USA have also been of importance. But more than 'pure' economics has been at work. The surge of investment has been led by ethnic Chinese within networks that stretch from South and East China all over Southeast Asia and around the 'Pacific Rim'. Common languages – Cantonese for Hong Kong and Guangdong, the Fujianese dialect for Taiwan and the closest provinces on the mainland – and a common culture ease the flow of money, managers and trade. Taiwan and China are officially still on a war footing, Taiwan being the refuge for the Nationalist government after the Communist revolution in 1949, and all economic relations are carried on through Hong Kong. But in 1992 China became the biggest single destination for Taiwanese foreign direct investment, displacing Malaysia into second place. Pressure is mounting on the Taiwan government to allow direct trade and investment with the mainland. With the 'spillover' of investment and trade to the mainland, China is following economically where Taiwan led (Andrews, 1992; Jones *et al.*, 1993).

The euphoria engendered by the recent economic transformation of coastal China should be tempered somewhat by recognizing a number of serious barriers to continuation of present trends. One is the poor condition of transport infrastructure, particularly the railways in the interior (China has one of the world's smallest railway networks relative to population and arable land), and the lack of investment in higher education. The development along the south and east coasts is creating a significant disparity in growth and incomes between coastal and interior regions that will grow greater as the advantaged regions build up their external links. The central government may well have to intervene to prevent spatial polarization from generating internal political conflict. Tensions already in evidence lead some commentators to see a future disintegration of China (Goodman and Segal, 1994). Be that as it may, poverty in China remains a major problem, with at least 25 per cent of the population living below the national poverty line. This is around 320 million people (*The Economist*, 12 October 1996). Their condition contrasts markedly with that of those who have cashed in on the opening up of the Chinese economy.

A second problem is the continuing drain upon central government revenues of state-owned firms (*The Economist*, 22 February 1997). In particular, coal and oil companies are important loss-makers because of government insistence upon subsidized energy prices. Third, local governments have acquired considerable autonomy in pushing credit expansion and investment (Yang, 1994). This has led to an overheating of the national economy as expanding credit chases a shrinking money supply controlled from the centre. Fourth, for all of the success in creating a private economy there is still a level of government regulation without a guaranteed rule of law that, among other things, restricts labour mobility, makes for an arbitrary application of rules, and encourages corruption among civil servants (Rohwer, 1992).

At the root of many of these problems is the lack of political change paralleling the economic change. Unlike the countries of Eastern Europe and the former Soviet Union which have undergone political change prior to economic reform, China has created an increasingly capitalist economy within a still formally socialist state. The consensus view among professional 'China watchers' is that the Chinese Communist Party is now largely irrelevant to economic life and, as the revolt of the movement for greater political democracy that was suppressed in Tiananmen Square in Beijing in 1989 showed, its monopoly of state power is unacceptable to many. As one commentator expresses it: 'Communism in China will probably end not with a bang but a whimper' (MacFarquhar, 1992, p. 28). The capitalist genie unleashed in 1978 is probably too well established now to put back in the lamp of communist autarchy.

Summary

According to the World Bank (1983), the aggregate rate of growth of both industrial (mineral resources plus manufacturing) and manufacturing output were over 3 per cent per annum for 34 low-income countries and over 6 per cent per annum for 59 middle-income countries over the period 1960–81. These rates were higher than those for the industrialized countries and, consequently, the share of the LDCs in the world's manufacturing output rose somewhat – from 17.6 per cent in 1960 to 18.9 per cent in 1981 (World Bank, 1983). Their combined share of world exports of manufactures rose from 3.9 per cent to 8.2 per cent in the same period. Or, from a slightly different perspective, the LDCs' share of the manufactured imports of all industrial countries rose from 5.3 per cent in 1962 to 13.1 per cent in 1978 (World Bank, 1982). At the same time the share of the GDP of the poorest (low-income) countries coming from the industrial sector rose from 25 per cent in 1960 to 34 per cent in 1981. For manufacturing the rise was only from 11 to 16 per cent. In the middle-income group the changes were from 30 to 38 per cent for industry, and from 20 to 22 per cent for manufacturing alone (World Bank, 1983). Much of this growth was concentrated in a relatively small group of NICs (Sutcliffe, 1984).

With some notable exceptions to the general trends, including China, the East Asian NICs and some Southeast Asian countries, the growth of manufacturing industry in the periphery and its penetration of DC markets did not deepen much in the 1980s and early 1990s, even though it spread somewhat beyond the older NICs. For example, annual rates of growth in manufacturing in 1980–94 averaged 4 per cent for all LDCs, with low-income countries having higher rates (5.2 per cent average if China is excluded, 9.2 per cent if it is included) than middle-income ones (including the NICs) (2.4 per cent) (World Bank, 1996). The OECD (major industrialized) countries averaged 3.3 per cent per annum through the decade of the 1980s. At the same time the LDCs' share of the manufactured imports of all industrial countries sank to 12.9 per cent in 1990 from 13.1 per cent in 1978 (World Bank, 1992).

These figures call into question the generally optimistic conclusion of Warren (1980) and others (see Chapter 8) that a massive industrialization of the periphery is under way. Indeed, this chapter has suggested that there are significant constraints upon the industrialization of the periphery and the 1970s provided a time period uniquely favourable to the type of development that did take place. Indeed, even in that time period a number of economies stagnated or deindustrialized (e.g. Argentina). The economic problems of the industrial core have not created inevitable advantages for industrialization in the global periphery. Any advantages have been created there rather than simply given from the outside.

Some of the major conclusions are:

- Industrialization plays a major role in national ideologies of modernization and is seen as a solution to various major practical economic and social problems.
- The spread and intensification of industrialization since the late 1960s coincided with the declining rate of profit in the industrial core.
- Much of the new industrialization is export-oriented rather than directed (as in import substitution) to domestic markets.
- There are limits to the development of this industrialization: such as improving profit rates in the industrial core, protectionist measures in export markets, technological changes that reduce the attractiveness of low-wage locations, incredible debt loads, and relatively limited employment effects. Most LDCs are either not industrializing or industrializing only very slowly.
- Export-processing zones represent one geographical form taken by the New International Division of Labour but 'offshore production' of components or assembly is not the only feature of peripheral industrialization. There are now important industries engaged in the production of locally created final products.
- The new industrialization has favoured existing metropolitan areas and coastal regions. Attempts at decentralization to growth centres have not met with much success.
- Three profiles, of Brazil, Cuba and China, illustrate three different paths to recent industrialization. Although obviously dissimilar in many respects (size of economy, role of foreign TNCs, form of government, etc.) there is a remarkable similarity between Brazil and Cuba in their reliance in the recent past on debt-led growth. Both now face problems precisely because of this previous emphasis. China's growth in the 1980s reflects a conscious political decision to re-engage with the world economy after a long period on its margins. As a geographical extension of the existing NICs of East Asia, China represents the main wave of new industrialization in the world economy in the 1980s and 1990s. This was anything but a global trend.

Key Sources and Suggested Reading

Abumere, S. I. 1982. Multinationals and industrialisation in a developing country: the case of Nigeria. In M. Taylor and N. Thrift (eds), *The Geography of Multinationals: Studies in the*

Spatial Development and Economic Consequences of Multinational Corporations. New York: St. Martin's Press.

Adelman, I. 1984. Beyond export-led growth. *World Development*, **12**, 937–49.

Agnew, J. A. and Grant, R. J. 1997. Falling out of the world economy? Theorizing Africa in world trade. In R. Lee and J. Wills (eds), *Geographies of Economies*. London: Edward Arnold.

Amin, A. and Thrift, N. 1992. Neo-Marshallian nodes in global networks. *International Journal of Urban and Regional Research*, **16**, 571–87.

Anderson, A. B. 1990. Smokestacks in the rainforest: industrial development and deforestation in the Amazon Basin. *World Development*, **18**, 1191–205.

Andrews, J. 1992. A change of face: a survey of Taiwan. *The Economist*, 10 October, survey.

Armstrong, W. and McGee, T. 1986. *Theatres of Accumulation: Studies in Asian and Latin American Urbanization*. London: Methuen.

Ayeni, B. 1981. Spatial dimensions of manufacturing activities in Nigeria. Unpublished paper, Department of Geography, University of Ibadan, Nigeria.

Ayres, R. U. and Miller, S. 1983. Robotics, CAM and industrial productivity. *National Productivity Review*, **1**: 1, 452–60.

Baer, W. 1983. *The Brazilian Economy: Growth and Development*. 2nd edn. New York: Praeger.

Baer, W. et al. 1987. Structural changes in Brazil's industrial economy, 1960–80. *World Development*, **15**, 275–86.

Biersteker, T. J. 1995. The 'triumph' of liberal economic ideas in the developing world. In B. Stallings (ed.), *Global Change, Regional Response: The New International Context of Development*. Cambridge: Cambridge University Press.

Bradford, C. I. 1987. Trade and structural change: NICs and next-tier NICs as transitional economies. *World Development*, **15**, 299–316.

Brundenius, C. 1984. *Revolutionary Cuba: The Challenge of Economic Growth with Equity*. Boulder, CO: Westview.

Cavalcanti, L. et al. 1981. Multinationals, the new international economic order and the spatial industrial structure of Brazil. In F. E. I. Hamilton and G. J. R. Linge (eds), *International Industrial Systems*. Chichester: Wiley.

Cline, W. R. 1982. Can the East Asian model of development be generalized? *World Development*, **10**, 81–90.

Cohen, R. B. 1981. The new international division of labor, multinational corporations and the urban hierarchy. In M. Dear and A. J. Scott (eds), *Urbanization and Urban Planning in Capitalist Society*. London: Methuen.

Corbridge, S. 1982. Urban bias, rural bias, and industrialisation: an appraisal of the work of Michael Lipton and Terry Byres. In J. Harriss (ed.), *Rural Development: Theories of Peasant Economic and Agrarian Change*. London: Hutchinson.

Corbridge, S. 1986. *Capitalist World Development: A Critique of Radical Development Geography*. London: Macmillan.

Crow, B. and Thomas, A. 1985. *Third World Atlas*. Milton Keynes: Open University Press.

Cumings, B. 1984. The origins and development of the Northeast Asian political economy: industrial sectors, product cycles and political consequences. *International Organization*, **38**, 1–40.

Cunningham, S. 1986. Multinationals and restructuring in Latin America. In C. J. Dixon et al. (eds), *Multinationals and the Third World*. Boulder, CO: Westview.

Daniels, P. 1985. *Service Industries: Growth and Location*. London: Methuen.

The Economist 27 June 1987. The rag trade: on the road to Mandalay. 67–8.

The Economist, 5 October 1991. The South China miracle: a great leap forward. 19–22.

The Economist, 20 February 1993. The final frontier. 63.

The Economist, 29 June 1996. Getting together. 42–3.

The Economist, 12 October 1996. How poor is China? 35–6.

The Economist, 9 November 1996. The buying and selling of Brazil Inc. 83–4.

The Economist, 22 February 1997. The last Emperor. 21–5.

The Economist, 1 March 1997. Time to roll out a new model. 71–2.

The Economist, 17 May 1997. Reforming Brazil. Is it for real? 38–40.

Edquist, C. 1985. *Capitalism, Socialism, and Technology: A Comparative Study of Cuba and Jamaica*. London: Zed Books.

Edwards, C. 1985. *The Fragmented World: Competing Perspectives on Trade, Money and Crisis*. London: Methuen.

Evans, D. and Alizadeh, P. 1984. Trade, industrialization and the visible hand. In R. Kaplinsky (ed.), *Third World Industrialisation in the 1980s: Open Economies in a Closing World*. London: Cass.

Evans, P. 1979. *Dependent Development: The Alliance of Multinational, State, and Local Capitalism in Brazil*. Princeton: Princeton University Press.

Evans, P. 1987. Dependency and the state in recent Korean development: some comparisons with Latin American NICs. In K. D. Kim (ed.), *Dependency Issues in Korean Development: Comparative Perspectives*. Seoul: Seoul National University Press.

Evans, P. and Timberlake, M. 1980. Dependence, inequality and growth in less developed countries. *American Sociological Review*, **45**, 531–52.

Fishlow, A. 1972. Brazilian size-distribution of income. *American Economic Review*, **62**, 391–402.

Forbes, D. 1986. Spatial aspects of Third World multinational corporations' direct investment in Indonesia. In M. Taylor and N. Thrift (eds), *Multinationals and the Restructuring of the World Economy*. Beckenham: Croom Helm.

Forbes, D. and Thrift, N. 1987. International impacts on the urbanization process in the Asian region: a review. In R. J. Fuchs *et al.*, *Urbanization and Urban Policies in Pacific Asia*. Boulder, CO: Westview.

Frankel, J. A. 1991. Is a yen bloc forming in Asia? *AMEX Bank Review*, November, 2–3.

Fuchs, R. J. and Pernia, E. M. 1987. External economic forces and national spatial development: Japanese direct investment in Pacific Asia. In R. J. Fuchs *et al.* (eds), *Urbanization and Urban Policies in Pacific Asia*. Boulder, CO: Westview.

Gereffi, G. 1995. Global production systems and third world development. In B. Stallings (ed.), *Global Change, Regional Response: The New International Context of Development*. Cambridge: Cambridge University Press.

Gibson, M. L. and Ward, M. D. 1992. Export orientation: pathway or artifact? *International Studies Quarterly*, **36**, 331–44.

Goldberg, M. 1987. *The Chinese Connection: Getting Plugged in to the Pacific Rim Real Estate, Trade and Capital Markets*. Vancouver: University of British Columbia Press.

Goldsbrough, D. 1985. Foreign direct investment in developing countries: trends, policy issues, and prospects. *Finance and Development*, **22**, 31–4.

Goodman, D. S. G. and Segal, G. (eds) 1994. *China Deconstructs: Politics, Trade and Regionalism*. London: Routledge.

Gore, C. 1984. *Regions in Question: Space, Development Theory and Regional Policy*. London: Methuen.

Griffith-Jones, S. 1980. The growth of multinational banking, the Eurocurrency markets and their effects on the developing countries. *Journal of Development Studies*, 16, 96–109.

Griffith-Jones, S. and Rodriguez, E. 1984. Private international finance and industrialization in LDCs. In R. Kaplinsky (ed.), *Third World Industrialization in the 1980s: Open Economies in a Closing World*. London: Cass.

Griffith-Jones, S. and Stallings, B. 1995. New global financial trends: implications for development. In B. Stallings (ed.), *Global Change, Regional Response: The New International Context of Development*. Cambridge: Cambridge University Press.

Gugler, J. 1980. A minimum of urbanism and a maximum of ruralism: the Cuban experience. *International Journal of Urban and Regional Research*, 4, 516–36.

Hamilton, C. 1983. Capitalist industrialisation in the four little tigers of East Asia. In P. Limqueco and B. McFarlane (eds), *Neo-Marxist Theories of Development*. Beckenham: Croom Helm.

Hansen, N. 1981. Mexico's border industry and the international division of labor. *Annals of Regional Science*, 15, 1–12.

Harris, N. 1986. *The End of the Third World: Newly Industrialising Countries and Decline of an Ideology*. London: Taurus.

Hirschman, A. 1992. Industrialization and its manifold discontents: West, East and South. *World Development*, 20, 1225–32.

Humphrey, J. 1984. Labor in the Brazilian motor vehicle industry. In R. Kronish and K. S. Mericle (eds), *The Political Economy of the Latin American Motor Vehicle Industry*. Cambridge, MA: MIT Press.

Jenkins, R. O. 1987. *Transnational Corporations and the Latin American Automobile Industry*. Pittsburgh: University of Pittsburgh Press.

Jenkins, R. O. 1988. *Transnational Corporations and Uneven Development: The Internationalization of Capital and the Third World*. London: Methuen.

Joekes, S. P. 1987. *Women and Development*. New York: Oxford University Press.

Jones, R. S. et al. 1993. Economic integration between Hong Kong, Taiwan and the coastal provinces of China. *OECD Economic Studies*, 20, 115–144.

Kaplinsky, R. 1984. *Automation: The Technology and Society*. London: Longman.

Kawai, M. 1990. Comments. In Y. Suzuki et al. (eds), *The Evolution of the International Monetary System: How Can Efficiency and Stability be Attained?* Tokyo: University of Tokyo Press.

Kirchner, J. 1983. Supercomputing seen as key to economic success. *Computerworld*, 3 October, 8.

Krueger, A. 1978. Alternative trade strategies and employment in developing countries. *American Economic Review*, 68, 523–36.

Lavrov, S. B. and Sdasnyk, G. V. 1982. The growth pole concept and the regional planning experience of developing countries. *Development Dialogue*, 3, 15–27.

Leung, C. K. and Chin, S. S. K. (eds) 1983. *China in Readjustment*. Hong Kong: Centre for Asian Studies, University of Hong Kong.

Lewis, P. 1988. Third World funds: wrong way flow. *New York Times*, 11 February, D1, D9.

Linge, G. J. R. and Hamilton, F. E. I. 1981. International industrial systems. In G. J. R. Linge and F. E. I. Hamilton (eds), *Spatial Analysis, Industry and the Industrial Environment*. Chichester: Wiley.

Locatelli, R. L. 1985. *Industrialização, Crescimento e Empregno: Uma Avaliação da Experiência Brasileira*. Rio de Janeiro: IPEA/INPES.

Loup, J. 1983. *Can the Third World Survive?* Baltimore: Johns Hopkins University Press.

MacFarquhar, R. 1992. Deng's last campaign. *New York Review of Books*, 17 December, 22–8.

Mattera, P. 1985. *Off the Books: The Rise of the Underground Economy*. New York: St. Martin's Press.

Menzel, U. and Senghaas, D. 1987. NICs defined: a proposal for indicators evaluating threshold countries. In K. D. Kim (ed.), *Dependency Issues in Korean Development: Comparative Perspectives*. Seoul: Seoul National University Press.

Mericle, K. S. 1984. The political economy of the Brazilian motor vehicle industry. In R. Kronish and K. S. Mericle (eds), *The Political Economy of the Latin American Motor Vehicle Industry*. Cambridge, MA: MIT Press.

Morello, T. 1983. Sweatshops in the sun? *Far Eastern Economic Review*, 15, September, 88–9.

Moudoud, E. 1986. *The Rise and Fall of the Growth Pole Approach*. Syracuse, NY: Syracuse University Department of Geography Discussion Paper No. 89.

Murphey, R. 1980. *The Fading of the Maoist Vision*. London: Methuen.

Neuman, S. G. 1984. International stratification and third world military industries. *International Organization*, **38**, 167–98.

OECD 1979. *The Impact of the NICs on Production and Trade in Manufactures*. Paris: Organization for Economic Cooperation and Development.

Peet, R. 1987. Industrial devolution, underconsumption and the third world debt crisis. *World Development*, **15**, 112–43.

Petras, J. 1984. Toward a theory of industrial development in the third world. *Journal of Contemporary Asia*, **14**, 182–203.

Rada, J. 1984. *International Division of Labor and Technology*. Geneva: ILO.

Rau, W. and Roncek, D. W. 1987. Industrialization and world inequality: the transformation of the division of labor in 59 nations, 1960–81. *American Sociological Review*, **52**, 359–69.

Reid, M. 1996. Mercosur: remapping South America. *The Economist*, 12 October, survey.

Reitsma, H. A. and Kleinpenning, J. M. G. 1985. *The Third World in Perspective*. Totowa, NJ: Rowman and Allanheld.

Rowher, J. 1991. Brazil: a survey. *The Economist*, 7 December, survey.

Rowher, J. 1992. When China wakes: a survey. *The Economist*, 28 November, survey.

Rowher, J. 1993. A billion consumers: a survey of Asia. *The Economist*, 30 October, survey.

Samuelson, P. 1964. *Economics*. New York: McGraw-Hill.

Sanders, R. 1987. Towards a geography of informal activity, *Socio-Economic Planning Sciences*, **21**, 229–37.

Santiago, C. E. 1987. The impact of foreign direct investment on employment structure and employment generation. *World Development*, **15**, 317–28.

Scott, A. J. 1987. The semi-conductor industry in Southeast Asia: organization, location, and the international division of labour. *Regional Studies*, **21**, 143–60.

Seiber, M. J. 1982. *International Borrowing by Developing Countries*. New York: Pergamon.

Shunzan, Y. 1987. Urban policies and urban housing programs in China. In R. J. Fuchs *et al.* (eds), *Urbanization and Urban Policies in Pacific Asia*. Boulder, CO: Westview.

South, R. B. 1990. Transnational 'maquiladora' location. *Annals of the Association of American Geographers*, **80**, 549–70.

Stallings, B. 1990. The role of foreign capital in economic development. In G. Gereffi and D. L. Wyman (eds), *Manufacturing Miracles: Paths of Industrialization in Latin America and East Asia*. Princeton, NJ: Princeton University Press.

Suhartono, F. X. 1987. Growth Centers in the Context of Indonesia's Urban and Regional Development Program. MA thesis, Social Science Program, Syracuse, NY: Syracuse University.

Sutcliffe, R. B. 1971. *Industry and underdevelopment*. London: Addison Wesley.

Sutcliffe, R. B. 1984. Industry and Underdevelopment re-examined. In R. Kaplinsky (ed.), *Third World Industrialization in the 1980s: Open Economies in a Closing World*. London: Cass.

Szczepanik, E. 1969. The size and efficiency of agricultural investment in selected developing countries. *FAO Monthly Bulletin of Agricultural Economics and Statistics*, December, 2.

Tempest, R. 1996. Barbie and the world economy. *Los Angeles Times*, 22 September, A1, A12.

Thomas, V. 1987. Differences in income and poverty within Brazil. *World Development*, **15**, 263–73.

Thrift, N. 1986. The geography of international economic disorder. In R. J. Johnston and P. J. Taylor (eds), *A World in Crisis? Geographical Perspectives*. Oxford: Blackwell.

Tunstall, J. 1986. *Communications Deregulation*. Oxford: Blackwell.

Tuong, H. D. and Yeats, A. J. 1981. Market disruption, the new protectionism, and developing countries: a note on empirical evidence from the US. *The Developing Economies*, **19**, 107–18.

Turits, R. 1987. Trade, debt, and the Cuban economy. *World Development*, **15**, 163–80.

Tyler, W. G. 1976. Manufactured exports and employment creation in developing countries: some empirical evidence. *Economic Development and Cultural Change*, **24**, 355–73.

Tyler, W. G. 1981. Growth and export expansion in developing countries: some empirical evidence. *Journal of Development Economics*, **9**, 121–30.

Tyler, W. G. 1986. Stabilization, external adjustment, and recession in Brazil: perspectives on the mid-1980s. *Studies in Comparative International Development*, **21**, 5–33.

Uchitelle, L. 1987. Corporate profitability rising, reversing 15-year downturn. *New York Times*, 30 November, A1, D11.

UNIDO 1981. *A Statistical Review of the World Industrial Situation*. Vienna: United Nations Industrial Development Organization.

UNIDO 1985. *Industry in the 1980s*. Vienna: United Nations Industrial Development Organization.

Warren, B. 1980. *Imperialism: Pioneer of Capitalism*. London: Verso.

Webber, M. J. and Rigby, D. L. 1996. *The Golden Age Illusion: Rethinking Postwar Capitalism*. New York: Guilford Press.

White, G. 1987. Cuban planning in the mid-1980s: centralization, decentralization and participation. *World Development*, **15**, 153–61.

Wong, K. Y. 1987. China's special economic zone experiment: an appraisal. *Geografiska Annaler, B*, **69**, 27–40.

World Bank 1979. *World Trade and Output of Manufactures*. Staff Working Paper, January.

World Bank 1982, 1983, 1992. *World Development Report*. New York: Oxford University Press.

World Bank 1996. *From Plan to Market: World Bank Development Report*. New York: Oxford University Press.

Yang, D. 1994. Reform and the restructuring of central-local relations. In D. S. G. Goodman and G. Segal (eds), *China Deconstructs: Politics, Trade and Regionalism*. London: Routledge.

365

Yoffie, D. B. 1983. *Power and Protectionism: Strategies of the Newly Industrializing Countries.* New York: Columbia University Press.

Zimbalist, A. 1987. Cuban industrial growth, 1965–84. *World Development*, **15**, 83–93.

Zimbalist, A. and Eckstein, S. 1987. Patterns of Cuban development. *World Development*, **15**, 5–22.

ADJUSTING TO A NEW GLOBAL ECONOMY

In this concluding part of the book we examine some of the reactions to the emergence of ever-larger and more powerful economic forces and the time–space compression that have come to characterize the world economy. In Chapter 11 we explore the changing role of nation states within the world economy, emphasizing the relationships between economic change and the new geopolitics and, in particular, the spatial consequences of transnational political and economic integration that have occurred in response to the increased scale, sophistication, and interdependence of the modern world economy. In Chapter 12 we examine the other side of the coin: decentralist reactions to the changing world economy. Here, the focus is on regionalism and regional policy, nationalism and separatism, and grassroots movements towards economic democracy. Finally, in Chapter 13 we review the key arguments that have shaped the book, emphasizing the dynamic interdependence of global and local change.

Picture credit: Martin Wyness/Still Pictures

Picture credit: Sanja Iskov/Still Pictures

CHAPTER 11

TRANSNATIONAL INTEGRATION

In this chapter we return to the theme of the relationship between economic development and the role of the state. As Parts 2 and 3 have shown, nation states have been crucial, both in the struggle for domination within the core and as peripheral and semi-peripheral economies have struggled to reduce their dependency on core economies. As the world economy has become more and more globalized, however, nation states throughout the world economy have had to explore cooperative strategies involving transnational political and economic integration of various kinds. This chapter outlines the rationale for these strategies, describes the scope of the major transnational organizations, and illustrates some of the more important spatial implications of transnational integration.

ECONOMIC CHANGE AND THE NEW GEOPOLITICS

In order to understand the emergence of transnational organizations we must first remind ourselves of the shifting economic and geopolitical foundations of the world economy since the Second World War (see Taylor, 1992). In the aftermath of the war, the capitalist economy was reordered as a more open system. It was a system without the economic barriers of the trading empires that had been set up in the 1930s. Instead, it was based on free-market capitalism with stable monetary relations and rapidly diminishing barriers to trade. This required, first of all, an *orderly* world, internally peaceful and secure from outside threats. Second, it required leadership in providing and furthering mechanisms for establishing a stable reserve standard for international currency exchange rates and for ensuring access to world trade markets. The one state that could provide military order – the United States – was also the only state economically strong enough to impose order on the economic system. The Soviet world-empire had turned inward in its attempt to restructure economy and society along different ideological lines, but its existence was extremely important because (until its dissolution in 1989) it served to mobilize an ideological reaction – anti-communism – that provided both an economic stimulus and political solidarity within the core economies.

In short, the world economy was characterized by the *hegemony* of the United States. Under US hegemony, as we saw in Chapter 5, the world economy came to be characterized by Fordism, the socio-economic system that links mass production with mass consumption. A tense but durable relationship between big business, big labour and big government enabled Fordism to provide the basis for the long postwar boom and unprecedented rise in living standards throughout much of the capitalist world. This boom was also crucially dependent on the massive expansion of world trade and international investment flows made possible under the umbrella of US financial and military power. Following the Bretton Woods agreement that made the US dollar the world's reserve currency (see p. 49), Fordism was implanted in Europe and Japan, either directly, during the occupation phase, or indirectly, through the Marshall Plan and foreign direct investment by US companies. The consequent opening-up of foreign trade, observes Harvey (1988, p. 4):

> permitted surplus productive capacity (and potentially surplus labour reserves) to be absorbed in the United States, while the progress of Fordism internationally meant the formation of global mass markets and the absorption of the mass of the world's population, outside the communist world, into the global dynamics of a new kind of capitalism. . . . At the input end, the opening up of foreign trade meant the global-ization of supply and often ever cheaper raw material. This new internationalism also brought a host of other activities in its wake – banking, insurance, hotels, airports, and, ultimately, tourism. It also meant a new international culture and a new global system of gathering and evaluating information.

The postwar period saw, therefore, the rise of a series of industries – autos, steel, petro-chemicals, rubber, etc. – that acted as the propulsive engines of economic growth, coordinated through the collective powers of big labour, big business and big government. And out of this there arose a series of grand production regions in the world economy – the Midwest of the United, States, the West Midlands of Britain, the Ruhr and the Tokyo–Yokohama production region – built on world financial and governmental centres, and reaching out to dominate an increasingly homogeneous world market.

The logic of Fordist production also fostered, as we have seen (Chapter 6), the emergence of transnational corporations with the capacity to move capital and technology rapidly from place to place, drawing opportunistically on resources, labour markets and consumer markets in different parts of the world. Transnational corporations have now gone far beyond the point where they can be seen simply as extensions of a specific national economy; and even some small firms have now acquired both the capability and the propensity to operate globally. The significance of this is that, *although private companies are by no means absolute masters of their own fate, they do have the ability* (as compared with governmental units) *to redefine their commitments and objectives in response to the changing opportun-ities presented by the globalization of the economy.*

Within this new context of political and economic interdependence, regional and international shifts in economic and political power began to occur, as we saw in Chapters 5 and 6: shifts that the policies of particular governments seemed powerless to prevent by normal means. The ascent of the NICs brought a new dimension to the world economy in the form of second-order economic powers that effectively created a hierarchical geopolitical system. Meanwhile, as the regional influence of NICs has grown, so they have come to exert an independent effect on the landscapes of the world's core economies:

> None has had a greater political/psychological effect on the major powers than the omnipresence of persons, symbols and signs in Europe's great cities, and in such American cities as New York, Miami, New Orleans and Los Angeles. In Europe, billboards advertising Asian, African and Latin American Airlines, store signs in Arabic script, national airline offices and ethnic food restaurants from three continents, a plethora of Arabic-language newspapers displayed prominently in kiosks and, above all, the businessman, tourist, shopper, student and adolescent youth from these newly-powerful countries demonstrate that the world has changed. They have joined the overseas symbols of American power – the Hilton, the Holiday Inn, Hertz, Avis, ESSO, Mobil, IBM, the English-language newspaper, the American bar and restaurant, and the tourist, student and businessman – to share the landscape of sight, sound and taste with Americans.
>
> (Cohen, 1982, p. 227)

In the core economies, meanwhile, the prosperity associated with Fordist regime had been replaced by uncertainty, destabilization and crisis resulting from the conjunction of an episode of stagflation and the 1973 OPEC oil embargo. This created national economic management problems that could not be solved without accelerating the inflation that had undermined the role of the US dollar as the international reserve currency.

At the same time, the role and relative power of nation states changed significantly. Economic circumstances reduced the ability of governments to deliver full employment as well as a full range of welfare services; and the growth of the global financial system blunted the power of individual nation states to pursue independent fiscal and monetary policies with any degree of success. In particular, the United States had to struggle hard to maintain its hegemony, running a mounting trade deficit and an enormous public debt, and having persistently to devalue the dollar in order to maintain competitiveness with Japan and Germany.

In the decentralized, restructured and consolidated world economy that emerged in the 1980s, new communications technologies, new forms of corporate organization, and new business services were intensifying 'time-space compression', decreasing the time horizons of both public and private decision-making and making it easier to spread those decisions over an ever-wider space. As we saw in Chapter 7, one result has been the acceleration of shifts in the patterning of uneven development as more flexible corporate organization and production systems have been able to quickly exploit particular local mixes of skills and resources. Another outcome is that local governments are being forced to be much more competitive with one another as they attempt not only to protect their economic base during a time of upheaval and transition but also to identify and exploit some comparative edge with which to lure the newly flexible flows of finance and production. This inter-governmental competition has bred so-called 'entrepreneurial' cities, whose governments have been drawn beyond questions of tax policies, infrastructure provision and service delivery to explore public–private partnerships, foster favourable 'business climates' and initiate controls on labour through contract negotiations with municipal workers.

TRANSNATIONAL INTEGRATION

It is within this changing economic and geopolitical context that we have to see the various attempts to adjust to the modern world economy through strategies of transnational economic and political integration. We should nevertheless remember that, as Parts 2 and 3 have shown, the dominant processes in both intra-core rivalry and in the struggle by the periphery to escape from dependency have been dominated by conflict and competition. Economic nationalism, whether drawing on practical examples (e.g. eighteenth-century Britain and nineteenth- and twentieth-century Japan – Chapter 5), political ideology (e.g. Juan Péron in Argentina and Getúlio Vargas in Brazil in the 1940s and 1950s) or development theory (e.g. the import-substituting industrialization espoused by Raoul Prebisch), continues to dominate global economics and geopolitics.

Having acknowledged this, however, we must also recognize the long-term trend among the world's space-economies towards the progressive integration and interdependence of local, regional and national economic systems (Gibb and Michalak, 1994). *What has happened is that the logic of the world economy has in many ways transcended the scale of nation states.* The logic and apparatus of statehood are not conducive to transnational integration, economic or political; but the outcomes of Neo-Fordism have forced many states to explore cooperative strategies of various kinds. As a result, the world's economic landscapes now bear the imprint, in a variety of ways, of transnational economic and political integration.

The logic of integration

The increased scale, sophistication and interdependence of the modern world economy would not have been feasible if it were not for the fact that new technologies and new forms of corporate organization gradually made it possible to conquer several of the frictions that tend to operate against hierarchical flows of production and consumption. In addition to the obvious – for geographers – friction of distance itself, these include the frictions associated with spatial variations in social organization and culture. As railroads, the telegraph, automobiles, aircraft, computer networks, satellite communications systems and fibre optics have successively 'shrunk' the globe, Fordist principles of mass production have brought about a convergence of patterns of social organization, and radio and television have undermined local and regional cultures and replaced them with an international culture characterized by the language and artifacts of consumerism: American Express, Benetton, Burger King, Coca Cola, Gucci, Laura Ashley, Marlboro, Mercedes-Benz, MTV, Rolex, Shiseido, Sony, Visa, and so on.

The framework of nation states, however, is a source of friction that has persisted. The main reason for this, of course, is that the functional logic of statehood hinges on reinforcing *differences between* nations while reinforcing *similarities within* nations. In order to establish the required feelings of common identity, even the oldest states have had to engage in the process of creating and diffusing a distinctive identity. Much of the ideology and symbolism of nation states in Europe, for example, has centred on the systematic mythologizing of history, reinforced by the stereotyping of outsiders. One very important outcome of this was the jingoism and xenophobia that set the context for the First World War, nurtured the ambitions of the Third Reich, and hampered postwar attempts to establish common economic and legal ground.

Among the more explicit functions of nation states that have contributed to the frictions affecting the world economy are those relating to national security and the promotion of homogeneous internal standards and conditions. The latter include controlling fiscal and monetary policy, upholding labour contracts, establishing standards for everything from education to ballbearings, and overseeing key industries such as telecommunications.

Once a significant amount of economic activity had spilled beyond national boundaries, however, nation states had to confront the need to rethink these activities in order not to become isolated or to become even more vulnerable to underdevelopment. In short, *it was the international trade system that provided the major impetus for nation states to be drawn into various forms of institutionalized integration.* For core nations, the objective was primarily to protect and consolidate existing advantages through increased international security, access to wider markets, investment opportunities and labour markets. For peripheral nations, the objective was primarily to minimize or reduce dependency through harnessing more resources and more investment potential. In addition, most nations were able to subscribe, in public at least, to the more lofty ideals of good international relations and a more equitable international economic order.

The particular *advantages* of formalized transnational integration include:

1. the potential for economies of scale, particularly for the smallest nation states and the weakest national economies;
2. the potential for creating multiplier effects from the existence of enlarged markets; and
3. the potential for strengthening regional interaction by easing the movement of labour, goods and capital.

The particular *disadvantages* of formalized transnational integration include:

1. the potential loss of sovereignty over a broad spectrum of issues; and
2. the potential for the intensification of internal inequalities as a wider geographical context makes for more pronounced processes of uneven development.

Types and levels of integration

Figure 11.1 summarizes the 'where' and 'when' of transnational economic integration since 1945. In practice, integration can be pursued in a variety of ways and at different levels. It can be *formal*, involving an institutionalized set of rules and procedures (e.g. United Nations Organization, European Union (EU; formerly the

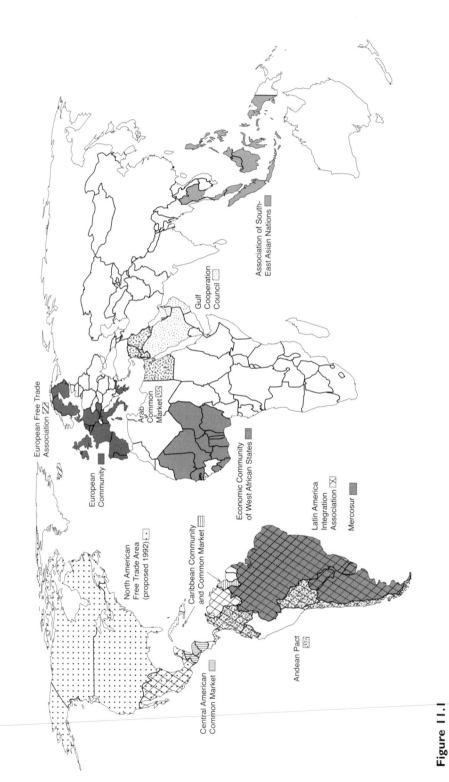

Figure 11.1

Regional integration in the world economy

Source: Updated from Edwards (1985), Fig. 8.6

European Community, or EC), General Agreement on Tariffs and Trade); or *informal*, involving coalitions of interests (UN voting blocs). It can be *trans*national, involving attempts to foster integration between nation states (NATO, OAU, WTO, British Commonwealth Association); or *supra*national, involving a commitment to an institutionalized body with certain powers over member states (EU). It can be *economically* oriented (WTO, the European Free Trade Association), *strategic* (NATO, the Warsaw Pact), *political* (UN voting blocs), *socio-cultural* (UNESCO), or *mixed* (EU, OAU) in orientation.

The GATT framework and the WTO

Our immediate concern here is with economically oriented integration schemes. Within the capitalist world, these have had to conform to the rules of the General Agreement on Tariffs and Trade (GATT), a transnational association of most of the world's trading nations formed in the aftermath of the Second World War to promote world-wide free trade and to untangle the complex trade restrictions that had accumulated. The original GATT agreement (in 1947) reduced the average tariff on goods from over 40 per cent to less than 30 per cent. Subsequent rounds of renegotiation have brought the average tariff level down to about 5 per cent (Table 11.1).

Yet the GATT became the victim of its own success. As more countries have joined the agreement and the world economy became increasingly globalized and interdependent, so trade issues became increasingly complex. The original agreement was written to deal primarily with trade in manufactured goods among developed countries, yet by 1990 only about 60 per cent of world export earnings came from manufactures. Services accounted for an increasing share of world trade; many of the NICs were not fully subject to GATT rules; and foreign direct investment by transnational corporations was beyond the scope of the GATT. And although tariffs on manufactured goods were successfully reduced through the GATT, substantial **non-tariff barriers** (e.g. import quotas, import licences, exchange rate manipulation, government subsidies to domestic industries, special labelling and packaging regulations, etc.) remained a problem. Thus, whereas the early rounds of GATT renegotiation took several months, more recent rounds took several years. The most recent round of renegotiation, the Uruguay Round, began in 1986 and was not concluded until December 1993. The chief obstacle was disagreement between the USA and the European Union over non-tariff barriers in the form of various subsidies that both were paying to their farmers. The crowning achievement of the Uruguay round was the creation of the World Trade Organization as a replacement for the GATT (which had become labelled by wags as the 'General Agreement to Talk and Talk'). Whereas the GATT had little ability to enforce its decisions, the WTO is a global body with both judicial and regulatory power. Its framework is a series of agreements 24 000 pages long, agreements which extend beyond trade in manufactured goods to cover investment, services, and intellectual property rights. In the words of the organization's director-general, Renato Ruggiero, the WTO 'is writing the constitution of a single global economy'. A significant step towards this was taken in February 1997 when the 68 original members of the WTO signed an agreement to free up their markets to international

non-tariff barriers
Policy instruments (other than import taxes) designed to protect domestic industry from foreign competition. Examples include special standards and regulations, quotas, and licensing requirements.

Table 11.1: Average tariff levels for selected countries, 1988 and 1996.

	1988	1996
Australia	15.6	5.0
Canada	3.7	1.6
Chile	19.9	11.0
China	39.5	23.0
European Union	5.7	3.6
Hong Kong	0.0	0.0
Indonesia	18.1	13.1
Japan	4.3	4.0
Malaysia	13.6	9.0
Mexico	10.5	9.8
New Zealand	14.9	5.7
Philippines	27.9	15.6
Singapore	0.3	0.0
South Korea	19.2	7.9
Taiwan	12.6	8.6
Thailand	31.2	17.0
United States	4.2	3.4

free trade association
A form of international economic integration that involves the elimination of some (but not necessarily all) trade barriers between member states, but where each member state continues to set its own tariffs and quotas as trade barriers to non-member states.

customs union
A form of international economic integration that involves the elimination of some (but not necessarily all) trade barriers between member states and the creation of a common set of trade barriers to non-member states.

competition in telecommunications. By mid-1997, the WTO had a membership of 130 nations, with a waiting list of 29 others, including Russia and China.

Many of the WTO's agreements are derived from GATT rulings, including the provision that each member state shall extend most-favoured status to all other member countries. (Thus if, say, the USA were to lower its import duty on textile products from Canada, it would immediately have to extend that same reduced rate to every other WTO member.) There is, however, an exception to this principle for free trade associations and customs unions, members of which may reduce their tariffs against each other without extending such concessions to remaining WTO members. It is this exception that has provided the basis for regional economic integration within the globalizing world economy.

The institutional forms of integration

In a **free trade association**, member countries eliminate tariff and quota barriers against trade from other member states, but each individual member continues to charge its regular duties on materials and products coming from outside the association. The only free trade association of any real significance has been the European Free Trade Association (EFTA), whose membership now comprises only Iceland, Lichtenstein, Norway, and Switzerland.

A **customs union** also involves the elimination of tariffs between member states, but has a common protective wall against non-members. Where, in addition, internal restrictions on the movement of capital, labour and enterprise are removed, the result is a **common market**. Most customs unions have gone at least some way towards common market status. Examples include the European Union (Austria, Belgium, Denmark, Finland, France, Germany, Greece, Ireland, Italy, Luxembourg, Netherlands, Portugal, Spain, Sweden, and the United Kingdom), the Central American Common Market or CACM (Costa Rica, El Salvador, Guatemala, Honduras, Nicaragua), the Arab Common Market (Egypt, Iraq, Syria, Jordan), the Andean Pact (Bolivia, Colombia, Ecuador, Panama, Venezuela), the Caribbean Community and Common Market, The Economic Community of the States of Central Africa, and the Economic Community of West African States.

A still higher form of integration is the **economic union**, which, in addition to the characteristics of a common market, provides for integrated economic policies among member states, as in the European Union. The highest form of integration possible would have to involve some form of **supranational political union**, with a single monetary system and a central bank, a unified fiscal system, a common foreign economic policy, and a supranational authority with executive, judicial and legislative branches.

Beyond the EU, however, many free trade associations and common markets have found it difficult to overcome the obstacles imposed by memberships that include nations at very different levels of development, and that involve enormous distances and poorly developed transportation networks. It was in response to such problems that the GATT authorized, in 1971, the waiver of the Article 1 most-favoured nation provision for developing nations offering concessions to other developing nations. As a result, Mexico, for example, could offer to reduce its duty on a product from Bolivia without having to extend the same lower rate to the United States. The GATT decision meant that developing countries were free to experiment with a variety of integration models without incorporating internal free trade as a legally binding obligation. The result has been the emergence of a series of **trade preference associations** such as the Associations of Southeast Asian Nations or ASEAN (Brunei, Indonesia, Malaysia, the Philippines, Singapore, Thailand) and the Latin American Integration Association, formerly the Latin American Free Trade Association (Argentina, Bolivia, Brazil, Chile, Colombia, Ecuador, Mexico, Paraguay, Peru, Uruguay, Venezuela).

The increasing globalization of the world economy has broadened and deepened the trend towards regional economic integration. The Andean Pact, for example, decided in 1990 to establish an Andean Common Market by 1996. The following year, the Pact approved more trade liberalization reforms than it had in the previous 22 years of the Pact's existence. In 1989, the ASEAN countries joined with Australia, Canada, China, Hong Kong, Japan, New Zealand, South Korea, Taiwan and the United States to form the Asia Pacific Economic Co-operation group, with the objective of promoting the liberalization of trade and promoting cooperation in trade and investment around the Pacific Rim. In 1992, EFTA and the European Union agreed to establish a unified free trade zone, the European Economic Area (EEA), with a combined market size of 379 million people. (The EEA took effect on

377

common market
A form of international economic integration that involves the elimination of tariffs and other trade barriers between member states, the removal of internal restrictions on the movement of factors of production, and the creation of a common set of trade agreements with non-member countries.

economic union
A form of international economic integration that involves the removal of all internal barriers to trade and the movement of factors of production, the creation of a common set of trade barriers and trade agreements with non-member states, and the coordination of integrated economic policies within the union.

supranational political union
A form of international economic integration that extends beyond economic union unified fiscal and monetary system controlled by a supranational authority with executive, judicial, and legislative powers.

trade preference associations
Loose forms of transnational economic integration that involve reduced trade barriers between member states.

1 January 1993, but without the Swiss, whose electorate had failed to ratify the agreement.)

Meanwhile, Canada, Mexico, and the United States established a trading zone of 360 million consumers in 1992 with the completion of the North American Free Trade Agreement (NAFTA). This was not only an unprecedented economic integration of core countries and a semi-peripheral country but also the first instrument of economic integration to liberalize trade in services. The NAFTA will phase out tariffs and other trade and investment barriers between the countries over 15 years and will inevitably lead to a further reorganization of the economic geography of North America. Many manufacturing jobs have already been switched from Canada and the USA to Mexico; while the growing Mexican market is now open to US and Canadian automobiles and automobile parts, telecommunications, and financial services. Eventually, there will be total access in agricultural markets, which will rewrite the agricultural geography of Mexico and significantly modify agricultural patterns in the southwestern United States.

SPATIAL OUTCOMES OF ECONOMIC INTEGRATION

It follows from the basic principles of economic geography that the enlargement of markets and the removal of artificial barriers to trade will result in a realignment of patterns of economic activity. Two main sets of effects can in fact be anticipated. The first relates to patterns of trade. With transnational integration, the *removal of trade barriers should lead to a more pronounced regional division of labour*, with each region in the larger association tending to specialize in those activities in which it has the greatest comparative advantage. In effect, production is thereby reallocated from high-cost to low-cost settings, and a great deal of trade is generated within the association. At the same time, lower costs can, theoretically, be passed on to consumers, thus contributing to improved levels of living. These effects of integration are generally referred to as **trade creation effects**.

trade creation effects
The positive effects of transnational economic integration, resulting from the free movement of factors of production and free trade, which allows each region to specialize according to its comparative advantage, thus leading to a greater overall productivity and internal trade.

Countries that do not belong to the association, however, tend to lose trade: the external tariff wall prevents them from competing effectively with higher-cost internal producers whose output is able to circulate duty-free within the association. To the extent that the old sources of supply were more efficient producers than the new ones, **trade diversion** will have taken place, with the result that consumption is shifted away from lower-cost external sources to higher-cost internal sources, consumers have to pay more for certain goods, and levels of living may be depressed.

trade diversion
The displacement of pre-existing trade flows as a result of transnational economic integration.

The extent to which trade creation might outweigh trade diversion depends on several factors, including the degree to which the range of goods produced in member states overlap and the degree of pre-integration reliance on trade with countries outside the association. If integration *is* successful in the long run in creating trade and accelerating economic growth, it is possible that consequent increases in demand for goods and raw materials will generate 'spread effects', thus creating a positive spillover effect for other economies.

The second set of effects relates to patterns of regional development. Because of the need to exploit new patterns of comparative advantage, a certain amount of relocation of production must take place, with related activities tending to cluster together in the most efficient settings. The corollary is the disinvestment that takes place as production is withdrawn from less efficient locations. Given the logic of cumulative causation, the net effect in terms of regional development within the association will clearly be a tendency for *spatial polarization* as a result of 'backwash effects' (see p. 250). Because of the political dimension inherent to integration, this in turn provides a powerful case for a strong *regional policy*.

Meanwhile, integration can also be expected to precipitate other changes in patterns of regional development. A reorganization of patterns of production may occur where changes in patterns of comparative advantage are not sufficient to write off past investment or to prompt relocation, but are sufficient to justify intra-industry specialization. Steel-producing regions, for example, may come to special-ize in certain kinds of steel products rather than producing a broad spectrum of steel products for a domestic national market; or agricultural regions may move from mixed farming to a more specialized set of outputs.

Another important consequence of integration is the stimulus that is provided for foreign direct investment. Excluded by high external tariff barriers, foreign sup-pliers are likely to seek to open branch plants inside the association in order to get access to its market. If successful, this not only makes for a drain of capital when profits are repatriated; it also makes for a degree of *external control* of some local labour markets. Finally, we must consider the implications of integration for patterns of regional development *outside* the association. The most striking effects in this context will be those related to the dislocations experienced by specialized regions whose exports are no longer competitive within the protected market of the association.

These same principles and tendencies mean that we should expect integration to reinforce the dominant core–periphery patterns in the world's economic landscapes at the macro scale. Patterns of trade between core economies, for instance, are already so strong that integration is able to draw on a good deal of momentum. At the same time, *it is relatively easy for core states to meet the political, social and cultural prerequisites for successful economic integration*. These include:

- similarity in power of units joining the association;
- complementarity of élite value systems;
- the existence of pluralistic power structures in member countries;
- positive perceptions concerning (a) the expected equity of the distribution of benefits from integration, and (b) the magnitude of the costs of integration;
- the compatibility of states' decision-making styles;
- the adaptability, administrative capacity and flexibility of member states' governments and bureaucracies.

The success of the European Union has dramatized how effective integration between core states can be. Between 1959 and 1971, trade between the six original member countries increased nearly sixfold; by 1993 the expanded Union of 15 nations accounted for 21.5 per cent of all world trade, excluding intra-Union trade.

In the case of peripheral economies, on the other hand, patterns of trade offer little realistic scope for the reallocation of output following the removal of trade barriers in trade preference organizations, common markets or free trade associations. As we have seen (Chapters 2 and 9), most peripheral nations produce primary commodities which are exported to the core economies rather than to each other, and most are so short of capital that even pooled resources are likely to be insufficient to trigger economies of scale of sufficient magnitude to be able to break free from their functional dependency on trade with core economies. Experience has shown, meanwhile, that it is difficult for peripheral states to meet the political, social and cultural preconditions for successful economic integration. ASEAN, for example, despite having generated a growing sense of regional identity, has been unable to progress beyond a preliminary stage of economic regionalism. Regional projects such as the Asian Highway and the Mekong River Project have been discussed and tentative national responsibilities and commitments planned; but it has not proved possible to foster a significant increase in inter-ASEAN trade. Indeed, inter-ASEAN trade actually fell, as a proportion of total ASEAN trade with the world, during the 1970s. Moreover, a large proportion of inter-ASEAN trade is accounted for by exports that are transshipped through Singapore with only marginal value added by processing or packaging. Quite simply, the ASEAN economies are more complementary to those of Japan, the United States and Western Europe than they are to one another: ASEAN itself cannot absorb all the primary products it produces, and it is still dependent on these core economies for capital, technology, and many consumer goods.

Even less successful is the Andean Common Market, where integration has been truly half-hearted. Members have been unwilling to build integration into their own economic planning and policy-making and are unable to reach agreement about the harmonization of policies with regard to foreign trade, industrial development or fiscal affairs. Although some progress has been made at diplomatic levels (on pronouncements in favour of human rights in Nicaragua, for instance) and some increase has been achieved in absolute levels of intra-market trade, the negative effects of spatial and socio-economic polarization, particularly in Bolivia and Ecuador, have led to new tensions. Meanwhile, there have been virtually no positive effects in terms of the promotion of new sectors of production or the strengthening of existing regions of production (Puyana de Palacios, 1982).

One important response to such problems has been the so-called 'North–South dialogue'. The most important platform for this dialogue has been the United Nations Conference on Trade and Development (UNCTAD), launched in Geneva in 1964. By the end of the Geneva meetings, a degree of political solidarity had emerged among developing countries. Under the banner of the 'Group of 77' they issued a declaration:

> The unity [of the developing countries in UNCTAD] has sprung out of the fact that facing the basic problems of development they have a common interest in a new policy for international trade and development. The developing countries have a strong conviction that there is a vital need to maintain, and further strengthen, this unity in the years ahead. It is an indispensable instrument for securing the adoption of new attitudes and new approaches in the international economic field.

By 1987, the Group of 77 had nearly 130 members and had succeeded in articulating demands for a 'New International Economic Order' (not to be confused with the New International Division of Labour). Central to the new order envisioned by the Group of 77 are demands for fundamental changes in the marketing conditions of world trade in primary commodities. These changes would require a variety of measures, including price and production agreements among producer countries, the creation of international buffer stocks of commodities financed by a common fund, multilateral long-term supply contracts, and the indexing of prices of primary commodities against the price of manufactured goods. Such changes have been at the centre of discussions in a series of UNCTAD conferences, special sessions of the United Nations General Assembly, meetings of a specially convened Conference on International Economic Cooperation, and successive meetings of the heads of state of the British Commonwealth. Throughout these discussions, however, the core countries in general and the United States in particular have been reluctant to do more than agree to general statements about the desirability of a new international economic order. As a result, Williams' observation (1981, p. 99) remains true: 'It is clear to everyone that so far the North–South dialogue has failed, and the New International Economic Order is still a dream.'

In practice, therefore, there have been two dominant sets of spatial outcomes of transnational economic integration. One has simply been the reinforcement of the dominant core–periphery structure of the world economy because of the relative success of economic integration between core states. The second has been the imprint of this success on particular regions. This imprint can be discerned: (1) in terms of the effects of trade creation, trade diversion, spatial polarization, regional policy and socio-spatial tensions within core associations, and (2) in terms of the dislocations experienced within non-member states. In the remainder of this chapter we illustrate the importance and complexity of the second of these sets of spatial outcomes – the consequences of the success of economic integration between core states – using the example of the European Union.

The imprint of the European Union

The European Union had its origins in pragmatic responses to the changed economic climate of postwar Europe. The objective was to recapture the core status within the world economy that Europe had forfeited as a result of the war. Although there was a good deal of popular concern over the dominance of US-based transnationals in Europe's postwar economic recovery, the crux of the problem was that the centre of technological advance had moved to the United States. As a result, 'The real challenge was to ensure that Europe did not remain dependent on imported capital goods, and that it began to generate its own research so as to preempt the United States in any future technological cycle of production' (George, 1991, p. 59).

The European Community was formed in 1967 by an amalgamation of three institutions which had been set up in the 1950s in order to promote progressive economic integration along particular lines: Euratom, the European Coal and Steel

Community (ECSC) and the European Economic Community (EEC). This amalgamation fostered the recovery of core status:

> by providing favourable conditions for multinational investment, so bringing jobs and prosperity back to Europe; by encouraging the emergence of European multinationals that had cultural reasons for situating their headquarters and research facilities in Europe; by providing the conditions in which capital accumulation could proceed to the point where research and development funds were available from profits; but also by an injection of public funding into the process, through Euratom, and through EEC industrial research programmes.

(George, 1991, p. 59)

Having expanded from its six original members – Belgium, France, Italy, Luxembourg, the Netherlands and West Germany – to include Denmark, the Republic of Ireland and the United Kingdom in 1972, Greece in 1981, Portugal and Spain in 1984, and Austria, Finland and Sweden in 1995, the European Community now boasts a population of over 370 million, with a combined GDP 10 per cent larger than that of the United States. It has developed into a sophisticated and powerful institution with a pervasive influence on patterns of economic and social well-being within its member states and a significant impact on certain aspects of economic development within some non-member countries.

The initial cornerstone of the Community was a compromise worked out between the strongest two of the original six members. West Germany wanted a larger but protected market for its industrial goods; France wanted to continue to protect its highly inefficient (but large and politically important) agricultural sector from overseas competition (Baldwin, 1994). The result was the creation of a tariff-free market within the Community, a common external tariff, and a Common Agricultural Policy (CAP) to bolster the Community's agricultural sector. Given the nature of this compromise, it should be no surprise that the European Community performed very unevenly.

Meanwhile, the rest of the world economy had changed significantly, intensifying the challenge to Europe. By the early 1980s, the US and Japanese economies, having accomplished a large measure of restructuring, were becoming increasingly interdependent and prosperous on the basis of globalized producer services and new, high-tech industries. This threatened much of Western Europe with the prospect of reverting once again to semi-peripheral status within the world economy. London's once pre-eminent financial services were losing ground to those of New York and Tokyo; and even West Germany, with the Community's healthiest economy, faced the prospect of being left as a producer of obsolescent capital goods and consumer goods. In response, the European Community relaunched itself, beginning in the mid-1980s with the ratification of the Single European Act (1985) that affirmed the ultimate aim of economic and political harmonization within a single supranational government (Wise and Gibb, 1993). The relaunch was completed in 1992 with the Treaty of European Union (the 'Maastricht Treaty'), which conferred on the Union most of the major functions of a sovereign nation state, including:

- the creation of a single currency;
- the coordination, supervision and enforcement of economic policies;

- the maintenance of a completely free internal market;
- the preservation of law and order;
- the protection of fundamental rights of individual citizens;
- the maintenance of equity and, where necessary, the redistribution of wealth between regions;
- the management of a common external policy covering all areas of foreign policy and a common defence policy.

This relaunching represents an impressive achievement, particularly since it had to be undertaken at a time when there were major distractions: having to manage a changing relationship with the United States through GATT renegotiations, having to cope with the reunification of Germany and the break-up of the former Soviet empire in Eastern Europe, and, not least, having to cope with a resurgence of nationalism (see Chapter 12). Ratification of the Maastricht Treaty was in fact achieved in 1993 only after last-minute manoeuvrings prompted by the concerns of Danish and UK voters over aspects of sovereignty. Despite such misgivings, however, the economic benefits of EU membership are widely recognized. Indeed, there is a growing list of countries seeking membership, including former Soviet satellites Bulgaria, the Czech Republic, Hungary, Poland, Romania and Slovenia.

Trade creation

The economic benefits of EC membership were soon felt. Even by 1970, trade between member countries was 40–50 per cent more, overall, than it would have been if the Community had not been formed; by 1980 the figure had risen to a gain of between 100 and 125 per cent. The net benefits of this increase are far from clear, however, since it is generally acknowledged that the overall increase in intra-Union trade has been the product of a high degree both of *trade creation* and of *trade diversion*.

It may seem somewhat surprising that there is little hard evidence as to the actual magnitude of these effects, given that the putative benefits of trade creation are fundamental to the Union's existence, but it must be acknowledged that it is very difficult to isolate the effects of Union membership from other effects, such as transnational corporate activity. What does seem clear, however, is that overall increases in intra-Union trade have generated *scale economies* for EU producers that have in turn stimulated further trade, accelerated changes in industrial structure and corporate organizations, and brought about efficiency gains in both importing and exporting countries. Owen (1983) estimated that these economic benefits could be more than half as great as the value of trade itself.

Spatial polarization

It is also clear that these benefits have been associated with a significant amount of regional change within the Union (Dawson, 1993; Williams, 1991), although once again it is difficult to isolate the effects of the common market from others. In overall terms, the removal of internal barriers to labour, capital and trade has worked to the clear disadvantage of peripheral regions within member states and in particular to the disadvantage of those furthest from the 'Golden Triangle' (between

Amsterdam, Brussels and Cologne) that is increasingly the 'centre of gravity' in terms of both production and consumption. At the same time, integration has accelerated and extended the processes of concentration and centralization, creating structural as well as spatial inequalities. As Holland put it:

> the market of the Community is essentially a capitalist market, uncommon and unequal in the record of who gains what, where, why and when. Its mechanisms have already disintegrated major industries and regions in the Community and threaten to realize an inner and outer Europe of rich and poor countries.

(1980, p. 8)

Slower-growing states with economies dominated by inefficient primary or manufacturing industries are, in short, in danger of becoming backward problem regions within a prosperous EU. Evidence on trends in personal incomes supports this prognosis. Although the overall range of incomes has remained more or less constant, there is clear evidence of a steady convergence of per capita income at the top end of the range in Belgium, Denmark, France, Luxembourg, the Netherlands and Germany. This has led in effect to a two-tier Community as far as income levels are concerned, with the lower tier consisting of Greece, Ireland, Italy, Portugal, Spain and the United Kingdom (Hadjimichalis and Sadler, 1995).

The effects of the CAP

The most striking changes in the regional geography of the EU, however, have been those related to the operation of the CAP. It is the CAP which dominates the EU budget. For a long time, it accounted for more than 70 per cent of the EU's total expenditures, and it still accounts for more than 40 per cent. Its operation has had a significant impact on rural economies, rural landscapes and rural levels of living; and has even influenced urban living through its effects on food prices.

The basis of the CAP was a system of support for farmers' incomes that was operated through the artificial support of wholesale prices for agricultural produce. While motivated mainly by political considerations, the CAP provided a relatively risk-free environment in which investment for farm modernization could be encouraged. At the same time, stable, guaranteed prices provided security and continuity of food supplies for consumers. Assured markets also allowed trends in product specialization and concentration by farm, region and country to proceed at a faster rate than might otherwise have occurred, as Bowler (1985) showed in his survey of the geography of agriculture under the CAP. Not all products have been subject to CAP support, however. While regions specializing in crops and livestock subject to price guarantees, intervention and market regulation have been able to intensify their specialization, other regions have been subject to Union-wide competition.

The overall result has been a *realignment of production patterns*, with a general withdrawal from mixed farming. Ireland, the United Kingdom and Denmark, for example, have increased their specialization in the production of wheat, barley, poultry and milk; while France and Germany have increased their specialization in the production of barley, maize and sugar beet. It is at regional and sub-regional scales that these changes have been most striking. CAP support for oilseeds, for example, made rapeseed a profitable break-crop in cereal-producing regions of the

United Kingdom, with the bright yellow flowers of the crop bringing a remarkable change to the summer landscapes of the countryside.

The reorganization of Europe's agricultural geography under the CAP also brought some *unwanted side effects*, however:

1. Environmental problems occurred because of the speed and scale of modernization, combined with farmers' desire to capitalize on generous levels of guaranteed prices for arable crops. In particular, moorlands, woodlands, wetlands and hedgerows have come under threat, and some 'vernacular' landscapes have been replaced by the prairie-style settings of specialized agribusiness.

2. Another serious problem with geographical implications concerns the large surpluses fostered by the price support system. Prices set to give a reasonable return to producers on small farms were so favourable to the modernized sector of European agriculture that 'mountains' of beef, butter, wheat, sugar and milk powder and 'lakes' of olive oil and wine had to be sold off at a loss to neighbouring countries, dumped on world markets, or 'denatured' (rendered unfit for human consumption) at a considerable cost.

3. A third set of problems arose from the income transfers caused by CAP policies. Price support mechanisms involve a transfer of income from taxpayers to producers and from consumers to producers. There is plenty of evidence to show that these transfers are regressive within member countries and inequitable between them. Expenditure on food generally accounts for a larger proportion of disposable income in poorer households than in better-off households. Producers, on the other hand, benefit from price support policies in proportion to their total production, so that the larger and more prosperous farmers receive a disproportionate share of the benefits. Spatial inequity arises because countries or regions that are major producers of price-supported products receive the major share of the benefits while the costs of price support are shared among member nations according to the overall size of their agricultural sector. Furthermore, the CAP pricing system made no concessions for a long time to the variety of agricultural systems practised on farms of different sizes and in different regions. As a result, areas with particularly large and/or intensive or specialized farm units (such as northern France and the Netherlands) benefited most, together with regions specializing in the most strongly supported crops (cereals, sugar beet and dairy products). Effectively, this has meant that the most prosperous agricultural regions have benefited most from the CAP, so that farm income differentials within member countries have been maintained, if not reinforced.

4. In addition to all this, the budgetary cost of the CAP escalated. By 1983, budgetary problems had become acute; but reform of the CAP was hampered by domestic political considerations in member countries that were the biggest beneficiaries of the CAP. The CAP became a source of serious disharmony, particularly in the United Kingdom, where, before EU membership, food policies had been progressive, subsidizing lower-income households. Embracing the CAP meant a higher and regressive system of food prices without any

compensatory benefits: peasant farming and inefficient agricultural practices had been purged from the UK economy long before.

Meanwhile, EC agricultural subsidies had become a serious issue in GATT renegotiations; and the relaunch of the EC in the mid-1980s required a more open and competitive approach to internal markets in every sector, including agriculture. Together with increasing awareness of the unwanted side-effects of the CAP, these considerations eventually led to a reform of the CAP. As a first step, the Union agreed in 1992 to cut guaranteed cereal prices by 29 per cent and guaranteed beef prices by 15 per cent over the following 3 years. At the same time, production quotas for dairy farmers were reduced.

Regional policy

The United Kingdom's accession to the Community in 1972 highlighted the lopsidedness of Community policy in favour of rural interests compared with those of industrial areas. As a result, the Community was persuaded to launch the European Regional Development Fund (ERDF). Although the Community had effectively operated 'regional' policies through the ECSC and the European Investment Bank (EIB) for some time, there had been no comprehensive, coordinated framework within which to operate. The ECSC was limited to the 'readaption' of workers and the 'conversion' of local economies in depressed coalmining and steel-producing regions. The EIB was a Community banking system designed to reduce intra-Community disparities in economic development by disbursing loans to selected projects in priority regions; but although it was particularly influential in sponsoring projects in marginal, trans-frontier regions, it was simply not equipped to deal with the casualties of regional economic restructuring within an expanding common market.

The entry of the United Kingdom to the Community not only made for a significant increase in the scope and intensity of regional restructuring processes but also brought a legacy of chronic regional problems and, with them, a certain political resolve. Following an examination of the issues (Commission of the European Communities, Thompson Report, 1973), the Community launched the ERDF in 1975 with a relatively modest budget (1300 million European Units of Account, compared with the 2250 million recommended by the Thompson Report).

The addition of Portugal and Spain to the Community in 1984 changed both the nature and the intensity of regional problems (Artis and Lee, 1994). The proportion of the EC population living in 'least favoured' regions (those where Gross Domestic Product is under half the EC average) doubled, with most of the increase being accounted for by depressed rural regions. At the same time, the relaunch of the EC, with more open internal markets, brought the probability of intensified spatial polarization. This was recognized by the Single European Act (SEA), which raised 'economic and social cohesion' to the status of a new policy objective within the Community. The SEA doubled the funding – in real terms – for grants for regional development assistance (from 7 billion ECU to 14 billion ECU at 1988 prices). This was further reinforced by the budget for 1993–97 (the so-called Delors II Package), which contained a real increase of 30 per cent in the Community budget, including

10 billion ECU over 5 years for a Cohesion Fund to help Greece, Ireland, Portugal, and Spain achieve comparable levels of economic development to the rest of the Community. By 1999, the Structural Funds will account for 33 per cent of the overall EU budget.

Meanwhile, the relaunch of the Union and the reform of the CAP provided the impetus for a reform of EU regional policy objectives (Sadler, 1992). The result was that six *Priority Objectives* have been established to guide the disbursement of regional development assistance grants through three *Structural Funds*. These are defined as follows:

Objective 1: to promote the development and structural adjustment of under-developed regions
Objective 2: to redevelop regions or local labour markets which are seriously affected by industrial decline
Objective 3: to combat long-term unemployment, to provide career prospects for young people, and to reintegrate persons at risk of being excluded from the labour market
Objective 4: to facilitate the adaptation of workers to industrial change and developments in production systems
Objective 5: to speed up the adaptation of production, processing, and marketing structures in agriculture and forestry and to help modernize and restructure the fisheries and aquaculture sectors (Objective 5a); and to promote the development of rural areas (Objective 5b)
Objective 6: to promote the development of the northern regions in the new member states in Scandinavia.

The main instrument for achieving Objectives 1, 2, 5b and 6 is the ERDF; objectives 3 and 4 are addressed mainly through the European Social Fund (ESF); and Objective 5b is addressed through the European Agricultural Guidance and Guarantee Fund (EAGGF). Objectives 1, 2, 5b and 6 have an explicit regional dimension (Fig. 11.2). Objective 1 covers those regions where capita GDP is less than 75 per cent of the EU average. These are predominantly peripheral regions with relatively little industry or where industry is threatened. They include the whole of Greece, Ireland, and Portugal, some 70 per cent of Spain, the Italian Mezzogiorno, Corsica, northeastern France, Northern Ireland, the Scottish Highlands, Merseyside, Burgenland (Austria), Flevoland (Netherlands), Hainaut (Belgium), and the whole of former East Germany. More than 25 per cent of the population of the EU lives in these regions.

Declining industrial regions (Objective 2) have a rate of unemployment above the EU average, higher than average levels of employment in manufacturing, and significant job losses in the manufacturing sector. About 58 million people, nearly 17 per cent of the total EU population, live in these areas. The areas covered by Objective 5b are selected on the basis of their level of socio-economic development, assessed in terms of per capita GDP; the proportion of jobs in agriculture; levels of agricultural incomes; low population density; and depopulation. They account for 28.5 million (8.2 per cent) of the EU population. The regions targeted by Objective 6 are the very sparsely populated areas of Finland and Sweden north of the 62nd parallel.

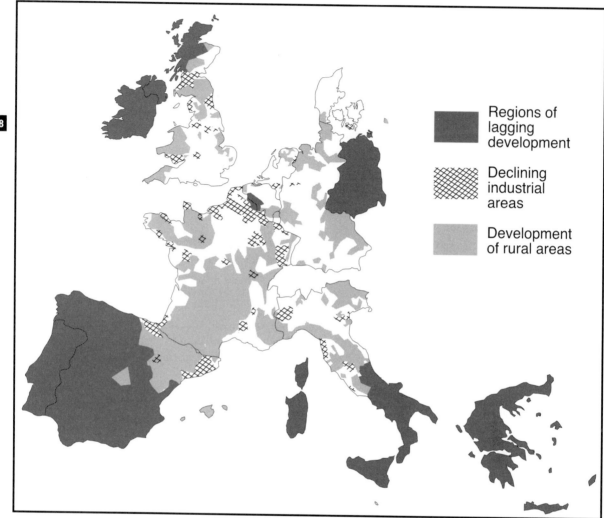

Regions of lagging development

Declining industrial areas

Development of rural areas

Figure 11.2
Regional policy in the
European Union

These policies clearly represent a serious response to the spatial implications of economic integration. It is too soon to tell how effective they will be in redressing the regional restructuring and spatial polarization that have accompanied the creation and enlargement of the common market. Yet it is debatable whether regional policies can in fact do so, particularly since the new reach and flexibility of multinational corporations can exploit cost advantages elsewhere in the world that EU incentives could never hope to match (Hardy *et al.*, 1995).

External effects of the EU

Meanwhile, the scale of the EU and its maintenance of a strongly protectionist agricultural policy as well as a protected common market have inevitably had a significant impact on non-member countries: diverting trade and creating complex new layers of interdependence. Much of this complexity relates to the 'pyramid of privilege' that has arisen from the Community's trade agreements with different groups of non-member countries. At the base of the pyramid is a Generalized System of Preferences negotiated through UNCTAD. This allows access to the Community market for a broad range of products from developing countries. Bilateral trade agreements also exist with some countries as a result of attempts by the EU to extend and diversify its trading patterns. The most favourable trading privileges are extended to a large group of countries in Africa, the Caribbean and the Pacific (the 'ACP states'), most of them former colonial territories of member states. Originally established at the Youndé Convention in 1963 and later extended at Lomé Conventions in 1975 and 1979, these privileges allow access to the Community market for tropical agricultural products without having to provide reciprocal privileges to EU members or abandon trading agreements with other developing countries. They also involve an export revenue stabilization scheme – STABEX – that covers nearly 50 key primary products and raw materials.

In detail, the mechanics of these privileges are complex, and it is very difficult to assess their impact on patterns of trade and development. It is clear, however, that the 'privileges' extended to non-members are essentially designed to enhance the position of the EU rather than to contribute to a New International Economic Order. 'Sensitive' products (i.e. those that compete directly with EU agricultural and industrial products), for example, are excluded from preferential treatment, or are subject to seasonal restrictions. Moreover, the net effect of the Lomé 'privileges' has been to increase the dependency of many countries on exports of a narrow range of primary produce to the EU market. Particular examples include Burundi (coffee), Uganda (coffee), Chad (cotton), Ghana (cocoa), Ivory Coast (groundnuts), Senegal (wood, groundnuts), Sudan (coffee), and Tonga (copra) (Bowler, 1985). As a result, EU relations with ACP countries have been interpreted as neo-imperialist, effectively extending the core–periphery structure of the world economy.

It is not only peripheral countries that have been affected by the EC, however. The EU represents an outcome of the struggle for economic power within the core and to preserve power in relation to the semi-periphery as much as it is an attempt to consolidate power in relation to the periphery. EU trade relations with the United States have been fractious, while the EU has been forced to mount a 'diplomatic assault' on Japan in an attempt to stem the impact of Japanese direct investment in sophisticated manufacturing industries (automobiles, electronics, etc.) within the EU.

Core and semi-periphery countries have also been directly affected by the trade-diverting effects of the EU's protection of 'temperate' agricultural products. Trade diversion effects are particularly evident where EU subsidies have produced large surpluses for export. EU exports of beef, for example, rose from 5 per cent of world trade in 1977 to over 20 per cent in 1980 and in so doing displaced Australian and Argentinian exports to Egypt and Uruguayan exports to Ghana (Bowler, 1985). Other striking cases of trade diversion have occurred as specialist producers of 'temperate' products with strong traditional ties to European markets found themselves largely excluded by the EU's external tariff wall. New Zealand is a good example. The United Kingdom used to take nearly all of New Zealand's butter, cheese and lamb, so that after the United Kingdom joined the EU, New Zealand agriculture had to be restructured, new products had to be developed (a notable success here being the kiwi fruit) and new markets had to be penetrated in Latin America, India and Japan – in the face of competition from the subsidized surpluses of dairy produce from the EU.

Summary

In this chapter we have shown how the imprint of transnational economic and political integration has begun to affect the world's economic landscapes as nation states have responded to the changing economic and geopolitical context of the world economy. This response has resulted in a variety of forms and levels of integration, but in practice the basic principles of economic geography have resulted in three main outcomes:

- the reinforcement of the dominant core–periphery structure of the world economy;
- the spatial reorganization of production as trade creation and trade diversion affect both member and non-member states;
- the creation and intensification of regional polarization as the economies of scale and multiplier effects in regions most favoured by integration create backwash effects elsewhere.

Yet, while it is important to acknowledge transnational integration as a response to the globalization of the world economy, it would be unwise to overstate its effects. Transnational and supranational organizations are not about to replace nation states, and we must recognize their limits as contributors to the constant rewriting of the world's economic landscapes. The further integration of the EU, for example, is seriously hampered by a number of issues that transcend its territory and jurisdiction, including its inability to curb inflation, to reduce unemployment, or to stem the 'brain drain'. Moreover, as Nairn pointed out, spatial polarization within the EU has 'sought out and found the buried fault lines of the area. ... Nationalism in the real sense is never a historical accident, or a mere invention. It reflects the latent fracture lines of human society under strain' (1977, p. 69). Indeed, nation alism and localism can be seen to be intensifying, not only in response to the backwash effects of transnational integration but also in response to the overall globalization of the economy, the internationalization of culture and society, and the insecurity and instability generated by the transition to advanced capitalism. In the next chapter of the book, therefore, we turn to an examination of decentralist reactions to the changing world economy.

Key Sources and Suggested Reading

Artis, M. and Lee, N. (eds) 1994. *The Economics of the European Union.* Oxford: Oxford University Press.

Baldwin, R. E. 1994. *Towards an Integrated Europe.* London: Centre for Economic Policy Research.

Bowler, I. 1985. *Agriculture under the Common Agricultural Policy.* Manchester: Manchester University Press.

Cohen, S. B. 1982. A new map of global geopolitical equilibrium: a developmental approach. *Political Geography Quarterly,* 1, 233–41.

Commission of the European Communities 1973. Report on the Regional Problem in the Enlarged Community (Thompson Report). Brussels: Commission of the European Communities.

Dawson, A. H. 1993. *A Geography of European Integration*. London: Belhaven Press.

Day, G. and Rees, G. (eds) 1991. *Regions, Nations, and European Integration: Remaking the Celtic Periphery*. Cardiff: University of Wales Press.

Edwards, C. 1985. *The Fragmented World. Competing Perspectives on Trade, Money, and Crisis*. London: Methuen.

George, S. 1991. European political cooperation: A World-Systems perspective. In M. Holland (ed.), *The Future of European Political Cooperation: Essays on Theory and Practice*. New York: St Martin's Press.

Gibb, R. and Michalak, W. (eds) 1994. *Continental Trading Blocs: The Growth of Regionalism in the World Economy*. Chichester: Wiley.

Hadjimichalis, C. and Sadler, D. (eds) 1995. *Europe at the Margins: New Mosaics of Inequality*. Chichester: Wiley and European Science Foundation.

Hardy, S., Hart, M., Albrechts, L. and Katos, A. (eds) 1995. *An Enlarged Europe: Regions in Competition?* London: Regional Studies Association.

Harvey, D. W. 1988. The geographical and geopolitical consequences of the transition from Fordist to Flexible Accumulation. In G. Sternlieb and J. W. Hughes (eds), *America's New Market Economy*. New Brunswick, NJ: Center for Urban Policy Research.

Hoekman, B. M. and Kostecki, M. M. 1996. *The Political Economy of the World Trading System: From GATT to WTO*. New York: Oxford University Press.

Holland, S. 1980. *UnCommon Market*. London: Macmillan.

Martin, R. 1986. Thatcherism and Britain's industrial landscape. In R. Martin and B. Rowthorne (eds), *The Geography of Deindustrialization*. London: Methuen.

Nairn, T. 1977. Super-power or failure? In T. Nairn (ed.), *Atlantic Europe?* Amsterdam: Transnational Institute, 68–77.

Owen, N. 1983. *Economies of Scale, Competitiveness, and Trade Patterns within the European Community*. Oxford: Clarendon Press.

Puyana de Palacios, A. 1982. *Economic Integration Among Unequal Partners: The Case of the Andean Group*. New York: Pergamon.

Sadler, D. 1992. Industrial policy of the European Community: strategic deficits and regional dilemmas. *Environment & Planning A*, **24**, 1711–30.

Taylor, P. J. (ed.) 1992. *The Political Geography of the Twentieth Century. A Global Analysis*. London: Belhaven.

Williams, A. M. 1991. *The European Community*. Oxford: Blackwell.

Williams, G. 1981. *Third World Political Organizations*. Montclair, NJ: Allenheld, Osmun.

Wise, M. and Gibb, R. 1993. *Single Market to Social Europe*. London: Longman.

Picture credit: Martin Wyness/Still Pictures

THE REASSERTION OF THE LOCAL IN THE AGE OF THE GLOBAL: REGIONS AND LOCALITIES WITHIN THE WORLD ECONOMY

A persisting theme of this book is the existence of trends towards ever more powerful states and ever-larger corporate structures. In the broader sweep of change within the world economy, these trends can appear to be inexorable and irreversible. Similarly, the increasing prominence of supranational institutions and initiatives (such as the EU, NAFTA, the WTO or the IMF) and transnational corporations can suggest a pervasive bureaucratization of modern life under the control of fewer and fewer organizations and individuals. Yet, while trends toward centralization, homogenization and standardization are real enough, there is also evidence for persisting and even increasing differentiation and decentralization: the peripheral industrialization that has come with the New International Division of Labour and the growth of the NICs, the apparent reversal of previously depressed or underdeveloped local economies (e.g. the Sunbelt phenomenon in the USA), and the revival or creation of regional identities (e.g. Ukraine, Quebec, Scotland, Catalonia, Lombardy, Punjab), for example.

The two sets of phenomena are often related. Thus, for example, it is the centralization of economic power in transnational corporations that has often led to a decentralization of their productive activities (as noted in Chapters 3, 6 and 10); and it is attempts at political and cultural homogenization through political unification that have generated resistance at the local or regional level (as noted in Chapters 3 and 8). The end of the Cold War and the collapse of the former Soviet Union have given an added fillip to economic decentralization and political fragmentation. Whether this signals for Eastern Europe and the former Soviet Union a permanent trend or a temporary hiatus prior to renewed political-economic centralization is open to question.

However, a general trend all over the world in the wake of the increased integration of the world economy has been an enhanced differentiation between places. So, even as the world has shrunk in real terms with respect to flows of goods, services and investments, small differences in economic characteristics and cultural practices have taken on greater significance. As a result, pressures towards a localizing of political decision-making power and political identities have increased. Three kinds of 'decentralist reaction' have been increasingly common since the 1970s:

First, national governments have had to satisfy local and regional constituencies that they represent their best interests. When faced by geographically differentiated patterns of economic growth and decline, regional policies and regional devolution have been important responses.

Second, many modern national states are internally divided along cultural lines with regional/geographical bases. This has sometimes led to national separatist movements directed towards achieving autonomy or independence for disaffected regions.

Third, and most generally, the growing globalization of the world economy has encouraged decentralization rather than centralization in the location of economic activities; whether in the form of the branch plants of big corporations or the localization and clustering of specialized small firms inherent in industrial districts. In particular, small-scale production has become of increasing importance (the 'batch production', etc. noted in Chapters 7 and 10). Ideas such as 'basic needs', 'appropriate technology' and 'local control' have become increasingly attractive in this context in framing the basic demands of new political movements calling for economic as well as political democracy. Though usually dismissed as 'utopian' such ideas have become especially attractive as global resource and pollution problems arising as by-products of constant increases in global production and consumption have attracted more attention.

REGIONALISM AND REGIONAL POLICY

As we have shown in previous chapters, processes of economic growth and decline are not geographically neutral in their impact. In particular, the locational require-ments of new profitable manufacturing production under the market-access regime and the growing service industries (such as finance) are likely to differ from those of established production (see Chapters 3, 6 and 10). It is in this context that appeals for governmental action arise to 'help' a particular region or set of regions either 'adjust' to a new economic situation or encourage compensating investment by means of fiscal measures such as tax breaks or relocation allowances.

In the 1950s and 1960s there was widespread acceptance in many core countries of the need to encourage regional 'balance' in economic growth at a time when established regional economies were beginning to experience challenges to their competitive advantage and 'poorer' regions (such as the Italian South or the US South) were seen as lagging behind other regions. It is no coincidence that this acceptance flourished at a time of relative prosperity: quite simply, affluent societies could afford to indulge in redistributive policies. In some countries this took the form of revitalizing or establishing lower 'regional' tiers of government. Regional governments were viewed as agents for maintaining or attracting private invest-ment. In some countries, such as Italy, Norway, and, to a lesser degree, Britain, regional authorities were introduced to encourage regional economic planning and foster local industrial regeneration. In many countries, especially those with federal political systems (such as the United States, Canada, Australia, Switzerland and Germany), lower-tier governments have traditionally played an important role in stimulating economic growth within their territories.

Two questions are especially pertinent with respect to the history of regional policy. One concerns the extent to which **regionalism**, or explicit commitment to spatial or regional planning, has inspired regional policy. The second involves the impact, if any, of regional and local development policies organized by lower tiers of government.

With regard to the first, some countries such as France, Germany and Italy have long established traditions of regionalism. In particular, French programmes of 'territorial management' and the Italian 'Cassa del Mezzogiorno' (Southern Development Agency, replaced in 1987 by several smaller agencies) provide well-known examples. Britain and the United States acquired formal regional policies in the 1930s but in neither case has there been the same political consensus in favour of such policies (or anything that smacks of formal 'planning') as in other European countries and elsewhere (e.g. Japan, Brazil, India). In both cases earlier initiatives have been largely abandoned since the late 1960s in favour of either very localized programmes, such as 'enterprise zones', or lower-level government rather than national-level policies.

This reflects in part the coming to national power of governments ill-disposed on ideological grounds to government intervention in the direction of economic activities to particular places. But it also reflects a negative appraisal of the effects of previous planning activities. If regional unemployment rates can be used as an indicator of regional 'economic well-being', the fact that such rates are highly correlated over time and across countries irrespective of the commitment to regional policies suggests that such policies do not make much difference (Chisholm, 1990, pp. 167–9). Incidentally, however, this also suggests a lack of evidence for the long-run spatial equilibrium in the distribution of economic activities assumed in most static models of regional development (discussed in Chapters 7 and 10) and, hence, for the ideology that has often inspired the abandonment of regional policies!

The trend throughout the core countries over the past 20 years has been away from formal regionalism sponsored by national governments towards the adoption of competitive spatial policies by regional and local governments. This has older roots, particularly in the United States where it dates back to the years immediately after the Second World War when southern states such as Tennessee and Mississippi began 'attracting' firms from the Northeast and Midwest with a mix of low production costs and fiscal advantages (especially low taxes). This approach spread widely in the USA with the onset of the massive restructuring of industry in the early 1970s. At the same time the narrow focus on attracting industry shifted to a broader concern with general local economic competitiveness, primarily through improving the overall 'climate' for business and creating a mix of incentives for stimulating 'new' industries with potential multiplier effects.

Most of the American states and many municipalities have now created economic development agencies to attract industry and foster indigenous economic development. Particularly conspicuous have been 'business–public sector partnerships' and offices established abroad, in London, Tokyo or Brussels, to entice

---

regionalism
Designation of regional units as the basis for allocating economic activities by central government.

foreign business to particular locations in the USA. This latter strategy has paid off handsomely for some states, such as Ohio and Kentucky, which succeeded in beating off other US states in attracting major Japanese auto assembly plants to their jurisdictions. Local government intervention of a similar type began to appear in Britain and other European countries in the 1980s, in part taking a leaf out of the American book but also reflecting the availability of funds from the EU to provide grants and cheap loans to prospective employers.

The overall effectiveness of these local development efforts remains in doubt. While 'success stories', such as that of Kentucky in the USA or the local government–small business linkages present in many parts of central and northeast Italy, are well known, evidence for the positive impact of these efforts in general is mixed. Frequently, local programmes of tax abatements and subsidized plant and equipment merely remove industry from another state or municipality rather than building fresh capacity and employment. Ironically, given the usual association of 'good' business climate with low taxes and limited public services, some evidence suggests that in the USA state and local education, training, and infrastructure expenditures are more beneficial in generating fresh investment and new industry than are subsidies to individual firms (Wasylenko, 1991). Systematic fiscal reform, in the sense of aiming for low tax rates and broad tax bases, combined with efficient service delivery seem to offer the best formula for successful local development efforts in the USA. Specific tax subsidies to firms (including so-called enterprise zone experiments) seem a much less successful route to job growth and overall economic development in local economies (Fisher, 1997; Wasylenko, 1997; Fisher and Peters, 1997). The theoretical framework outlined in Chapter 3 would suggest that the best local policies would be those that assist clustering by firms so as to increase transactional, learning, work-training and other linkages. Otherwise competition between states and localities might encourage a 'race to the bottom' with jurisdictions such as Mississippi becoming the norm against which states with long traditions of active government intervention to regulate workplaces and encourage economic development based on high quality public services would find themselves at a disadvantage in the scramble to attract inward investment (Harrison, 1997).

A basic dilemma remains unresolved, however. When local policies successfully promote economic development, both the capital and the labour (when educated or skilled) that benefited from the policies will probably act to undermine what the policies initially achieved. Capital will do this through takeovers and moving investment elsewhere, labour by migration to higher wage areas. This is the paradox of planning regional and local initiatives in the context of a world economy that is dynamic and 'placeless' in its orientation to securing improvements in rates of return on investment. Yet, at the same time, rather than 'rooting capital' in weak as well as strong local economies, local policies will, as the model of regional cumulative causation described in Chapter 7 might suggest, produce deepening spatial inequalities. Richer localities will have advantages in revenues and infrastructure that poor ones lack. The end result of this will be greater geographical concentration of productive economic activities (see, e.g. Eisenschitz and Gough, 1996).

Whatever the strength of this logic, however, after the downturn of the world economy beginning in the mid-1970s and extending into the 1990s local governments in many countries did became relatively focused on economic development efforts. What they were able to achieve, however, was constrained by their relative autonomy and by the need for national governments to curb public spending and reduce taxation. In Britain the absence of a formal regional tier of government has been a particular drawback to local initiative. Localities, such as city governments, are often too small to create effective economic development policies. Their control over physical (land use) planning is likewise too parochial when the environmental and labour market impacts of 'new' industries extend beyond jurisdictional boundaries. Indeed, from one point of view Britain has the worst of all worlds because of the absence of a regional tier of government: the national government monopolizes most controls over economic development and local government exercises control over physical planning. There is no intersection of authority or coordination of powers (Cheshire *et al.*, 1992). This disadvantages Britain as a whole in a context of increased international competition in which regions elsewhere are able to offer 'packages' of advantages not available in Britain. But even in Italy, usually presented as a 'showcase' of successful economic regionalism, the regions are relatively weak institutions with limited powers and a low popular profile. The Italian administrative regions have served mainly as spending agencies for national government policies. They also vary substantially in size, competence and legitimacy; with those in the South particularly disadvantaged (Bellini, 1996). Strong 'regional motors' (such as those in central and northern Italy), therefore, do not necessarily generate a parallel strong model of regional or local governance.

One trend that goes against the tendency for enhanced competition between regions and localities is the emergence of 'compacts' or agreements between regions in different countries with respect to technology licensing and plant establishment (see e.g. Cooke, 1992). Thus, such regions as Baden-Wurttenberg in Germany and Wales in Britain have arranged contacts to encourage the flow of technology and investment from the former to the latter. Of course, such cooperation can be regarded as simply a way of encouraging competitive advantage for the regions concerned rather than something totally different from more typical competitive strategies.

The 'fiscal crises' experienced by many national governments beginning in the 1970s, increasing expenditures on 'entitlements' (social security, healthcare) and defence were not matched by corresponding increases in revenues, led to increased pressure on local governments to provide compensatory spending on education and other public services. This has at one and the same time increased the fiscal problems of local governments, especially apparent in the USA, and reduced their ability to negotiate special 'deals' with increasingly mobile businesses. Indeed, a strong case can be made that there has been a rolling back of government in many countries irrespective of the level at which it operates. Regions and localities are as caught up in the pressures from the market-access regime as are national governments. The primacy of market forces leads to an emphasis on entrepreneurialism, the imperative of flexibility and a 'business localism' in which there is little or no popular oversight or control (see, e.g. Lloyd and Meegan, 1996).

Direct central government intervention has still remained an important dimension of regional policy, if in some countries more than others. Substantial regional variations in economic development (especially when they are reflected in high regional unemployment rates) have been widely viewed as creating a national political problem, threatening the social and economic cohesion of the state itself. As a result, and beginning in the 1930s, national governments have felt it necessary to intervene in the economic geography of their territories by manipulating the costs of production.

The policies adopted have varied over time and by country. Some have sought to reduce the costs of fixed capital in declining regions by undertaking government investment in infrastructure, providing subsidies for private investment, etc. (German and Italian policies have emphasized such strategies). Others have sought to reduce the costs of capital in declining regions through subsidies for labour costs in those regions or through artificially increasing costs in 'overheated' regions (in the 1960s French and British policies tended in this direction). In other words,

> The goal has been both to direct employment-generating activities into the depressed regions and to stimulate growth generally. Conventional wisdom suggested that the former should occur in any case, that the geographical disparities in wages should stimulate the migration of workers to the higher-wage areas and the migration of employers to low-wage areas. However, the problems of the immobility of labour, of abandoning expensive fixed investments, and of inertia prevented a new equilibrium situation occurring naturally and quickly.
>
> (Johnston, 1986, p. 270)

The increased globalization of the world economy, however, has challenged the relevance of such conventional regional policies. In the first place, the largest firms no longer choose sites from among a single national set. There is no longer much identity between the scale of economic-locational decision-making and the spatial scale over which governments can exert their fiscal powers. Indeed, in this context 'the difficulties of guaranteeing full employment nationally mean that the state has increasingly to focus its attention on the national rather than the regional crisis; regional policy is in large part irrelevant' (Johnston, 1986, p. 274).

In the second place, the cost of subsidies raises government spending and produces a macroeconomic environment that is unattractive to global capital. Regional policy is then viewed as both an expensive luxury and an increasing liability. This was very much the view adopted by the Thatcher/Major governments in Britain and the Reagan/Bush administrations in the United States. Reducing state spending on regional policy and other 'welfare' programmes is seen as a necessity for improving national competitiveness in a global economy. In Britain between 1979 and 1985, for example, regional aid was cut from 842 to 560 million pounds; a cut in real terms of exactly one-third. Moreover, the areas eligible for aid were 'rolled back' considerably (Fig. 12.1).

Third, and finally, as Doreen Massey (1984, p. 298) pointed out: 'No longer is there really a "regional" problem in the old sense. No longer is there a fairly straightforward twofold division between central prosperous areas and a decaying periphery.' Rather, because 'it is not regions which interrelate, but the social relations of production which take place over space' (p. 122), a new spatial division

Figure 12.1

The Thatcher government's rolling back of regional aid, 1979–84

Source: Martin (1986), Fig. 8.4, p. 273

of labour based upon spatial division of a firm activities has given rise to a new and more localized pattern of spatial inequality. Her summary argument is as follows (p. 295):

> The old spatial division of labour based on sector, on contrasts between industries, has gone into accelerated decline and in its place has arisen to dominance a spatial division of labour in which a more important component is the inter-regional spatial structuring of production within individual industries. Relations between economic activity in different parts of the country are now a function less of market relations between firms and rather more of planned relations within them.

From this perspective regional policy no longer engages with economic reality. Rather than discrete regional economies in competition with one another, 'a new spatial division of labour made possible by new information and communication technologies and new non-spatial scale economies has created a localized pattern of restructuring' (Agnew, 1988, p. 131). In a world in which exogenous links have become central to local development, area-based policies for endogenous development often can appear less and less fruitful.

As decentralist reactions, therefore, regional policy and, especially, regionalism appear increasingly problematic. As Martin and Hodge (1983, p. 319) have argued with respect to Britain: 'however much regional policies of the conventional type are strengthened, if they are pursued against a background of continued unfavoura-ble macro-economic conditions, their "social" role will be largely limited to simply spreading the misery of mass unemployment more "fairly" around the country while providing little boost to total economic activity or employment'.

It may be too early to write off regional policy entirely, even if regionalism, the commitment to planning in a nationally coordinated fashion the economic develop-ment of fixed regional units, is largely in retreat. For one thing, to counterbalance its commitment to improving the overall efficiency of European manufacturing industry in global competition through enhanced European competition, the EU has placed renewed emphasis on regional incentives to compensate for losses of industries and employment (see Chapter 11). Regional disparities in employment and production increased throughout the EU from the mid-1970s to the late 1980s after declining steadily from the late 1950s. In response, the European Commission has concentrated its spending on infrastructural projects and vocational training. There has been no attempt to direct industries to particular areas. Whether this is sufficient to rebalance regional disparities is questionable given the continued high rates of job loss in less developed regions (Dunford, 1992). Second, some emerging industries enjoy economies of agglomeration (particularly localization economies). Once established in specific areas such industries have a local dependence that generates competitive advantage. Unfortunately, not everywhere can benefit from this. This has been the problem with proliferating 'growth poles', trying to mass-produce the dynamism associated with initial advantage at great public expense but without much of a return on investment. There can only be a limited number of 'Silicon Valleys'. Hence, there is an economic return to early organization of specialized complexes. If a national economy is going to share in the potential for national economic development represented by these, then a national government will require a regional policy to encourage early response.

For example, Japanese regional policy is now centrally concerned with creating the conditions for early response in growing sectors such as computing and biotechnology. There is a network of 170 regional centres that channel support for innovation and research into companies with fewer than 300 employees. Three-quarters of the staff of these *kohsetsushi* centres are engineers who carry out applied research and offer training and advice. The centres encourage small firms to collaborate with one another and with the large firms they often supply with components and services. The regional system provides an organizational frame-work for Japanese economic innovation (Coghlan, 1993). Countries such as Britain and the United States have suffered from the absence of such government 'priming' of the pump of innovation while waiting for firms to do it themselves. Firms, however, have had no incentive to look beyond their own short-run interests to the interest of the country as a whole. From one point of view, that is what govern-ments are for.

NATIONALIST SEPARATISM

The growth of industrial capitalism in the nineteenth century was accompanied by promotion of the nation state and the growth of nationalism. The conviction grew among élites and populations at large that each state should be clearly bounded geographically; it should be organized as an economy; and it should be as linguistically and culturally homogeneous as possible. To many in Western Europe, North America and Australasia the blessings of material abundance and personal freedom became associated with the interrelated development of capitalism and the nation state.

In the twentieth century, however, nationalism has regularly been perverted into fascism. Dreadful wars have been fought. National independence has been no guarantee of national prosperity. Even in the original or 'founding' states of Europe such as Spain, France, and Britain, political movements rejecting established national claims and asserting political and economic rights for regional and ethnic populations have become widespread (e.g. Harvie, 1994; Buzzetti 1996).

This last trend reflects the fact that state-making and nationalism have never redounded equally to the benefit of all the nominal citizens of a country. In particular, since their earliest formation modern national states have contained a diversity of cultural groups within their boundaries. In the social science literature the term ethnicity has come to signify the organization of cultural diversity within modern states. Thus, an ethnic group can be defined as 'a collectivity of people who share some pattern of normative behaviour, or culture, and who form a part of a larger population, interacting within the framework of a common social system.' (Cohen, 1974, p. 92)

But ethnic groups are not simply primordial groupings, even though they usually draw on myths of common ancestry and cultural distinctiveness. They are differentiated from one another and integrated internally through such mechanisms as a cultural division of labour, political favouritism and historically created economic roles. Indeed, ethnicity can be viewed as a mechanism for allocating wealth and power within states that have not inherited or successfully imposed a unifying set of cultural practices and symbols upon their populations. A clear example of the use of ethnicity in this way would be Northern Ireland between 1920 and 1972 where a dominant élite of Protestant landowners and businessmen maintained their hegemony over the region through a web of mutual obligations, customs, duties and economic favours that bound Protestant workers and small farmers to them while excluding the Catholic population (MacLaughlin and Agnew, 1986). This process of ethnic competition for control over the fruits of economic growth and government policy is extremely widespread the world over; the more so the greater the number of ethnic groups and the weaker the alternative means of political mobilization (e.g. social class).

Students of ethnicity have noted that in recent years the level and intensity of conflicts between ethnic groups have been on the increase. Writing of Indonesia, Clifford Geertz (1973, pp. 244–5) has provided a particularly vivid description:

> Up until the third decade of this century, the several ingredient traditions – Indic, Sinitic, Islamic, Christian, Polynesian – were suspended in a kind of half-solution in which contrasting, even opposed styles of life and world outlook managed to coexist, if not wholly without tension, or even without violence, at least in some sort of workable, to-each-his-own sort of arrangement. This modus vivendi began to show signs of strain as early as the mid nineteenth century, but its dissolution got genuinely under way only with the rise, from 1912 on, of nationalism; its collapse, which is still

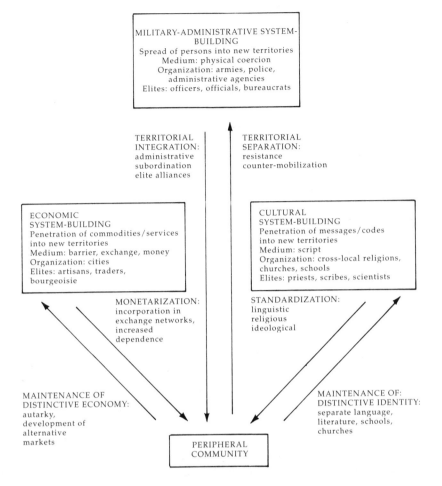

Figure 12.2

An abstract model of processes of interaction and resistance within large-scale territorial systems

Source: Rokkan and Urwin (1983), Fig. 1.3, p. 15

not complete, only in the revolutionary and post-revolutionary periods [1945 on]. For then what had been parallel traditionalisms became competing definitions of the essence of the New Indonesia. What was once, to employ a term I have used elsewhere, a kind of 'cultural balance of power' became an ideological war of a peculiarly implacable sort.

Some, for example Kedourie (1960), have seen this ethnic 'schismogenesis' as a world-wide process associated with the diffusion of the idea of nationalism from Europe along with colonialism. Others have emphasized modernization or industrialization as universal processes laying the material foundations for the politics of nationalism (e.g. Gellner, 1964). In fact, ethnic nationalism seems to have developed in different ways and with different causes in different parts of the world. Rokkan and Urwin (1983), for example, argue that processes of economic, military–administrative, and cultural 'system building' have combined in different ways to produce different effects in different European localities (Fig. 12.2). In turn, ethnic nationalism has been both encouraged in some settings (such as the Celtic fringe of the British Isles, the Basque provinces of Spain and the Balkans) and discouraged elsewhere (for example in Alsace, France; and the South Tyrol, Italy).

Whatever its precise origins in particular cases, however, ethnic conflict and nationalist separatism (when ethnic groups are geographically concentrated) have become increasingly marked features of the contemporary world. From Ireland to Yugoslavia to Lebanon to India to Sri Lanka to Canada, to name only a few of the best-known cases, ethnic groups and ethnic conflict have become major elements in national political life. Four factors seem to be especially important in this trend. One of these is the increased economic-geographical differentiation within states and its relationship to ethnic divisions. It is not that ethnic conflict always involves increasingly poor regions rebelling against more affluent ones. It is difference *per se* generating a sense of deprivation or exploitation. In Spain, for example, it is the Basque and Catalan regions, the most prosperous in the country that are the most rebellious. Likewise in the former Yugoslavia, where the relatively well-off Slovenians and Croatians demonstrated their impatience with 'subsidizing' the ethnic groups (Serbs, Albanians, etc.) which occupy other regions, by agitating for and achieving political independence. Yet within the now Serb-dominated 'rump' of Yugoslavia it is the poorer Albanians in the province of Kosovo who are the most likely candidates for the next round of nationalist separatism.

The break-up of the former Yugoslavia illustrates a second factor of singular importance in the explosion of nationalist separatism in 1989–93: the collapse of the Soviet Union, the exhaustion of state socialism and the end of the Cold War. The demise of strong central governments and the exhaustion of state socialism as an ideology have opened the way for a re-emergence of political identities based upon ethnic divisions. Formerly communist states such as Yugoslavia and the Soviet Union were organized administratively around geographical units that reproduced ethnic cleavages. Even though some groups, such as Russians in the Soviet case, are to be found scattered in considerable numbers outside of their own 'republics', the dominant identity of particular administrative units remained that of the historically dominant ethnic group.

Within Russia itself the absence of the distinct social groups that underpin the political divisions in other countries, such as organized labour or religious traditions, has led to political organization by entrenched vested interests from the bureaucracies and by ethnic group. One of the early results of the break-up of the Soviet Union within Russia was the shift in power from the centre to the regions. Local governments were formerly the instruments of central rule but by 1992 they had become major protagonists in political-economic development (Parker, 1992). Some of the 27 million non-Russians in Russia have even declared sovereignty within their local government units. These units have different economies, some are raw-materials producers and others are industrial, and thus have different interests in terms of pricing and macroeconomic policies. So there is the possibility within Russia of ethnic and economic differences becoming mutually reinforcing as they did within the former Soviet Union as a whole (Fig. 12.3).

At the same time, the 'freezing' of political boundaries that both sides in the Cold War had quietly accepted no longer can be tied to an overriding global conflict. Each and every territorial dispute is no longer a potential spark for World War III. This opens the way for a possible proliferation of nationalist separatisms (and also

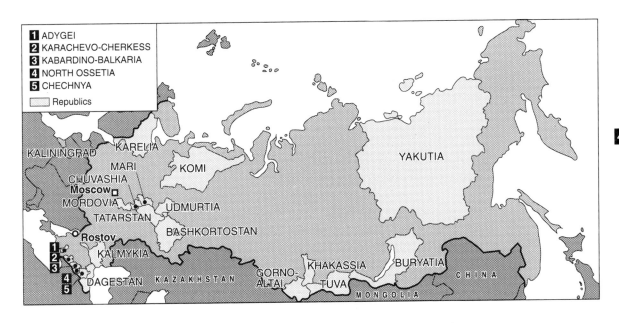

Figure 12.3
The governmental decentralization of Russia, 1993

Source: The Economist, 5 December 1992, Survey: Russian reborn, p. 25

expansionist claims by existing states) in all world regions as established political boundaries lose their previous inviolability.

A third factor is the increased bureaucratization of the state and the growth of the welfare state. Ethnic identity can be a basis for collective action against the intrusiveness of the modern state and its destruction of particularity. Fox (1981) note, for example, how political mobilization in Western Europe increasingly takes regional and cultural rather than social class forms. But the welfare state in its increased provision of social services has also reduced the appeal of class-based politics and stimulated the growth of ethnic politics.

The fourth factor is the growing globalization of political and economic activity. The shift of power and control over local economies to ever more distant locations provides an incentive for regional counter-mobilization. The development of the EC in Europe may have been one stimulus, the growing importance of transnational corporations may have been another. At the same time the increase in the flow of international migrants, especially into Western Europe, has introduced new ethnic groups, such as Indians in Britain, Turks and East Europeans in Germany, and Algerians in France, which both stimulate the demands of indigenous ethnic groups and provide new 'out-groups' for new rounds of ethnic conflict and nationalist politics.

With the exception of the former Soviet Union and some parts of Eastern Europe (Yugoslavia, Czechoslovakia) outright separatist movements have not met with success, at least as measured in terms of political independence. In both Europe and peripheral countries the ethnic–territorial status quo has been largely maintained. During the Cold War the world superpowers generally refused to back separatist movements, perhaps for fear of stimulating them at home or within their own spheres of influence. Often political changes short of outright independence have proved satisfactory responses to regional–ethnic revolt. These include federalism, regional devolution, and consociationalism (power-sharing among ethnic groups as in Switzerland and the Netherlands). What is clear, however, irrespective of the prospects for nationalist separatism as such, is that 'there is little likelihood of an abatement of ethnic nationalism in the near future' (Williams, 1982, p. 36).

GRASS-ROOTS REACTIONS

A peculiar paradox of the growing globalization of the world economy has been the stimulus it has provided to the destruction of some specialized local/regional economies and the decentralization/localization of single plants or use of sub-contractors at disparate locations. New information and transportation technologies have made it possible to decentralize production operations to 'cheaper' locations (lower wage bills, etc.) or ones with special advantages in terms of access to technology, labour skills or markets at the same time that central corporate control is maintained or enhanced. This process has been brought about by increased competitive pressures upon large firms from the appearance of foreign competitors. Many large firms, especially in Europe and the United States, now face a global market-place far more competitive than the more geographically restricted ones they had known for the previous 40 years. Of course, as argued in general in Chapters 1 and 3 and then repeatedly in subsequent chapters, branch-plant industrialization is only one among a number of strategies for re-establishing firm competitiveness. And, it does not signify that there are not continuing and new pressures for the clustering of production facilities. The point to be made here is more that today there are greater incentives and technological possibilities for firms – even small firms – maintaining control over production at a distance than was the case in the past.

Coincidentally, union–management conflicts in established production facilities and changes in market conditions, especially the increased demand in many developed countries for customized rather than mass-produced goods, have also encouraged decentralization of production. In Italy, for example, one can see evidence of both causes. Large firms in Piedmont (around Turin) and Lombardy (around Milan) have increasingly contracted-out to small firms for parts and services that used to be provided on-site at large factories. Moving production in this way both undermines the power of workers in large factories, where solidarity is more easily achieved than in scattered small factories, and protects the large firm from the need to shed labour cyclically. In Emilia-Romagna (around Bologna) and elsewhere in Central Italy many small firms provide customized products (both industrial and consumer goods) to domestic and export markets that are better served by the flexible response to shifts in fashion that small firms can provide (Brusco and Sabel, 1981; Mattera, 1985).

In the United States both types of decentralization are also increasingly common, although the small firm as supplier to the plants of a large firm was most characteristic from the late 1960s until recently (Sabel, 1982; Ettlinger 1997).

Companies with fewer than 500 workers added 1.2 million jobs in the United States between 1976 and 1984, while larger companies lost 300 000 jobs. By 1992 companies with fewer than 250 employees accounted for more than half of the US's manufacturing jobs, up from 42 per cent in 1988 (Scott, 1992). Paralleling this, production facilities have also shrunk in size. Plants in the USA built before 1970 and still operating in 1979 had on average more than twice the number of workers of a plant built in the 1970s or 1980s (Fig. 12.4). These data cast doubt on the claim that not much has changed in either the scale of production or the average-ownership size in US manufacturing industries (e.g. O hUallacháin, 1996). What perhaps has not changed in the USA is the *directing* role of large firms, only now in relation to a variety of production arrangements rather than solely in terms of vertical integration of all stages of production within one firm (see Chapter 3; also see Harrison, 1994).

A similar fragmentation of production has been noted in Britain. Since the late 1970s average employment in plants of firms employing more than 999 people has shrunk and the average number of plants operated by these firms has increased significantly. In 1973 average employment per plant was 459 and there were 12.4 plants per firm. The comparable figures for 1982 were 338 and 15.0 (Shutt and Whittington, 1987). Yet, at the same time there was a decrease in the number of firms in the category of 999+ employees, indicating an increasing concentration of ownership. In the British case reduction in the size of plants has not involved a

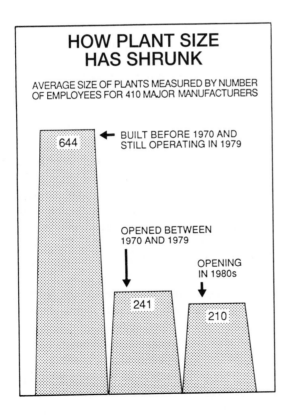

Figure 12.4
How plant size in the United States has shrunk

Source: *Business Week*, 22 October 1984, p. 156

reduction in the concentration of ownership of manufacturing industry. However, some of the decrease in the average size of all workplaces (as opposed to just those of the larger firms) is due to the expansion of employment in small firms. Although in this respect Britain appears to lag well behind other countries such as Italy, Japan and the United States (Allen, 1988: Hirst and Zeitlin, 1989).

The new 'decentralized economy', however, is incredibly volatile. Patterns of new firm creation/destruction are extremely sensitive to minor cyclical fluctuations in demand from larger firms and final markets. Many firms have tiny inventories and limited equity. Though advantageous financially this introduces major limitations in terms of employee security and long-term commitment to local economies. In this context local rather than aggregate national conditions are of increased significance for the welfare of populations. Political parties such as the Labour Party in Britain have proposed decentralization of governmental powers over economic development as a reaction (in part, at least) to the increased localization of economic activities and perhaps as a response to their exclusion from national government. In Italy, the Communist Party (and now that part of it called the Democratic Party of the Left), excluded from national governments on a permanent basis during the Cold War, has been a major sponsor of both economic and political decentralization in the central regions (Emilia-Romagna, Tuscany and Umbria) where it dominates local government. Without supportive national government policies, however, local economic initiatives organized on a geographical basis are problematic. Above all, they face the dilemma of protecting current employment while attempting to create local economies that can capture the benefits of 'emerging' industries (Eisenschitz and Gough, 1992).

Perhaps local initiative and control or 'small is beautiful' can only work when decision-making power is no longer vested in giant transnational corporations, as it still is even when small firms proliferate to serve them (as is the case even in many industrial districts; Scott, 1992; Amin and Thrift, 1992). For example, the strategy for procuring components used by Ford in Europe involves the increased use of widely scattered and expendable suppliers. This approach is hardly amenable to either government controls or local economic development strategies geared towards long-term stability in employment (Wells and Rawlinson, 1992). The lack of a 'natural evolution' of economies towards greater security of employment and increased participation in the benefits of economic growth has stimulated the revival of thinking about the possibility of wider popular participation in economic decision-making. The theme of 'economic democracy' is especially strong in some recent proposals for reviving the US economy. Economic democracy refers to an egalitarian form of political-economic structure in which a serious attempt is made to democratize the economic sphere in general and workplaces in particular. The major point is to challenge the political and economic position of global capital and the commitment to it and its international role by the major states (the USA, Britain, etc.). It builds on the view characteristic of movements for participatory democracy that, to be more than a sham, democracy should be extended beyond episodic political activities such as voting into the economic sphere (Boyte, 1980).

Economic democracy differs from democratic capitalism (in both *laissez-faire* and welfare state manifestations), in which democracy is limited to periodic involvement in electoral politics and the means of production are largely privately owned. It also differs from conventional state socialism, especially of the now-defunct Soviet variety, in which markets are prohibited (in public), there is little meaningful electoral politics and the means of production are owned by the state.

There are a number of reasons why interest in both the theory and practice of economic democracy has increased in recent years. First, as the heavy 'smokestack' industries in the United States and other developed countries have become less profitable, there have been numerous attempts by their employees to save their jobs by buying failing factories. Changes in tax laws have also made employee stock option plans (ESOPs) more attractive to businesses. Between 1976 and 1983 the number of ESOPs grew in the United States from less than 300 to over 5 000 (English, 1983). While owning stock is hardly the same as control, ESOPs do raise the question of where employees' participation should stop. But it is probably the competitive environment for businesses and the prospect of bankruptcy that do most to encourage talk about and proposals for producer cooperatives and worker self-management (Hancock et al., 1991).

Second, interest in economic democracy reflects consideration of actual practices in a number of countries which have had high levels of economic growth and high standards of living. In West Germany, 'co-determination' allocates positions on corporate boards to employees. In Sweden, after much controversy, a tax on corporate profits will be used to purchase corporate assets. In addition, the success of several large-scale prototypes has given advocates of economic democracy a ready reply to those who assert that self-management by employees is inherently utopian. The extensive network of Basque producer cooperatives in Spain, 'Mondragon', employs over 10 000 workers and was for many years more efficient than its more conventional competitors (Thomas and Logan, 1982; Hancock et al., 1991).

Third, and finally, especially in the LDCs, economic democracy can be seen as an alternative to both American-style corporate capitalism and socialist-style central planning. Under nationalistic pressure to avoid becoming satellites of the major world powers, the rhetoric and sometimes the substance of economic democracy arise. But the pressure is also immediately practical. Considerable evidence suggests that rapid economic growth in LDCs does not necessarily improve the welfare of large numbers of their people. Markets, based on effective demand, which means the given distribution of income, have generally failed to allocate resources to the basic human needs of the poor in many LDCs: mass poverty, unemployment and malnutrition are the consequences (Yapa, 1980). Directing attention to local production for 'basic needs' has been one response, based on the use of indigenous knowledge and traditions of production as opposed to imported ones. Discussion of 'appropriate technology' and 'alternative' development strategies, however, are both inspired by similar concerns. They also reflect increasing concern over the limits to the growth of the world economy. Can the world's natural resources and increasingly fragile physical environment support the levels of production that it would take for the entire population of the world to enjoy American or Swiss levels of consumption?

There are of course a range of possible 'grassroots reactions' within the general confines of economic democracy. Among advocates of economic democracy are some who are committed to a vision of large firms with powerful, central councils, while others look to smaller, decentralized firms embedded in non-state networks of social association. Ellen Comisso (1979) compared this difference to that between federalists (such as Hamilton) and Jeffersonians at the time of the founding of the United States. To take one example, Paul Hirst (1994) has argued for what he calls 'associative democracy' resting on two major principles: a decentralized economy based on cooperation and mutuality and a system of governance based on self-governing associations of mutual interests. State power is to be weakened through a pluralizing and federalizing of political authority and economic power is to be redistributed to local economies. Hirst draws upon the examples of manufacturing success in Italian industrial districts, German regions and Japanese firms to propose an associative regional economy linked into others by federal channels of authority. Hirst claims that such a model would both stimulate economic growth and redistribute its fruits in a more egalitarian fashion than is the case with the present world economy.

One important objection to economic democracy of all types, but especially the more decentralized ones, is that they are utopian dreams since those in power will not hear of them. The world is now bureaucratized; bureaucracies will not seriously consider participatory democracy (see, for example, Ross, 1982). In particular, most states are run by oligarchies determined to fix the best deals for themselves from transnational corporations and the most powerful states. William Appleman Williams (1981), however, has defended the need for utopias. He argues that

> the purpose of a radical utopia is to create a tension in our souls. . . . We must imagine something better. That defines us as people who offer our fellow citizens a meaningful choice about how we can define and live our lives . . . . Radicals must confront centralized nationalism and internationalism and begin to shake it apart, break it down, and imagine a humane and socially responsible alternative. It simply will not do to define radicalism as changing the guard of the existing system.
>
> (pp. 95, 98)

In essence this is what grassroots reactions inspired by a vision of economic democracy are really all about.

A second objection is more by way of a critique of possibilities of successful long-term local development in the absence of a relatively strong state presence in economic regulation. Ash Amin (1996, 309–10), for example, is critical of proponents of local 'associative democracy' (particularly Paul Hirst) for 'failing to distinguish between different forms of state economic intervention and state practice' and for undervaluing 'the strategic and developmental role played by the state in some of the most successful economies in the world'. In a somewhat different vein John Donahue (1997) has claimed that a state defines something of a 'commons' in which externalities across local and regional boundaries are so intense as to vitiate against the possibility of ever successfully separating out groups of people into discrete geographical communities that can be run as if the others did not exist. Federalism is about achieving a balance between the common and the particular. For Donahue (p. 42), the 'devil in devolution' is that in the American case the dominant consensus

> in favor of letting Washington [the US Federal Government] fade while the states take the lead is badly timed. The public sector's current trajectory – the devolution of

welfare and other programs, legislative and judicial action circumscribing Washington's authority, and the federal government's retreat to domestic role largely defined by writing checks to entitlement claimants, creditors, and state and local governments – would make sense if economic and cultural ties reaching across state lines were *weakening* over time. But state borders are becoming more, not less permeable.

Summary

In this chapter three types of decentralist reaction to the impact of the world economy have been described in the context of the trend towards globalization of economic activities.

Regionalism and regional policy under state sponsorship were common up until the 1960s but their relevance has been increasingly questioned in the face of the changed relationship between national and world economies. Regional and local tiers of government have tended to displace the regionalism carried out under the sponsorship of national governments.

Nationalist separatism challenges both existing states, supranational organizations and the existing distribution of economic activities. But most separatist movements will usually settle for something less than complete independence.

Finally, recent trends in the world economy have generated renewed interest in the possibility of economic democracy. Disillusionment with both American-style corporate capitalism and Soviet-style socialism in the face of an increasingly volatile world economy has directed attention to the possibility of people taking control of their economic activities and putting them to work for them.

Whether or not decentralist reactions increase in importance depends in part on whether the world economy recovers from its present problems, especially the debt crisis, the slowing of growth in world trade and the lack of congruence between global production and global consumption, and whether or not the 'free market' ideologies antithetical to many of these reactions and associated for many years with the Reagan/Bush administrations in the USA, and the Thatcher/Major governments in Britain continue to find support around the world and in international organizations such as the World Bank and the IMF. Without some dramatic rebalancing of global patterns of production and consumption and the geographical spread of the benefits of globalization, the trend towards decentralist reactions may prove inexorable as people begin to take their fate into their own hands.

Key Sources and Suggested Reading

Agnew, J. A. 1988. Beyond core and periphery: the myth of regional political-economic restructuring and a new sectionalism in American politics. *Political Geography Quarterly*, **7**, 127–39.

Allen, J. 1988. Fragmented firms, disorganised labour? In J. Allen and D. Massey (eds), *Restructuring Britain: The Economy in Question*. London: Sage.

Amin A. 1996. Beyond associative democracy. *New Political Economy*, **1**, 309–33.

Amin A. and Thrift, N. 1992. Neo-Marshallian nodes in global networks. *International Journal of Urban and Regional Research*, **16**, 571–87.

Bellini, N. 1996. Regional economic policies and the non-linearity of history. *European Planning Studies*, **4**, 63–73.

Boyte, H. 1980. *The Backyard Revolution: Understanding the New Citizen Movement*. Philadelphia: Temple University Press.

Brusco, S. and Sabel, C. 1981. Artisan production and economic growth. In F. Wilkinson (ed.), *The Dynamics of Labour Market Segmentation*. London: Academic Press.

Buzzetti, L. 1996. Efforts to reorganize the state within the new international framework. In A. Vallega *et al. The Geography of Disequilibrium: Global Issues and Restructuring in Italy*. Rome: Società Geografica Italiana.

Cheshire, P. C. *et al.* 1992. Purpose built for failure? Local, regional and national government in Britain. *Environment and Planning C: Government and Policy*, **10**, 355–69.

Chisholm, M. 1991. *Regions in Recession and Resurgence*. London: Unwin Hyman.

Coghlan, A. 1993. Dying for innovation. *New Scientist*, **137**, 9 January, 12–14.

Cohen, A. 1974. *Two-Dimensional Man: An Essay on the Anthropology of Power and Symbolism in Complex Society*. Berkeley: University of California Press.

Comisso, E. 1979. *Workers' Control Under Plan and Market*. New Haven: Yale University Press.

Cooke, P. 1992. Regional innovation systems: competitive regulation in the new Europe. *Geoforum*, **23**, 65–82.

Donahue, J. D. 1997. The devil in devolution. *The American Prospect*, **32**, 42–7.

Dunford, M. 1992. *Socio-economic trajectories, European integration and regional development in the EC*. Brighton: University of Sussex Research Papers in Geography.

Eisenschitz, A. and Gough, J. 1992. *The Politics of Local Economic Policy: The Problems and Possibilities of Local Initiative*. London: Macmillan.

Eisenschitz, A. and Gough, J. 1996. The contradictions of neo-Keynesian local economic strategy. *Review of International Political Economy*, **3**, 434–58.

English, C. W. 1983. When workers take over the plant. *US News and World Report*, 18 April, 89–90.

Ettlinger, N. 1997. An assessment of the small-firm debate in the United States. *Environment and Planning A*, **29**, 419–42.

Fisher, P. S. and Peters, A. S. 1997. Tax and spending incentives and Enterprise Zones. *New England Economic Review*, March/April, 109–30.

Fisher, R. C. 1997. The effects of state and local services on economic development. *New England Economic Review*, March/April, 53–67.

Fox, R. G. 1981. Ethnic nationalism and the welfare state. In C. F. Keyes (ed.), *Ethnic Change*. Seattle: University of Washington Press.

Geertz, C. 1973. *The Interpretation of Cultures*. New York: Basic Books.

Gellner, E. 1964. *Thought and Change*. London: Weidenfeld and Nicolson.

Hancock, M. D. *et al.* 1991. *Managing Modern Capitalism: Industrial Renewal and Workplace Democracy in the United States and Europe*. Westport, CT: Praeger.

Harrison, B. 1994. *Lean and Mean*. New York: Basic Books.

Harrison, B. 1997. Comments on 'The Effects of State and Local Public Policies on Economic Development.' *New England Economic Review*, March/April, 140–1.

Harvie, C. 1994. *The Rise of Regional Europe*. London: Routledge.

Hirst, P. 1994. *Associative Democracy*. Cambridge: Polity Press.

Hirst, P. and Zeitlin, J. (ed.) 1989. *Reversing Economic Decline: Industrial Structure and Policy in Britain and Her Competitors*. Oxford: Berg.

Johnston, R. J. 1986. The state, the region, and the division of labor. In A. J. Scott and M. Storper (eds), *Production, Work, Territory: The Geographical Anatomy of Industrial Capitalism*. Boston: Allen and Unwin.

Kedourie, E. 1960. *Nationalism*. London: Hutchinson.

Lloyd, P. and Meegan, R. 1996. Contested governance: European exposure in the English regions. *European Planning Studies*, **4**, 75–97.

MacLaughlin, J. G. and Agnew, J. A. 1986. Hegemony and the regional question: the political geography of regional industrial policy in Northern Ireland 1945–1972. *Annals of the Association of American Geographers*, **76**, 247–61.

Martin, R. L. and Hodge, J. S. C. 1983. The reconstruction of Britain's regional policy, 2: towards a new agenda. *Environment and Planning C; Government and Policy*, **1**, 317–40.

Massey, D. 1984. *Spatial Divisions of Labour*. London: Methuen.

Mattera, P. 1985. *Off the Books: The Rise of the Underground Economy*. New York: St. Martin's Press.

O hUallachàin, B. 1996. Vertical integration in American manufacturing: evidence for the 1980s. *The Professional Geographer*, **48**, 343–56.

Parker, J. 1992. Survey: Russia reborn. *The Economist*, 5 December, 58.

Rokkan, S. and Urwin, D. 1983. *Economy, Territory, Identity: Politics of West European Peripheries*. London: Sage.

Ross, R. 1982. Regional illusion, capitalist reality. *Democracy*, **2**, 93–9.

Sabel, C. F. 1982. *Work and Politics: The Division of Labor in Industry*. New York: Cambridge University Press.

Scott, A. J. 1992. The role of large producers in industrial districts: a case study of high technology systems houses in Southern California. *Regional Studies*, **26**, 265–75.

Shutt, J. and Whittington, R. 1987. Fragmentation strategies and the rise of small units: cases from the North West. *Regional Studies*, **21**, 13–23.

Thomas, H. and Logan, C. 1982. *Mondragon: An Economic Analysis*. London: Allen and Unwin.

Wasylenko, M. 1991. Empirical evidence on interregional business location decisions and the role of fiscal incentives in economic development. In H. W. Herzog and A. W. Schlottmann (eds), *Industry Location and Public Policy*. Knoxville: University of Tennessee Press.

Wasylenko, M. 1997. Taxation and economic development: the state of the economic literature. *New England Economic Review*, March/April, 37–52.

Wells, P. and Rawlinson, M. 1992. New procurement regimes and the spatial distribution of suppliers: the case of Ford in Europe. *Area*, **24**, 380–90.

Williams, C. H. (ed.) 1982. *National Separatism*. Cardiff: University of Wales Press.

Williams, W. A. 1981. Radicals and regionalism. *Democracy*, **1**, 87–98.

Yapa, L. 1980. The concept of the basic goods economy. In R. L. Singh et al. (eds), *Rural Habitat Transformation in World Frontiers*. Varanasi: National Geographical Society of India.

Picture credit: Thomas Raupach/Still Pictures

CHAPTER 13

CONCLUSION

The task facing economic geography is to make sense of the geographical pattern of economic activities around the world. In this book the task has been pursued using the geographical metaphor of core and periphery to provide a basic model or way of seeing, against which actual patterns of economic activities at a range of scales can be compared. A number of claims about the processes creating and recreating economic landscapes have been investigated using a historical–geographical framework deriving from an evolutionary perspective on the development of the modern world economy. In this short concluding chapter the key elements in the argument are drawn together as a way of summarizing the perspective and pointing the reader back to the themes of the introductory chapters after reading the detailed empirical studies that form the body of the book. A number of controversial issues at the heart of contemporary debates in economic geography are also introduced to highlight the open-ended and exciting character of the field.

First, the modern world economy is an open system that evolves over time. Although there is an obvious path dependence to its motion, as illustrated by the return to initial advantage produced through increasing returns to scale, there are also locational reversals and shifts in the way it works under different organizational–political and technological conditions. Thus, although the core of the world economy was (by definition) the first to industrialize on a massive scale, some other parts of the world have recently experienced a substantial expansion in industrial activities. This reflects the shifts in the operation of the world economy detailed in Chapters 4–10, especially the recent slow disintegration of the Fordist/ organized capitalism that was important in the core for much of the twentieth century. The workings and outcomes of the world economy, therefore, are not set in stone but evolve and change over time. In particular, the evolution of the world economy is not best thought of as a cyclical repetition of what has happened previously only with different technologies and countries but with much the same process of capital accumulation and territorial imperialism driving the long-wave cycle. The mechanisms driving the world economy have also changed. Today, for example, it is misleading to suppose, as some scholars still do, that the globalization of production and finance is essentially indistinguishable from the territorialization of production and trade within empires that characterized the world economy in the late nineteenth and early twentieth centuries. These are different processes and should be seen as such (see, e.g., Agnew and Grant, 1996).

Second, this dynamic understanding of the world economy allows us to combine a focus on general economic forces with a concern for the local variability which characterizes the world's economic geography. The expansion of the world economy has incorporated regions with distinctive economic histories that are themselves changed in different ways as they engage with the interests and influences emanating from organizations that span ever-larger geographical areas. The growth of the world economy has produced difference rather than homogenization. The uniqueness of different places and their economies is the result of interaction over space rather than of a singularity produced by isolation. This approach challenges the assertion that globalization of production and finance portend the demise of uneven development or a progressive reduction of economic differences between places. The 'end of geography' is nowhere in sight. Indeed, recent trends indicate a deepening of differences between regions and localities within countries as well as between countries at a global scale. These reflect not only the impact of decisions by transnational corporations and macroeconomic differences between countries but also the relative success of localities and regions in inserting themselves into and protecting themselves from the circuits of global capitalism (Cox, 1997). At the same time, however, some parts of the world have become largely excluded from the trend towards a more integrated world economy. Africa, for example, finds itself on the margins of globalization, without even the exploitative trade relationships of the colonial period to fuel economic development (Grant and Agnew, 1996). From one point of view, this is to Africa's advantage, since it can now 'develop' in its own way without external interference. From another viewpoint, however, this is nothing short of disastrous, because it means that Africa is closed off from the technological and lifestyle choices that can come with increased exposure to the modern world economy.

Third, the evolution of the world economy has followed a number of long-term cyclical fluctuations that correlate highly with the emergence of distinctive technological systems. Of particular importance are the so-called Kondratiev cycles describing distinctive epochs of economic and political development since the late eighteenth century. A vital dimension of change bringing about these shifts has been the transformation of the nature of capitalism; more especially, change in its mode of regulation. There have been two phases so far and now the world economy appears to be entering a third. The earliest phase, of competitive capitalism, lasted from the eighteenth century until the end of the nineteenth century. The second phase, of organized capitalism based upon close coordination between business, government and labour, dominated throughout much of the twentieth century. However, recently this mode of political-economic organization has begun to be replaced by a disorganized capitalism in which the geographical coincidence between production and consumption that characterized the previous phase has started to unravel. Rather than a sudden shift, however, the transition has been gradual with some 'old' and many more 'emerging' sectors representing most clearly the new modes of organization and production.

The focus on cycles can lead to the overemphasis on sharp breaks or ruptures and the exaggeration of the suddenness of change (see, e.g., Sayer, 1989). The idea of total change is open to doubt, as argued in the latter part of Chapter 3 where the new 'regional motors' of the world economy are placed in the context of a range of locational outcomes depending on the mix of externalities and spatial transaction costs associated with different economic sectors. Mass production is still of great importance, particularly in relation to branch-plant industrialization such as that in EPZs and *maquiladoras*. But it is mistaken to identify the continuance of some mass production with the persistence of Fordism or organized capitalism *tout court* (the association of mass production with high mass consumption by workers) or with the absence of *any* break with the past in the essential structure of industrial organization (e.g. Williams *et al.*, 1987; Gough, 1996). What is also clear, however, is that the present trend towards flexible modes of accumulation in the most industrialized countries (see Chapters 3 and 7) does not necessarily portend a stable and irreversible pattern that will extend indefinitely into the future (Gordon, 1988).

Fourth, as the world economy has evolved, so have the economic and locational principles that govern its operation. The perfect competition, transportation costs, factor endowments and comparative advantage that were important in the phase of competitive capitalism and, to a lesser degree, with organized capitalism, have faded in relative significance. Today, monopolistic competition (sectors dominated by small numbers of large firms), macroeconomic regulation, the market-access regime of international relations, position in the global urban hierarchy, global financial networks, regional motors, industrial districts, economies of scope and coordination, and competitive advantage are all more important (Chapters 3–10). This means that the models of economic activities that identify, for example, transportation costs as *the* key factor in determining location (characteristic of many economic geography textbooks) are potentially misleading in contemporary circumstances. Such static models need to be adjusted or supplemented by more dynamic models of the transition from organized to disorganized capitalism that now appears under way. But static models based on the presumed eternal significance of this or that factor are deaf to history. Part of the problem here involves the continuing preference for simple over more complex accounts in contemporary social science. Once locked in, simple models tend to be perpetuated irrespective of whether they actually offer much purchase on contemporary economic geography. Sad to say, in both economics and economic geography simplicity often appears to be preferred on aesthetic grounds more than because of its theoretical or empirical adequacy.

Fifth, state or political regulation of economic activities is of fundamental importance in understanding world economic geography yet is typically neglected or ignored in conventional economic geography. The achievement of comparative-competitive advantage by firms and places rests importantly upon political organization and coordination. The real world economy is not one of *laissez-faire* economics but one of political economy. Macroeconomic, agricultural and industrial policies have significant and often determining impacts on economic landscapes. The mobility of different national economies within the world economy both historically and as manifested today by the NICs reflects in large part the organizational and mobilization capabilities of governments. Financial systems are particularly important in mediating between states and the investment decisions of firms that are so important in bringing about patterns of economic location (Chapter 3). Today states face the challenge of coping with trends towards globalization and localization that make economic management much more problematic than it was in the past. One reaction on their part, creating trading blocs such as the EU (Chapter 11), threatens to both undermine the present relatively open world trading system and through its centralization of decision-making in distant seats of power further stimulate the decentralist politics of various types already underway around the world (Chapter 12). Success in many sectors of the modern world economy also seems related to certain social and cultural attributes, such as high levels of social trust and well-developed social networks supporting (for example) the links between dominant firms and sub-contractors. This strongly suggests that economic geography can no longer remain isolated from consideration of the sociological and institutional bases of economic activity (see, e.g., Lee and Wills, 1997).

As we have tried to show throughout this book, the world economy is a complex and changing inter-meshing of institutions and markets producing different effects in different places, rather than a closed global isotropic plain governed by the same invariant determinants everywhere and always. This is the central message we hope you take away from reading this book.

Key Sources and Suggested Reading

Agnew, J. A. and Grant, R. J. 1997. Falling out of the world economy? Theorizing 'Africa' in world trade. In R. Lee and J. Wills (eds), *Geographies of Economies*. London: Edward Arnold.

Cox, K. R. (ed.) 1997. *Spaces of Globalization: Reasserting the Power of the Local*. New York: Guilford Press.

Gordon, D. 1988. The global economy: new edifice or crumbling foundations? *New Left Review*, **68**, 24–64.

Gough, J. 1996. Not flexible accumulation – contradictions of value in contemporary economic geography: 1. Workplace and interfirm relations. *Environment and Planning A*, **28**, 2063–79.

Grant, R. J. and Agnew, J. A. 1996. Representing Africa: the geography of Africa in world trade, 1960–1992. *Annals of the Association of American Geographers*, **86**, 729–744.

Lee, R. and Wills, J. (eds.) 1997. *Geographies of Economies*. London: Edward Arnold.

Sayer, A. 1989. Post-Fordism in question. *International Journal of Urban and Regional Research*, **13**, 666–95.

Williams, K. *et al.* 1987. The end of mass production? *Economy and Society*, **16**, 405–39.

INDEX

Bold type indicates terms with marginal definitions.

420